Results and Problems in Cell Differentiation

Series Editors:
W. Hennig, L. Nover, U. Scheer

31

Springer-Verlag Berlin Heidelberg GmbH

M. Elizabeth Fini (Ed.)

Vertebrate Eye Development

With 37 Figures

 Springer

M. Elizabeth Fini, Ph. D.
Vision Research Laboratories
New England Eye Center
Tufts University School of Medicine
750 Washington Street, Box 450
Boston, MA 02111
USA

ISBN 978-3-642-53678-6   ISBN 978-3-540-46826-4 (eBook)
DOI 10.1007/978-3-540-46826-4

Library of Congress Cataloging-in-Publication Data
Vertebrate eye development/M. Elizabeth Fini.
    p. cm – (Results and Problems in Cell Differentiation; 31)
    Includes bibliographical references (p.).

    1. Eye–Growth. 2. Eye–Differentiation. 3. Eye–Cytology. 4. Developmental biology.
1. Fini, M. Elizabeth. II. Series.
QP475.V472000
573.8'816–dc21

© Springer-Verlag Berlin Heidelberg 2000
Originally published by Springer-Verlag Berlin Heidelberg New York in 2000
Softcover reprint of the hardcover 1st edition 2000

Cover design: Meta Design, Berlin
Typesetting: Scientific Publishing Services (P) Ltd, Madras
SPIN: 10697053   39/3136 - 5 4 3 2 1 0 - Printed on acid-free paper

Acknowledgements

The Editor wishes to acknowledge two of her previous fellows, Judith West-Mays, PhD and Katherine Strissel, PhD, for their significant contribution to her education in eye development, and expresses gratitude to her chairman, Carmen A. Puliafito, MD, MBA, for creating a situation that made this education possible. Thanks are also extended to Ms Laoti Russo for her expert editorial assistance.

This project was supported by grants from the National Eye Institute (EY09828 and EY12651) and from the National Institute of Arthritis, Musculoskeletal, and Skin Diseases (AR42981). The Editor is a Jules and Doris Stein Research to Prevent Blindness Professor.

Boston, January 2000
M. Elizabeth Fini

Preface

*"Who would believe that so small a space could contain
the images of all the universe?"*

Leonardo da Vinci

The last years of the 20th century have found the discipline of Developmental
Biology returning to its original position at the forefront of biological research. This progress can be attributed to the burgeoning knowledge base on
molecules and gene families, and to the power of the molecular genetic approach. Topping the list of organ systems which have provided the most
significant advances would have to be the eye. The vertebrate eye was one of
the classic embryologic models, used to demonstrate many important principles, including the concepts of inductive tissue interactions first put forth in
the early 1900s. Within the last decade of this century, a return to some of the
old questions with the new approaches has put eye development back into the
limelight. I find this a highly appropriate topic for a book which aims to
spark research for the new millennium.

We begin with a chapter that discusses the anatomy of eye development,
providing the basic reference information for the chapters that follow. A
novel aspect of this introduction is the connection made between developmental strategies and the eye's optical function. What also emerges from this
chapter is the number of important eye structures that have barely been
touched by the modern developmental biologist. Work on cornea and anterior chamber development has lagged behind lens and retina. Also, essentially
nothing is known on the molecular level regarding development of the glands
associated with the anterior of the eye. I predict that these structures will
provide the next important area of discovery.

Even though this book is about the vertebrate eye, so much of what is
known has been derived from studies in flies that it seemed essential to
include chapters on *Drosophila* eye development. The *Drosophila* eye is
structurally and functionally quite different from the vertebrate eye and the
two organs were thought to have arisen separately during evolution. A major
paradigm switch occurred recently, however, when the *Drosophila eyeless*
gene was determined to be the homologue of the vertebrate gene, *Pax6*. Of
equal significance is the role of this regulatory molecule at the very top of the
developmental hierarchy, with the capacity to switch on an entire genetic
program for organ formation. Ectopic expression of *eyeless* caused the appearance of fully formed and functional compound eyes to appear on the
wings or antennae, sparking the imagination of scientists and nonscientists
alike. It is not surprising, therefore, that *eyeless/Pax6* was chosen as a runner-

up for molecule of the year by the journal, *Science*, in 1995. *Eyeless* is covered in the fly chapters and a chapter on the role of *Pax6* in early vertebrate eye development is also included.

The explosion of new information over the last decade has been primarily about molecules involved in patterning and determination, and most of the regulators identified have been transcription factors and other signaling molecules. Two chapters on early eye development provide a good overview of this area, and two additional chapters review the cell fate specification in the retina. A chapter on induction of the lens updates to the molecular level the classic work on embryonic induction done at the turn of the last century. In addition, two chapters are included that describe work that has received much less attention, but that is likely to provide new directions in the future, i.e., the role of reciprocal interactions between cells and their extracellular matrix in controlling eye development.

Unique about this book is a broader definition of eye development than is usually applied. We include a chapter on the formation of connections between the eye and the brain, another old area of research that has taken off in molecular directions in the past few years. Also included are two chapters on regeneration in the eye, a field that is just starting to break through into the molecular era and holds much promise for advances important to medicine.

We conclude this volume with two chapters on the major genetic models for eye development. The use of transgenic and knockout gene technology, coupled with the availability of new positional cloning methods, has been responsible for the recent surge of progress in the mouse genetic model. The newer zebrafish genetic model, however, offers additional advantages which make possible large-scale genetic screens comparable to the *Drosophila* model. Flies have been invaluable for elucidating many aspects of vertebrate eye development, however, the position of this genus on the evolutionary scale precedes the neural crest, an embryonic structure from which much of the anterior of the vertebrate eye is formed. The zebrafish model overcomes this deficiency and lends itself to addressing an area that is underdeveloped. Continued work with genetic models in the next millennium will be instrumental in elucidating the function of genes involved in human-inherited disorders of the eye.

I am grateful to all the authors who have contributed their efforts and ideas to the chapters herein. Special thanks to Ms. Laoti Russo, for her help in coordinating and editing this volume. I hope readers will find this to be an informative and current overview of the field, and a useful resource for teaching and research.

Boston, January 2000
M. Elizabeth Fini

Contents

Embryonic Induction

Retinal Differentiation

Cell Fate Specification in the *Drosophila* Retina
Justin P. Kumar and Kevin Moses

Roles of the Extracellular Matrix in Retinal Development
and Maintenance
Richard T. Libby, William J. Brunken, and Dale D. Hunter

Adhesive Events in Retinal Development and Function:
The Role of Integrin Receptors
Dennis O. Clegg, Linda H. Mullick, Kevin L. Wingerd, Hai Lin,
Jason W. Atienza, Amy D. Bradshaw, Dennis B. Gervin,
and Gordon M. Cann

Formation of Neural Pathways for Vision

Connecting the Eye with the Brain:
The Formation of the Retinotectal Pathway
Karl G. Johnson and William A. Harris

Regeneration

Regeneration of the Lens in Amphibians
Panagiotis A. Tsonis

How the Neural Retina Regenerates
Pamela A. Raymond and Peter F. Hitchcock

Genetic Models

Mouse Mutants for Eye Development
Jochen Graw

Genetic Analysis of Eye Development in Zebrafish
Jarema Malicki

Vertebrate Eye Development and Refractive Function: An Overview

Barbara Sivak and Jacob Sivak[1]

Introduction and General Eye Development

The vertebrate eye is a sophisticated neurosensory/optical instrument capable of detecting minute quantities of light and providing high-resolution ability. It is made up of a variety of cellular and noncellular components derived from ectodermal and mesodermal germinal sources related to two primary functions: retinal image formation and retinal image processing. The overview which follows will describe the sequence of ocular development in general terms.

The optical characteristics of the eye are primarily determined by whether it is used in air or water (or both), or whether it is used under diurnal or nocturnal conditions (Walls 1942; Sivak 1980). While general overviews of ocular development abound, little attention has been directed to the relationship between developmental strategies and the optical performance of the eye. Because optical performance is a unique and primary feature of the eye, this chapter will give prominence to the development of optical structures and to the ramifications resulting from the developmental strategies that have evolved. It will also describe research carried out over the past two decades which explores the extent to which postembryonic optical development of the eye can be influenced by early visual experience.

1.1
Early Embryogenesis

A zygote is the single cell resulting from the fertilization of an ovum by a sperm. The first week of prenatal human development is devoted to cleavage. Zygotic cleavage results in the development of a hollow sphere structure, the blastocyst. The blastocyst differentiates at one pole into an inner cell mass from which the embryo will form. During the second week of embryonic development, the inner cell mass organizes into the primary germ layers: the

---

[1] School of Optometry, University of Waterloo, Waterloo, Ontario, Canada N2L 3G1

Results and Problems in Cell Differentiation, Vol. 31
M. E. Fini (Ed.): Vertebrate Eye Development
© Springer-Verlag Berlin Heidelberg 2000

ectoderm, mesoderm, and endoderm. Further differentiation of the primary germ layers will produce the major tissue types. In the development of the eye, the ectoderm (surface, neural, and neural crest cells) play important roles (Bron et al 1997; Jakobiec and Ozanics 1982).

The eye is an extension of the central nervous system (Mann 1969). In humans, the central nervous system differentiates from the ectoderm on the dorsal surface of the embryo. The ectoderm thickens along the middorsal region of the embryo to form the neural plate. By a process of neural induction, the neural plate begins to fold at its lateral edges to form a neural groove. The signals for the induction process are believed to include signals from the underlying mesoderm. The neural folds continue to converge toward each other. Fusion of the lateral folds results in a hollow structure, the neural tube. Finally, the neural tube (neural ectoderm) is separated from the surface (Fig. 1).

The cavity of the neural tube will become the ventricular system of the central nervous system, while the epithelial cells that line the walls of the tube will produce the neurons and neuroglial cells of the central nervous system. Proliferation of the cells at the rostral end of the tube leads to the formation of the earliest components of the brain, the forebrain (prosencephalon), midbrain (mesencephalon), and hindbrain (rhombencephalon). In addition, neuroepithelial cells also give rise to a special group of neural crest cells from the dorsal region of the neural tube soon after the neural folds have fused. Because there are no mesodermal somites in the head region, the cranial neural crest cells contribute to the formation of the connective tissues of the eye and its adnexal structures, in addition to peripheral nervous system structures associated with the head (Noden 1982).

1.2
Formation of the Optic Cup

The neural ectoderm components of the eye develop from the bilateral outgrowths of the forebrain (Duke-Elder 1958; Mann 1969). Even before the neural tube is completely closed, two depressions (the optic pits) are visible in the developing forebrain. With further rapid growth, these pits become vesicles (pouches). Later, with still more growth, the outer wall of each vesicle collapses inward, or invaginates, to form the optic cup (Fig. 1). This is the first indication of the spherical shape that is associated with ocular development. The embryonic eye is at this stage connected to the forebrain by way of an optic stalk. The invagination of the optic vesicle to form the cup results in the formation of two layers, one on top of the other. A ventral slit in the cup, the embryonic fissure, permits vascular contact with the inner structures of the cup; later, the fissure disappears.

The two layers of the cup eventually form all of the retina, parts of the ciliary body (excluding the musculature system that serves to control the

NEURAL ECTODERM

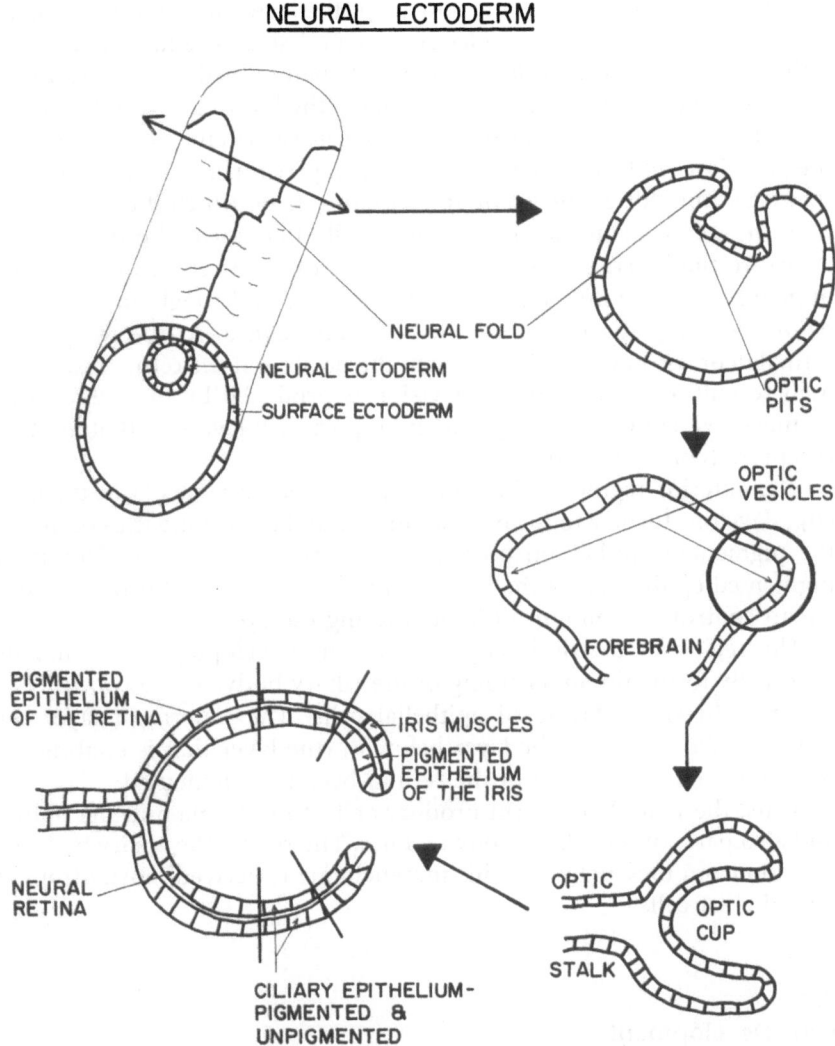

Fig. 1. Schematic sequence showing the development of the dorsal hollow nerve cord and the optic cup from neural ectoderm and the contribution of the optic cup to the structures of the eye. (Sivak and Bobier 1990)

shape or position of the lens), and a portion of the iris. The outer layer of the optic cup (the layer closest to the optic stalk) remains a relatively simple structure consisting of a single layer of epithelial cells, the retinal pigmented epithelium. As its name suggests, this layer is pigmented: it helps to form the dark chamber needed for image formation. The retinal pigmented epithelium also plays a role in the turnover of visual pigment of the specialized photo-receptor cells of the retina, the rods and cones.

Part of the inner layer of the optic cup (the layer closest to the center of the developing eye) becomes an elaborate neural structure known as the neural retina. This layer differentiates extensively before birth into a region of three main cell layers; a layer of rods and cones (the layer closest to the pigmented epithelium); a middle layer, the bipolar cell layer; and the innermost layer, the ganglion cell layer. These three neural layers form an interneural network that is as complex as the brain, the structure from which it develops directly. It is the nerve axons of the ganglion cells that form the optic nerve, the structure that carries visual information from the retina to the brain. The human retina is an inverted one. Light must travel through much of the retina before reaching the receptors, the rods, and cones. At the fovea, a central retinal depression that coincides with the portion of the retina responsible for high-resolution vision and color vision is produced by the lateral (i.e., horizontal) diversion of the ganglion and bipolar cell layers so that light hits the receptors (cones) directly.

The anterior region of the optic cup, the region adjacent to the lip of the cup, develops further to form the sphincter and dilator muscles of the iris and the pigmented epithelium on the inner surface of the iris. This layer, the pigmented epithelium of the iris, is, in part, responsible for the ability of the iris to control the amount of light entering the eye.

The region of the optic cup between the developing retina and the iris eventually forms the inner lining of the ciliary body. These two layers of the cup remain single layers of epithelial cells. The innermost layer remains unpigmented, whereas, the layer below it, (the layer that is continuous with the pigmented epithelium of the retina) becomes pigmented.

Thus, the neural ectoderm produces all of the retina and the optic nerve and also parts of the ciliary body and iris. The rest of the ciliary body and iris develop from mesenchyme. This mesenchyme is derived mostly from cranial neural crest cells.

1.3
Lens Development

It should be noted that all ocular structures develop simultaneously. Descriptions that specify the development of one or other germinal tissues at one time do so for convenience only. Thus, while the optic cup is developing from the forebrain, a similar change is taking place in the surface ectoderm opposite the open lips of the developing optic cups. First, the cells of the surface ectoderm thicken to form a structure known as the lens placode. Rapid growth in this region results in the invagination of the surface cells to form a lens vesicle, which ultimately separates from the surface to form a hollow spherical vesicle. The walls of this vesicle consist of a single layer of epithelial cells. Because of the inward invagination, the basement membrane of these cells is found around the outside of the vesicle. The lumen of the lens vesicle is slowly filled by elongation of cells that form the posterior hemi-sphere of the original vesicle. These become identified as the primary lens fibers and form the embryonic nucleus of the lens. A single layer of cells, known as the epithelium of the lens, remains on the front (anterior) surface. Subsequent growth takes place at a mitotic zone around the equator of the embryonic nucleus. New cells grow over the old fibers and under the epi-thelium to form the secondary lens fibers. The basement membrane thickens and becomes the lens capsule. This structure is, at least until later years, an acellular elastic body that plays an important role in determining the shape of the human lens during accommodation. Thus, the entire lens originates from the surface ectoderm (Fig. 2).

1.4
Mesenchyme

The mesenchymal structures of the eye are formed mainly around the de-veloping optic cup, as noted earlier. First, a vascular and pigmented layer of tissue is laid down to produce a continuous vascular region known as the uvea. The uveal tract consists of the choroid, the vascular and muscular part of the ciliary body, and the anterior vascular region of the iris. A second mesenchymal layer of development overlies these vascular zones to form the fibrous outer tunic of the eye, the sclera and cornea. Eventually, mesodermal/mesenchymal development is responsible for the production of other struc-tures of the orbit, such as the extraocular muscles, the orbital bones, and the bulk of the eyelids.

A temporary embryonic vascular network, the hyaloid circulation, enters the optic cup from the mesenchyme around it through the embryonic fissure. This circulation supplies the early developing lens. Ultimately, the hyaloid vessels are resorbed, often leaving behind debris. With new growth it be-comes the definitive retinal circulation. As the embryonic fissure closes, in the distal to proximal direction, the hyaloid vessels are pushed to the region of the optic stalk. The growth of retinal ganglion cell axons, during the formation of the optic nerve, traps the major artery and vein. These become known as the central retinal vessels.

1.5
Development of the Sclera, Choroid, Ciliary Body, and Iris

As stated above, the development of the sclera, choroid, ciliary body and iris are intimately related to the condensation of mesenchyme cells in layers around the developing optic cup (Jakobiec and Ozanics 1982; Tripathi et al.; Bron et al. 1997). The first condensation of mesenchyme cells around the optic cup will differentiate into the choroidal stroma. Future choroidal dif-ferentiation is directed to the development of the choroidal vasculature. The

SURFACE ECTODERM

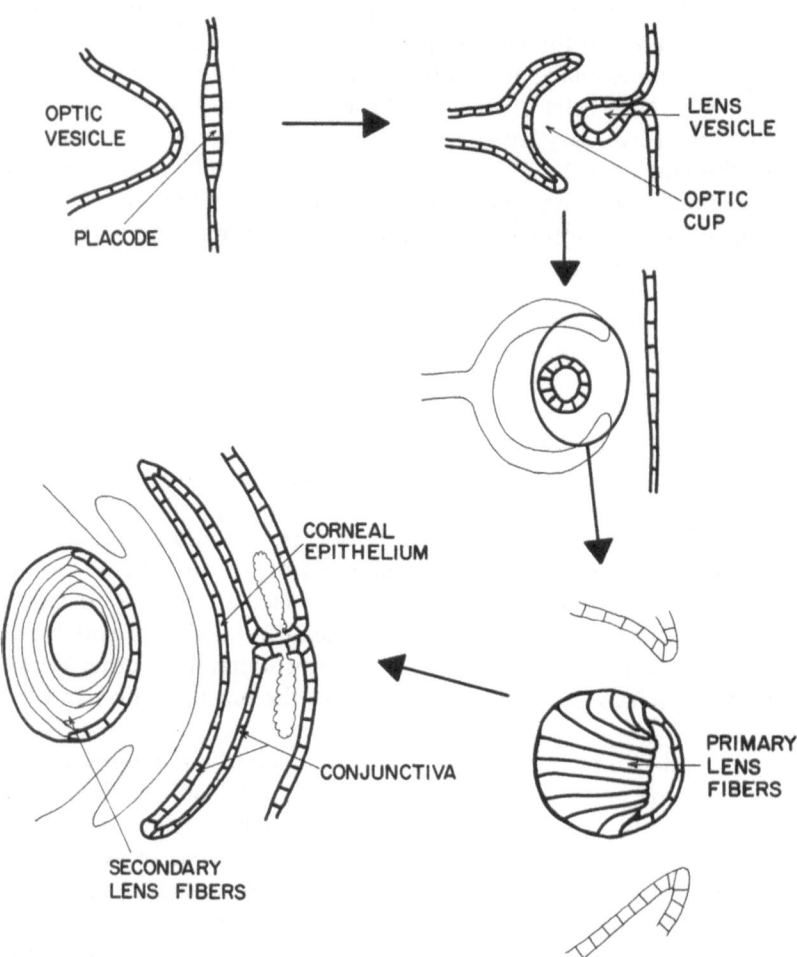

Fig. 2. Developmental sequence showing the formation of the lens, corneal epithelium and conjunctiva from the surface ectoderm (Sivak and Bobier 1990)

sclera forms from mesenchymal cells that condense around the optic cup, beginning in the area of the future limbus and continuing in a posterior direction, thereby surrounding the future choroid.

In contrast to the choroid and sclera, where mainly mesenchymal cells are involved in embryological development, the ciliary body and iris contains neuroectodermal elements from the developing optic cup. The future ciliary epithelium differentiates from the advancing margins of the two layers of neuroectoderm of the optic cup. Further layering of mesenchyme will contribute to the formation of ciliary stroma and muscle. As pointed out earlier, differentiation of the neuroectodermal component of the iris will form the future sphincter, dilator, and iris epithelial cells. The development of the iris stroma is closely associated with changes in the anterior portion of the tunica lentis vasculosa (primitive circulatory supply to the developing lens). The remnants of the atrophying tunica lentis vasculosa on the anterior surface of the lens will fuse with mesenchyme condensing along the anterior part of the optic cup to form the pupillary membrane. This layer of mesenchyme is positioned on an already existing layer of mesenchyme that is being pulled forward by the anterior growth of the edges of the optic cup. The vascular framework of the iris develops along with the iris stroma. Subsequent central atrophy of the pupillary membrane leads to the formation of the collarette between the ciliary and pupillary portions of the iris.

1.6
Retina

Further differentiation of the inner and outer parts of the optic cup will form the functioning retina. The inner part of the optic cup develops into the future nervous retina (Jakobiec and Ozanics 1982; Bron et al. 1997). The primitive neural retina differentiates into two zones: a cellular primitive zone (lying outermost, in close proximity to the outer layer of the optic cup); and an acellular marginal zone (lying innermost to the vitreous humor). The sequence of retinal maturation begins at the posterior pole and proceeds to the periphery. Cells in the primitive zone proliferate and migrate to form inner and outer neuroblastic layers. Between them is a narrow acellular layer, the transient fiber layer of Chievitz. Ganglion cells develop from the innermost part of the inner neuroblastic layer and migrate into the marginal zone of the nervous retina. Cells of the inner neuroblastic layer will also generate amacrine and Muller cells. Cells in the outer neuroblastic layer differentiate into photoreceptors, bipolar cells, and horizontal cell types. The bipolar and horizontal cells migrate toward the inner neuroblastic layer and take a position near the Muller and amacrine cells. The transient fiber layer of Chievitz disappears with the migration of bipolar and horizontal cells.

At this point in development, the cellular layers of the neural retina are in position for connections to be made between the layers. The ganglion cells send out fibers (axons) which grow toward the optic stalk and toward the lateral geniculate body in the thalamus of the brain. Further modification with respect to the movement of ganglion and bipolar cells continues after birth to allow for the direct stimulation of the photoreceptors in the area of the macula and fovea.

The pigmented epithelium of the retina develops from the outer part of the optic cup. Pigmentation occurs earlier in the pigmented epithelium of the retina than in the developing iris, ciliary body, or choroid. The pigmented epithelium of the retina remains one cell layer thick throughout its differentiation.

1.7
Corneal Development

The last significant ocular development involves further rapid growth and invagination of the surface ectoderm (Fig. 2). This results in the formation of the epithelium of the cornea (which fuses to the mesenchymal portions), the epithelium of the conjunctiva (which lines the inner eyelids and is reflected back on to the surface of the globe to continue as the epithelium of the cornea), and the glands of the eyelids.

The embryonic cornea starts out as a layer of cells of the surface ectoderm which reforms over the developing optic cup when the surface attachments of the lens vesicle have disappeared. These cells are joined by an internal layer of mesenchymal material which develops in concert with the condensation of mesenchymal tissue around the developing optic cup. This tissue, Descemet's mesothelium, will lead to the formation of Descemet's membrane and the endothelium of the cornea. As the surface ectodermal cells become stratified, a second mesenchymal layer develops below them to form the future stroma of the cornea.

2
Eye Development and Optical Function

2.1
Cornea

In the human (terrestrial) eye, both the cornea and lens are responsible for focusing an image on to the retina. Both have the same basic problem: how to maintain adequate physiological conditions for living tissue, while at the same time providing the image quality of a good optical device? Thus, both structures have evolved metabolic processes adequate for avascularity. Anatomically, both the lens and the cornea exhibit adaptations designed to minimize light scatter. Nevertheless, despite these and other similarities, differences in development (and location) have far-reaching consequences.

The cornea can be described as a collagen sandwich with epithelial tissue on both sides of a collagen core (the stroma) (Duke Elder 1958; Hogan et al. 1971). The external epithelium consists of stratified squamous epithelial cells, about five cell layers thick. A mitotic basal layer of collumnar cells rests on a basement membrane while the superficial squamous layer is subject to a process of erosion, or sloughing. The full cycle of regeneration, from bottom to top, takes about 7 days. That the corneal epithelial cells adhere tightly to one another by means of desmosomes and other junctional devices is an important physiological factor that also minimizes light scatter. Unlike the epidermis, the superficial and deep epithelial surfaces are relatively smooth and parallel to one another.

The innermost layer of the cornea, the endothelium, consists of a single row of epithelial cells. In the adult eye, these cells do not exhibit a measurable regenerative cycle. The rest of the cornea (Bowman's membrane, stroma and Descemet's membrane) is made up of collagen fibers. The few cells, mainly fibroblasts, are few and far between. In the stroma, the collagen fibers are of uniform thickness and arranged parallel to the corneal surface. The characteristic regularity of interfiber spacing is believed to be necessary to produce destructive interference, a prerequisite for corneal transparency. The corneal stroma has a very slow rate of turnover, if any (Rodrigues et al. 1982).

The cornea is an external ocular structure which represents the optical interface between the eye and the external environment (Fig. 3). The external and internal optical surfaces are roughly parallel, so the refractive function of the cornea is a result of the difference in refractive index between the medium in front (air or water) and the medium behind (aqueous humor). Since the refractive index of aqueous humor is 1.335 and that of water is 1.333, corneal refractive power is virtually nil when the eye is in water (Sivak 1980, 1990). This is true despite the fact that the overall refractive index of the cornea (1.376) is appreciably greater than that of water.

In view of the aquatic habitat of early vertebrates, it is obvious that the corneas of early species were simply transparent windows. This is reflected in the fact the lenses of fishes and aquatic mammals, the only refractive elements of these eyes, are spherical in shape and very high in overall refractive index (1.65). When the eye is in air, the refractive contribution of the cornea is considerable, due to the difference in refractive indices of air and aqueous humor. Thus, the refractive power of the human cornea (in air) is about double that of the lens (Bennett and Francis 1962).

The external location of the cornea is also important because the older cells of the only part to undergo appreciable regeneration, the epithelium, can be lost and are not retained. Thus, basal collumnar cells reach the surface as

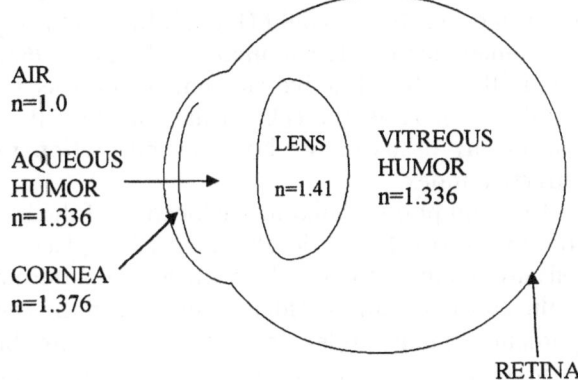

Fig. 3. Schematic representation of the human eye emphasizing the relative locations of the lens and cornea, their refractive indices (*n*) and those of the media which these two refractive elements contact. The external location of the cornea gives it a refractive advantage over the internally located lens

AIR
n=1.0

AQUEOUS
HUMOR
n=1.336

CORNEA
n=1.376

LENS
n=1.41

VITREOUS
HUMOR
n=1.336

RETINA

squamous cells in roughly 7 days and are sloughed off to the external environment.

2.2
Lens

Like the cornea, lens morphology is also characterized by regularity of structure (Hogan et al. 1971), but the lens is cellular. As already noted, a single layer of epithelium covers its anterior surface, and very elongated cells (fibers) fill the interior, extending to the posterior surface. The fibers are arranged in concentric cells of increasing diameter (Worgul 1982). Their size and shape determine the ultimate size and shape of the lens (Sivak 1985).

The lens fibers contain relatively few organelles and those in the central core lack nuclei. These adaptations are important for the maintenance of transparency. The fibers are interconnected by means of numerous ball-and-socket articulations, which help produce a highly ordered geometric pattern. The tight articulations between fibers reduce intercellular space and minimize light scatter.

In many species, particularly in mammals, the secondary lens fibers do not taper sufficiently to meet at points at the anterior and posterior poles. Rather, the fiber ends form lines or sutures of varying complexity, extending in depth into the lens, depending on the species or age of the mammal. For example, the sutures of the early postembryo human lens are Y-shaped, erect anteriorly, and inverted posteriorly. The adult human lens can exhibit a more complex nine branched suture arrangement. Paradoxically, the sutures are located along the optical axis of the lens, where they can most affect the optical quality of the eye, as shown in a series of correlative morphological and optical studies (Kuszak et al. 1991, 1994; Sivak et al. 1994).

The fact that the lens is located within the eye and that it is made up of cells of surface origin that continue to multiply through life creates a unique set of circumstances: because the lens is surrounded by the humors of the eye, it must have a refractive index that is substantially greater than that of water in order to focus light (Fig. 3). In fact, the equivalent refractive index of the human lens is 1.41 (Bennett and Francis 1962), while in fishes, where the lens is the only refractive element of the eye, the index is as high as 1.65 (Sivak 1990). Thus, the cells of the lens have the highest protein concentration of any tissue of the body, an adaptation to the need for an elevated refractive index.

A second point related to the location of the lens, as well as to the fact that continued growth and development takes place peripherally, is that old lens cells are retained. In fact, the growth ring pattern of lens development results in the concentration of older tissue toward the center and the formation of a gradient refractive index, the index at the center being higher than that at the periphery. This gradient plays an important optical role in reducing the optical aberrations of the eye, particularly spherical aberration (Bennett and Francis 1962; Sivak 1990).

A final point to be mentioned is that in many species the lens provides the eye with a variable focus mechanism, accommodation. (In at least some birds the cornea is also involved in accommodation.) In humans, accommodative change in lens shape takes place in response to neural directives given to the ciliary muscle/zonular apparatus of the eye. Presbyopia, the loss of accommodation with age, is a consequence, at least in part, of continued lens development.

3
Visual Environment and Eye Development

The optical properties of the eye are determined by the size and shape of the ocular globe, the curvature of the cornea, and the focal properties and position of the lens. In spite of over 100 years of intense study, the specific cause(s) of refractive errors are still unknown. While it is commonly assumed that postnatal ocular development is directed by both genetic and environmental factors, their relative importance is not clear. Research using animal models of eye development carried out during the past two decades have shown that early visual experience is of critical importance, at least in birds and mammals. This work has been carried out with a variety of species, including monkeys, cats, tree shrews, rabbits, etc. (Irving et al. 1995). However, the majority of studies involve the use of young chickens because the chick eye is large and relatively easy to measure, ocular development is very rapid (with most of the eye's growth taking place within 2 weeks of hatching) and, since chickens are precocial birds, the manipulation of the visual environment can begin as soon as they hatch.

The earlier experiments generally involved depriving the retina of the eye of a clear image, either by suturing the eyelids together or by applying some kind of transluscent or light-scattering device over the eye during early posthatching development (e.g, Pickett-Seltner et al. 1987). This form of treatment invariably leads to varying degrees of myopia, or nearsightedness, and has been referred to as form deprivation or experimental myopia. The myopia is a result, primarily, of excessive eye growth. Myopia can also be produced by raising chicks under constant light, while hyperopia, or farsightedness, can result from reduced growth (and reduced axial diameter) associated with constant darkness.

More recently, work has shown that it is possible to induce nearsightedness and farsightedness in young chicks by defocusing the retinal image with hood-mounted convex and concave spectacle lenses (Schaeffel et al. 1988). The range of induced refractive errors was increased through the use of a lightweight goggle/lens apparatus mounted over the cornea of the eye

(Irving et al. 1992, 1995). By making the young eye artificially nearsighted or farsighted with convex or concave lenses (Fig. 4), the eye grows faster or slower to compensate. In only 1 week or less (from the first day of hatching) the eye (with the goggle/lens removed) becomes myopic or hyperopic by an amount equal to the power of the inducing lens, up to limits of about plus or minus 20 diopters. Removal of the inducing apparatus results in a return to little or no refractive error after 1 or 2 weeks. The choroid can play a short-term role by rapidly decreasing the axial length of the eye when eye growth is slowed in response to a convex inducing lens or when the eye is recovering from induced myopia. While these refractive changes are primarily caused by change in the vitreous chamber depth of the eye, corneal flattening is found associated with higher levels of hyperopia (Irving et al. 1995) and subtle changes in lens focal properties (spherical aberration) have been noted with hyperopia and myopia (Priolo et al. 1999). Astigmatic refractive effects can also be induced by using cylindrical convex and concave inducing lenses. While the size of the blur caused by the inducing lenses provides the information necessary to correct for the size of the defocus, there is as yet no adequate explanation for how the eye compensates for the sign of the defocus.

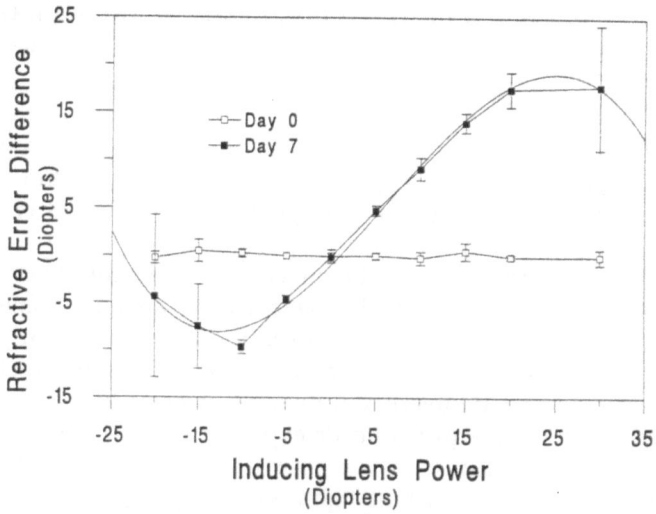

Fig. 4. Effect, in diopters, of 7 days of varying amounts of unilateral convex (convergent) or concave (divergent) defocus on the difference in refractive error of the treated and untreated eyes of a total 69 young chicks . The *open symbols* refer to the refractive difference on the day of hatching (day 0), while the *filled symbols* represent the refractive difference measured after 7 days of unilateral defocus. The *error bars* represent the variability as standard deviation. The relationship between inducing defocus and induced refractive error is described as a sigmoid function given by the third order polynomial $y = -0.968 + 0.975x + 0.019x^2 - 0.001x^3$, $r = 0.994$ (Irving et al. 1992)

   Additional studies have concentrated on determining the retinal mechanisms and retinal chemistry associated with induced refractive errors. Dopamine and its associated compounds are common retinal transmitter chemicals that have been implicated in refractive development from studies such as that of Guo et al. (1995), who reported that retinal dopamine levels are reduced, relative to control retinas, in myopic chick eyes, and elevated in hyperopic ones. More recently, attention has been drawn to the potential involvement of the retinal amacrine cells in controlling ocular refractive development, at least in chicks, and to the possible involvement of additional transmitter substances such as seratonin (Wong 1998).

   These studies are but a few of the very many that have followed the early work that highlighted the refractive plasticity of the eye during early development (Wallman et al. 1978). While it is still too early to see practical results from these efforts, the continuing increase in the incidence of ocular refractive problems in humans, particularly the alarming increase in the percentage of myopes among Asian populations, will ensure a high level of continued interest, particularly in the relationship between genetics and environment in controlling the refractive development of the eye.

References

Bennett AG, Francis JL (1962) The eye as an optical system. In: Davson H (ed) Visual optics and the optical space sense. The Eye vol 4. Academic Press, New York, pp 103–132

Bron AJ, Tripathi RC Tripathi BJ (1997) Wolff's anatomy of the eye and orbit, 8th edn. Chapman and Hall, London

Duke-Elder S (1958) System of opthalmology. vol II. The anatomy of the visual system. Henry Kimpton, London

Guo SS, Sivak JG, Callender MG, Diehl-Jones W (1995) Retinal dopamine and lens induced refractive errors in chicks. Curr Eye Res 14:385–389

Hogan MJ, Alvarado JA, Weddell JE (1971) Histology of the human eye. WB Saunders, Philadelphia

Irving EL, Sivak JG, Callender MG (1992) Refractive plasticity of the developing chick eye. Ophthalmic Physiol Opt 12:448–456

Irving L, Callender MG, Sivak JG (1995) Inducing ametropias in hatchling chicks by defocus-aperture effects and cylindrical lenses. Vision Res 9:1165–1174

Kuszak JR, Sivak JG, Weerheim JA (1991) Lens optical quality is a direct function of lens sutural architecture. Invest Ophthalmol Visual Sci 32:2119–2129

Kuszak JR, Peterson KL, Sivak JG, Herbert KL (1994) The interrelationship of lens anatomy and optical quality. II. primate lenses. Exp Eye Research 59:521–535

Mann I (1969) The development of the human eye. Grune and Stratton, New York

Noden DM (1982) Periocular mesenchyme: neural crest and mesodermal interactions. In: Jakobiec FA (ed) Ocular anatomy, embryology, and teratology. Harper and Row, Philadelphia, pp 97–119

Jakobiec FA, Ozanics V (1982) Prenatal development of the eye and its adnexa. In: Jakobiec FA (ed) Ocular anatomy, embryology and tertology. Harper and Row, Philadelphia, pp 11–96

Jakobiec FA, Ozanics V (1982) General topographic anatomy of the eye. Arch. Ophthalmol, 1, 1

Pickett-Seltner RL, Weerheim J, Sivak JG, Pasternak JJ (1987) Experimentally induced myopia does not affect post-hatching development of the chick lens. Vision Res 27:1779–1782

Priolo S, Sivak JG, Irving EL, Callender MG, Moore SE (1999) Effect of age and experimentally induced ametropia on the optics and morphology of the avian crystalline lens. Vision science and it's applications, Optical Society of America, Technical Digest Series, vol 1, pp 88–91

Rodrigues MM, Warring GO III, Hackett J, Donohoo P (1982) Cornea. In: Jakobiec FA (ed) Ocular anatomy, embryology and teratology. Harper and Row, Philadelphia pp 153–165

Schaeffel F, Glasser A, Howland HC (1988) Accommodation, refractive error and eye growth in chickens. Vision Res 28:639–657

Sivak JG (1980) Accommodation in vertebrates: a contemporary survey. In: Davson H, Zadunaisky J (eds) Current topics in eye research, vol. 3. Academic Press, New York, pp 281–330

Sivak JG (1985) Environmental influence on shape of the crystalline lens: the amphibian example. Exp Biol 44:29–40

Sivak JG (1990) Optical variability of the fish lens. In: Douglas RH, Djamgoz MBA (eds) The visual system of fish. Chapman and Hall, London, pp 63–80

Sivak JG, Bobier WR (1990) Optical components of the eye. In: Rosenbloom AA, Morgan MW (eds) Principles and practice of pediatric optometry. JB Lippencott, Philadelphia, pp 31–45

Sivak JG, Herbert KL, Peterson KL Kuszak JR (1994) The interrelationship of lens anatomy and optical quality. 1. Non-primate lenses. Exp Eye Res 59:505–520

Tripathi BJ, Tripathi RC, Wisdom J (1995) Embryology of the anterior segment of the human eye. In: Rich R, Schields MB, Krupin T (eds) The glaucoma. 2nd edn. Mosby, St Louis

Wallman J, Turkel J, Trachtman J (1978) Extreme myopia produced by modest changes in visual experience. Science 201:1249–1251

Walls GL (1942) The vertebrate eye and its adaptive radiation. Cranbrook Institute of Science, Bloomfield Hills, Michigan

Wong STY (1998) Effects of induced myopia and myperopia on dopaminergic and seratonergic amcrine neurons in chick retina. Doctoral Dissertation, University of Waterloo, Waterloo, Ontario, 167 pp

Worgul BV (1982) The lens. In: Jakobiec F A (ed) Ocular anatomy, embryology and tertology. Harper and Row, Philadelphia, pp 355–389

Pax6 and the Genetic Control of Early Eye Development

Stefan Wawersik[1], Patricia Purcell[1], and Richard L. Maas[1]

1
Introduction

While lens induction in vertebrates has been the subject of numerous embryological experiments, very little is known about the molecular identities of the regulators involved. Until recently, one of the few exceptions was the transcription factor *Pax6*, a member of the vertebrate paired box family, which is required in the head ectoderm for lens induction to proceed. Initially identified in mouse (Walther and Gruss 1991), *Pax6* homologues were subsequently isolated in human (Ton et al. 1991), rat (Matsuo et al. 1993), chick (Li et al. 1994), *Xenopus* (Altman et al. 1997), zebra fish (Krauss et al. 1991), ascidians (Glardon et al. 1997), cephalopods (Tomarev et al. 1997), *C. elegans* (Chisholm and Horvitz 1995; Zhang and Emmons 1995), and *Drosophila* (Quiring et al. 1994). In all of these organisms, the amino acid sequence of the Pax6 protein is highly conserved, arguing for functional importance across metazoan phyla.

While molecular biological and biochemical studies have offered insights into *Pax6* function and regulation, it is the gene's evolutionary conservation that has best illuminated the biological role of *Pax6* in eye development. Experiments in *Drosophila* have begun to elucidate the *Pax6*-dependent genetic pathway controlling eye formation, and homologues of fly optic development genes have been identified in vertebrates. The theme emerging from this work is that not only *Pax6*, but an entire genetic cascade has been highly conserved in the regulation of eye development.

Here, we will review the function of the *Pax6* gene in vertebrate oculogenesis. Initially, we will concentrate on studies that have led to an understanding of the role played by *Pax6* in vertebrate eye formation, the mechanism by which *Pax6* activates transcription, and how the *Pax6* gene itself is regulated. We will then focus on *eyeless*, one of two fly *Pax6* homologues, in eye development and how studies in *Drosophila* are providing

[1] Division of Genetics and Department of Medicine, Brigham and Women's Hospital and Graduate Program in Biological and Biomedical Sciences, Harvard Medical School, Boston, Massachusetts 02115, USA

Results and Problems in Cell Differentiation, Vol. 31
M. E. Fini (Ed.): Vertebrate Eye Development
© Springer-Verlag Berlin Heidelberg 2000

information for the examination of the *Pax6* genetic cascade in vertebrates. These studies demonstrate how the identification of *Pax6* as a pivotal regulator of eye development has allowed the analysis of lens induction to progress to the level of gene regulation, and has thus fueled an understanding of the molecular events underlying ocular development.

2
Pax6 Plays a Critical Role in Vertebrate Ocular Development

In mammals, the *Pax* gene family consists of nine known members unified by the presence of a paired domain. The paired domain is a 128-amino acid DNA-binding domain named after the prototypical *Drosophila* segment polarity gene *paired* in which it was first identified (Bopp et al. 1986; Burri et al. 1989; Walther and Gruss 1991). Haploinsufficiency for *PAX6* function in humans results in aniridia, a heritable panocular disorder characterized by iris and foveal hypoplasia which can also be accompanied by cataracts, corneal opacification, and progressive glaucoma (reviewed by Glaser et al. 1995). Homozygous *PAX6* mutations result in anophthalmia, nasal hypoplasia, and central nervous system defects (Glaser et al. 1994). In mouse, a naturally occurring mutation, *Small eye* (*Sey*), results from mutations in the murine *Pax6* gene (Hill et al. 1991). Like aniridia, *Sey* is inherited in a semidominant fashion, with *Sey/+* heterozygotes exhibiting corneal and lenticular abnormalities and *Sey/Sey* homozygotes lacking eyes entirely (Hogan et al. 1986). The close similarities in phenotype and mode of inheritance between aniridia and *Small eye* suggest that *Pax6* functions equivalently in both human and murine ocular development.

Development of the vertebrate eye begins with the formation of an outpouching of the diencephalon known as the optic vesicle (Fig. 1A). The optic vesicle eventually comes into close proximity with the surface ectoderm of the head and this interaction leads to the induction of a pseudostratified thickening of the ectoderm known as the lens placode (from Greek *plakos*, meaning flat; Fig. 1B, C) (Pei and Rhodin 1970; Zwaan et al. 1969; Zwaan and Pearce 1971). The lens placode subsequently invaginates to form the lens vesicle (Fig. 1D, E), while the optic vesicle invaginates around it to form the double-walled optic cup (Fig. 1D). The inner layer of the optic cup forms the neural retina, while the outer layer eventually forms the pigmented epithelial layer (Fig. 1E, F).

Examination of E8.5-10 *Sey/Sey* mouse embryos reveals that the lens and nasal placodes fail to form, the optic vesicle fails to constrict and subsequently degenerates, and mesenchymal cells become interposed between the surface ectoderm and the optic vesicle (Hogan et al. 1988). During mouse development, *Pax6* is expressed in both the head ectoderm and the optic vesicle at E8.5, prior to lens induction (Grindley et al. 1995). In the chick,

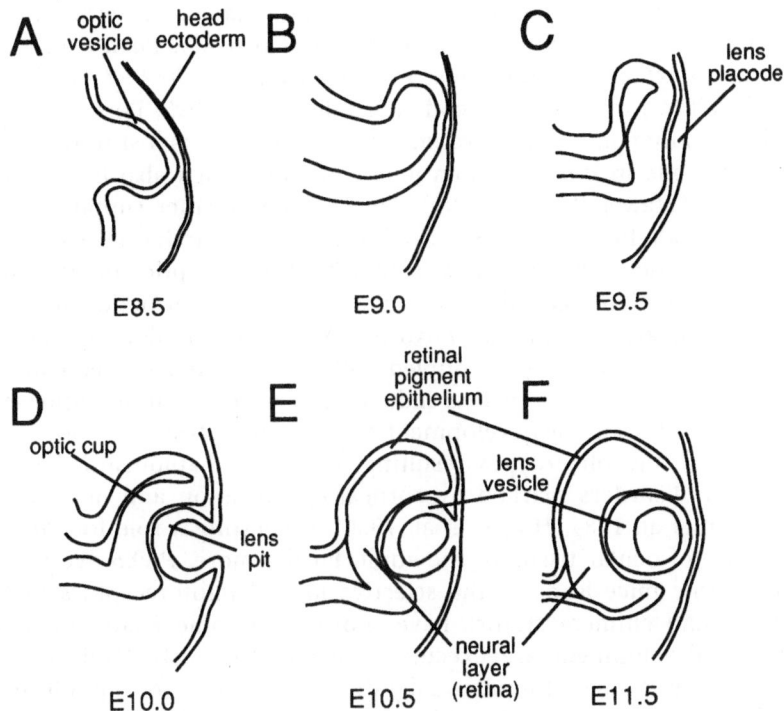

Fig. 1A–F. Induction of the murine lens. The optic vesicle forms as an outpouching of the forebrain at E8.5 (**A**). The optic vesicle comes in contact with the ectoderm of the head at E9.0 (**B**), and signals from the optic vesicle induce the lens placode to form by E9.5 (**C**). At E10.0, the lens placode invaginates to form a lens pit, while the optic vesicle folds into itself to create the optic cup (**D**). The invagination of the lens pit to form the lens vesicle is complete by E10.5, at which time the vesicle begins to separate from the overlying ectoderm (**E**). Differentiation of the two layers of the optic cup continues, with the inner layer forming the neural layer of the retina and the outer layer comprising the retinal pigment epithelium (**E, F**)

similar head ectodermal expression of *Pax6* is detected in the early embryo, and this expression is independent of contact with the optic vesicle (Li et al. 1994). Experiments using a rat *Small eye* mutation (r*Sey*) indicate a requirement for *Pax6* only in the surface ectoderm and not the optic vesicle for lens induction: tissue recombinations of r*Sey*/r*Sey* optic vesicle with wild-type ectoderm in organ culture form lenses at wild-type frequency, while the reciprocal combination fails (Fujiwara et al. 1994). In addition, injections of *Pax6* RNA into *Xenopus* embryos induce ectopic lenses, the majority without associated neural tissue (Altmann et al. 1997; Chow et al. 1999). These results suggest that *Pax6* function is required in the prospective lens ectoderm for lens induction, where it appears to function in a tissue-autonomous fashion.

At E9.5, prior to thickening of the head ectoderm to form the lens placode, *Pax6* expression becomes restricted to the presumptive placodal region, suggesting that *Pax6* expression is regulated in response to inductive signaling from the optic vesicle. This regulation may be due either to repression of nonplacodal *Pax6* or to maintenance of placodal *Pax6* expression. Ectodermal *Pax6* expression is lost in *Sey/Sey* and in *Lhx2 −/−* embryos, which both fail to form a lens placode (Grindley et al. 1995; Porter et al. 1997). In addition, though initially present, *Pax6* expression is lost in the presumptive lens placode in *Bmp7* homozygous embryos, which also fail to form a lens placode (Wawersik et al. 1999). These results further support the idea that lens placode formation is required to maintain ectodermal *Pax6* expression.

In later development, *Pax6* is expressed in the optic cup as well as in the invaginating lens placode. While the lack of lens formation in *Pax6* homozygotes makes it clear that *Pax6* is critical for lens development, the early nature of this phenotype coupled with the fact that subsequent eye development is dependent on lens induction precludes determination of the role played by *Pax6* in the development of other optic structures. Both the aniridia and *Small eye* phenotypes resulting from *Pax6* haploinsufficiency are progressive disorders marked by corneal opacification and neovascularization (Nelson et al. 1984; Hogan et al. 1986), suggesting a role for *Pax6* in maintenance of ocular tissue in the adult. Furthermore, *PAX6* overexpression in transgenic mice leads to the selective loss of photoreceptors (Schedl et al. 1996), and chimera studies have shown that some *Pax6* functions during retinal development are executed cell autonomously (Quinn et al. 1996). These results suggest a potentially direct role for *Pax6* in photoreceptor cell differentiation. Finally, *Pax6* has been implicated in the regulation of lens crystallins (reviewed by Cvekl and Piatigorsky 1996), suggesting that, in addition to a role in early lens patterning, *Pax6* may also function in later stages of lens development.

3
Molecular Biology of the *Pax6* Gene Product

Both the human and mouse *Pax6* genes encode identical 422 amino acid proteins with four distinct domains. An N-terminal paired domain (PD) and a centrally located homeodomain (HD) both function in DNA binding, and are separated by an intervening glycine-rich linker (Epstein et al. 1994a; Wilson et al. 1993, 1995). In addition, at the C-terminus of the protein is a proline-, serine-, and threonine-rich (PST) domain that has been shown to activate transcription, consistent with studies demonstrating that *Pax6* can act as a transcriptional regulator (Epstein et al. 1994b; Glaser et al. 1994).

The 128-amino acid paired domain (PD) is shared by all nine members of the Pax gene family and was originally identified in mouse through its high homology to the *Drosophila* segmentation gene *paired* (Walther and Gruss 1991). Both the *Drosophila* and mouse PDs are organized into distinct N- and C-terminal subdomains (also termed PAI and RED, respectively) each consisting of 3 α-helices arranged in a helix-turn-helix motif (Xu et al. 1995; Xu et al. 1999). Each PD subdomain encodes a sequence specific DNA binding activity (Treisman et al. 1991; Czerny et al. 1993; Epstein et al. 1994b; Jun and Desplan 1996). For fly *paired*, though both the N- and C-terminal subdomains are conserved, only the N-terminal domain is required for DNA binding (Treisman et al. 1991; Xu et al. 1995). However, mutational analysis of the vertebrate Pax genes suggests that their DNA binding involves both the N- and C-terminal subdomains (Czerny et al. 1993; Epstein et al. 1994b; Azuma et al. 1996; Tang et al. 1997). A recent cocrystal of the *PAX6* PD with a 26-bp DNA recognition site at 2.3 Å resolution confirms the essential details of this bipartite model, showing that both the N- and C-terminal subdomains make contacts with DNA (Xu et al. 1999).

Several mutations have been identified which affect only the PD (see http: //hgu.mrc.ac.uk/Softdata/PAX6/ for a summary of *PAX6* mutations in human). One of these is a missense mutation that creates a cryptic splice site, leading to the deletion of 36 amino acids from the PD (Hanson et al. 1993). This eliminates most of the C terminal subdomain but leaves the rest of the protein intact. The phenotypic outcome of heterozygosity for this mutation is aniridia, equivalent to deletion of the entire protein. Several other missense mutations have been identified which alter the PAX6 PD. Heterozygosity for these mutations leads to aniridia or to slightly milder phenotypes (Hanson et al. 1993, 1994; Azuma et al. 1996; Tang et al. 1997), indicating that the PD is critical to *Pax6* function.

An additional PD mutation helped corroborate the bipartite nature of the PD. *Pax6* contains an alternatively spliced exon known as exon 5a which introduces a 14-amino acid insertion into the PD, radically altering its DNA-binding specificity. By eliminating the ability of the PD N-terminal subdomain to bind DNA, the exon 5a insertion unmasks a latent DNA-binding activity of the C-subdomain, changing the sequence specificity of Pax6 (Epstein et al. 1994b). In the pedigree mentioned above, mutation of the exon 5a splice acceptor site alters the ratio of *Pax6* to *Pax6-5a* transcripts, leading to an aniridia-like phenotype marked by mild iris hypoplasia, glaucoma, and cataracts (Epstein et al. 1994b). In addition, RT-PCR analyses suggest that relative levels of *Pax6* vs. *Pax6-5a* transcripts are regulated in a tissue- and temporal-specific manner (Richardson et al. 1995; Koroma et al. 1997). A functionally similar insertion in the PD has also been observed in the *Pax8* gene product, and it has been suggested that these insertions may act as molecular toggles, in effect increasing the diversity of target genes which can be recognized (Epstein et al. 1994b; Kozmik et al. 1997).

Despite its importance in *Pax6* function, the high degree of amino acid conservation throughout the protein suggests that the PD is not the only domain required for functional activity. In addition to the PD, four of the nine known Pax genes also encode a DNA-binding homeodomain (HD, reviewed by Callaerts et al 1997). The 60-amino acid paired-class HD is characterized by a serine residue at position 50 and binds DNA of the palindromic sequence 5'-TAAT(N)$_{2-3}$ATTA-3' through cooperative dimerization mediated by the HD (Wilson et al. 1993, 1995). In both *C. elegans* and quail, *Pax6* transcripts have been identified which encode the HD and not the PD, and the *C. elegans* PD-less proteins have been shown to be important for normal development (Carrière et al. 1995; Chisholm and Horvitz 1995; Zhang and Emmons 1995).

Though not well studied in the context of vertebrate *Pax6*, the ability of the HD to cooperate with the PD in DNA binding has been examined in other members of the Pax family and in the *Drosophila paired* gene (Bertuccioli et al. 1996; Jun and Desplan 1996; Miskiewicz et al. 1996; Fortin et al. 1997; Jun et al. 1998). In the fly, both the HD and the PD's N-terminal subdomain are required for normal *paired* function (Bertuccioli et al. 1996; Miskiewicz et al. 1996), though *paired* mutations abolishing the ability of either of these domains to bind DNA do not affect the DNA-binding function of the other domain in vitro (Treisman et al. 1991). However, using an in vitro site-selection method, Jun and Desplan (1996) have identified several composite binding sites for the PD and HD which suggest that they may interact to varying degrees during DNA binding. Such a composite site has been identified in an enhancer of the *even-skipped* gene which is regulated by *paired* (Fujioka et al. 1996). Mutations in either the PD or the HD sites or in the spacing between them lead to dramatic increases in the expression of *even-skipped*, suggesting a functional interaction between the PD and the HD in vivo. Analysis of mutations in the mouse *Pax3* PD and HD also indicate that the two domains influence each others' DNA-binding activity (Underhill and Gros 1995, 1997; Fortin et al. 1997), and it appears that the interaction between these domains primarily involves the N-terminal PD subdomain, particularly the second helix (Fortin et al. 1998). These studies also show that the linker domain between the PD and the HD is important in DNA binding, as alterations of this region influence the functional interaction between the PD and the HD.

Whether the Pax6 PD and HD interact in DNA binding remains to be determined. A missense mutation leading to an amino acid substitution two residues before the start of the homeodomain suggests that the HD is important in Pax6 binding (Hanson et al. 1993). The result of this mutation is the disruption of a short amino acid motif (LKRK) which is conserved in all known proteins containing both a PD and an HD. Though the role of this conserved region is uncertain, it has been speculated that it is an extension of the N-terminal arm of the homeodomain and may provide sequence-specific contacts between the HD and its DNA target. Alternatively, this mutation may affect the linker domain between the PD and HD in a fashion analogous to that seen in *Pax3* (Fortin et al. 1998), or it may alter a nuclear localization signal, as it has also been shown to affect Pax6 nuclear localization (Glaser et al. 1995). In addition, a mutation in the mouse *Pax6* HD has recently been identified which leads to a hypomorphic *Small* eye phenotype (J. Favor, pers. comm.). Clearly, though these mutations suggest the involvement of the HD in Pax6 function, a full understanding of the role of the *Pax6* HD requires further investigation.

While the PD and HD are both DNA-binding domains (Epstein et al. 1994a; Wilson et al. 1993, 1995), the C-terminal end of the *Pax6* gene product encodes a proline-, serine-, and threonine-rich region known as the PST domain. Concordant with the fact that this domain is similar to those found in other known transcriptional activators (Mermod et al. 1989; Theill et al. 1989), the *Pax6* PST region can activate transcription of a reporter when fused to a GAL4 DNA-binding domain (Glaser et al. 1994). This is consistent with the fact that both Pax6 and Pax6-5a activate transcription through their respective binding sites (Epstein et al. 1994a, b), and indicates that the PST domain is sufficient for *Pax6*-mediated transcriptional activity.

The mechanism of transcriptional activation by the *Pax6* PST region is unknown. Molecular dissection studies indicate that the transactivation function is spread across the four constituent exons comprising the PST domain, rather than being localized to specific amino acid residues (Tang et al. 1998). The domains encoded by these four exons appear to act synergistically to produce the full activity of the PST domain. Thus, mutations truncating all or part of the PST domain lead to aniridia in humans and *Small eye* in mouse (Hill et al. 1991; Glaser et al. 1994). Interestingly, truncations of the PST domain appear to increase the binding affinity of the *Pax6* PD and HD with DNA and can therefore act as dominant negatives, perhaps due to changes in protein folding (Singh et al. 1998). However, this dominant negative effect cannot fully explain the haploinsufficiency seen in *Pax6* mutants, as truncations of the PST domain do not lead to a more severe phenotype than mutations which affect only the PD or HD (reviewed by Prosser and Van Heyningen 1998).

4
Genomic Structure and Transcriptional Regulation of *Pax6*

Much less is known about the regulation of the *Pax6* gene than about its DNA-binding and transactivation properties. In human, mouse, and quail, *Pax6* mRNA comprises 15 exons, the first four of which are noncoding (Glaser et al. 1992; Plaza et al. 1995a; Xu et al. 1999a). Several of these constituent exons are alternatively spliced: at least five different *Pax6* products have been characterized in quail (Carrière et al. 1995), including several without a PD. In addition, as was previously discussed, the inclusion of exon 5a in *Pax6* transcripts alters the DNA-binding properties of the PD (Epstein et al. 1994b). Though the transactivation activities and expression patterns of most of these *Pax6* isoforms have been examined, their biological functions are as yet unclear.

Pax6 transcripts are initiated from one of two promoters, termed P0 and P1, which were initially identified in the quail homologue (Plaza et al. 1995a).

Transactivation from P0 gives transcripts that include exon 0 as their initial exon, while P1-initiated transcripts start with exon 1. Subsequently, similar regulation by multiple promoters has been identified in the mouse and human *Pax6* genes (Xu et al. 1999a; Kammandel et al. 1999; Okladnova et al. 1998). RNAse protection analysis of quail P0- and P1-initiated mRNAs indicates temporally distinct expression of the two transcripts in the developing eye (Plaza et al. 1995a). In situ hybridizations with P0- and P1-specific riboprobes show that these transcripts are differentially expressed during mouse oculogenesis. Early in development, P0 transcripts are detected primarily in the lens placode, while P1 transcripts are found in both the lens placode and the optic vesicle. At later stages, both transcripts are present in lens and retina (Xu et al. 1999a). Within the P0 promoter, a 527 bp DNA fragment with a highly conserved 341 bp core element has been identified which is necessary and sufficient to drive reporter gene expression in the lens placode and lens starting at E8.75 of development (Williams et al. 1998), and this fragment has subsequently been narrowed to 107 bp (Kammandel et al. 1999). In addition to the P0 lens-specific element, a second region in the P0 promoter directs expression in a subset of retinal progenitor cells which becomes detectable at E11.5 (Xu et al. 1999a). P1-initiated expression appears to be involved in later stages of *Pax6* expression, as the P1 promoter drives expression in a subpopulation of retinal cells starting at E12.5. A 200 bp fragment of P1 is necessary for this expression, though it is not clear whether this element is sufficient for such regulation (Xu et al. 1999a). Together, these findings suggest that while both the P0 and P1 promoters are involved in the regulation of ocular *Pax6* expression, it is the P0 promoter that is critically important for early lens-specific expression.

Plaza et al. (1995b) have also identified a phylogenetically conserved intronic enhancer located 7.5 kb downstream of the P0 promoter. This enhancer functions in a position- and orientation-independent manner to drive reporter expression from a heterologous promoter in quail neuroretina cells. A potentially homologous 500 bp fragment from intron 4 of the mouse *Pax6* gene containing this 216 bp enhancer conferred expression in amacrine cells, ciliary body, and iris when placed upstream of either the P0 or P1 promoters. These expression patterns were identical with either promoter but distinct from those of either P0 or P1 alone (Kammandel et al. 1999; Xu et al. 1999a). These results show that the intron 4 region can function as an eye-specific enhancer and highlight the modular organization of *Pax6* regulatory elements.

While the factors that directly bind and regulate *Pax6* are as yet unknown, several in vivo effectors of *Pax6* expression have been identified. Overexpression of *Sonic hedgehog (Shh)* constricts the *Pax6* expression domain in the zebra fish diencephalic forebrain and optic vesicle (Ekker et al. 1995; MacDonald et al. 1995). However, it is unclear whether this regulation is direct, as it occurs in conjunction with expansion of the adjacent *Pax2*-positive domain. *Shh* concentrations have also been correlated with the graded expression of *Pax6* in ventral spinal cord, though again, the effect may be indirect (Ericson et al. 1997). Potential regulation by members of the TGFβ family of growth factors is suggested by the finding that Activin βA can activate *Pax6* expression in ventral spinal cord explants and that BMP7 is required to maintain *Pax6* expression in the lens placode (Pituello et al. 1995; Wawersik et al. 1999). Lastly, Fickenscher et al. (1993) and Gajovic et al. (1997) have shown that *Pax2* and *Pax6* are regulable by retinoic acid (RA) in P19 embryonic carcinoma cells and in ES cells, respectively, and further evidence indicates that both *Pax2* and *Pax6* are repressed by RA in the early zebra fish retina (Hyatt et al. 1996). These results suggest that *Pax6* may be subject to regulation by a diverse array of soluble factors. However, the relative importance of any of these regulators to the tissue-specific expression of *Pax6* in the developing eye is unknown.

5
Pax6 in *Drosophila*: Master Regulator or Network Manager?

As mentioned previously, *Pax6* homologues have been identified in numerous organisms across metazoan phyla. None of these homologues, however, has yet been as instructive in the examination of the *Pax6* genetic pathway as those in *Drosophila*. The surprising finding that two *Pax6* homologues, *ey* (*eyeless*) and *twin of eyeless* (*toy*), also function in fly eye development has led to the realization that a much higher degree of molecular conservation exists between mammalian and invertebrate oculogenesis than was originally appreciated (Zuker 1994; Quiring et al. 1994; Czerny et al. 1999). This apparent similarity in the control of early eye development has allowed the use of *Drosophila* genetics to identify additional genes regulating eye formation in vertebrates.

The *Drosophila* eye consists of a hexagonal array of approximately 750 repeated units called ommatidia. Each omatidium contains eight photoreceptor cells and a set of non-neuronal accessory cells which include lens-secreting cone cells, pigmented cells, and interommatidial bristles (reviewed by Wolff and Ready 1993). These ommatidial clusters form during the third instar of larval development from the columnar epithelium of the eye imaginal disc. Following a proliferative phase, a depression known as the morphogenetic furrow (MF) forms on the apical surface of the disc. As it advances across the eye disc, cells anterior to the MF are undifferentiated and their cell cycles unsynchronized. Cells posterior to the furrow become synchronized in G1 as they are sequentially recruited into ommatidial clusters and begin the process of differentiation into eye-specific cell types (Tomlinson and Ready 1987; reviewed in: Wolff and Ready 1993; Zipursky and Rubin 1994; Bonini and Choi 1995).

ey is expressed in *Drosophila* eye anlagen as early as they can be detected in the embryo. During the larval stages of development, *ey* expression continues in the eye disc, initially throughout the disc, subsequently becoming restricted to cells anterior to the MF (Quiring et al. 1994). *ey* is clearly a critical determinant in fly eye specification, as mutation of *ey* disrupts *Drosophila* eye formation (Ransom 1979). Furthermore, misexpression of *ey* in wing, antennal, and leg imaginal discs is sufficient to induce the development of morphologically normal ectopic eyes. One conclusion drawn from these experiments is that *ey* functions as a "master regulator" of eye development, acting to initiate a genetic cascade leading to eye formation (Halder et al. 1995).

The concept of a master regulator implies a high degree of specificity: such a signal would impose the identity of "eye" on cells in which it is expressed and should therefore be found only in optic tissue. However, *ey* is expressed at multiple sites in the fly, and its complete loss of function is lethal (Ransom 1979). This suggests that *ey* has additional functions auxiliary to eye development and that it must act in concert with other genes to specify optic fate. In fact, recent work has shown that the induction of ectopic eyes occurs only at sites where the BMP homologue *decapentaplegic (dpp)* is also expressed. Furthermore, misexpression of *dpp* together with *ey* dramatically increases the area over which ectopic eyes are induced, suggesting that the two genes act in concert to initiate eye development (Chen et al. 1999).

In addition to *ey*, *sine oculis (so)*, *eyes absent (eya)*, and *dachshund (dac)* are *Drosophila* genes in which mutations lead to complete loss of eye structures (Cheyette et al. 1994; Bonini et al. 1993; Mardon et al. 1994). *ey* expression is preserved in *so*, *eya*, and *dac* mutant eye discs, while expression of *so*, *eya*, and *dac* is lost in *ey* mutants (Shen and Mardon 1997; Halder et al. 1998), suggesting that *eya*, *so*, and *dac* each function downstream of *ey*. Indeed, Niimi et al. (1999) have shown that EY protein directly regulates *so* transcription through an eye-specific enhancer. Nevertheless, both *eya* and *dac* are weakly capable of imposing an eye fate upon nonocular tissues (Bonini et al. 1997; Shen and Mardon 1997).

Analysis of the epistatic relationship between *so*, *eya*, and *dac* reveals that *so* and *eya* regulate each others' expression while *dac* is downstream of both *so* and *eya* (Chen et al. 1997; Pignoni et al. 1997; Halder et al. 1998). As mentioned above, *eya* and *dac* alone each induce ectopic eyes when misexpressed, yet the penetrance of this effect is low and the induced eyes are small. However, like *ey*, *eya* and *dac* also show synergism with *dpp* in inducing ectopic eyes, suggesting a role for *dpp* in their function (Chen et al. 1999). Furthermore, when *eya* and *dac* are expressed together, or when *eya* and *so* are coexpressed, a much more dramatic induction of ectopic eyes is observed, apparently as a result of protein–protein interactions between eya and either so or dac (Chen et al. 1997; Pignoni et al. 1997). Together, these data suggest a complex network of interactions leading to eye development.

ey function is required for the induction of ectopic eyes by any of these downstream genes (Chen et al 1997; Pignoni et al. 1997; Halder et al. 1998), further calling into question the role of *ey* as the master regulator atop this genetic hierarchy. Rather than acting as a master regulator, *ey* appears more likely to serve as an entry point into a genetic network controlling eye formation, acting with *dpp* to initiate eye-specific expression of *eya*, *so*, and *dac* (Fig. 2). In turn, *eya*, *so*, and *dac* are capable of subsequent regulation of *ey* and *dpp* (Chen et al. 1997; Pignoni et al. 1997; Hazelett et al. 1998). The significance of this feedback in development of endogenous eyes is unclear, however, as *ey* expression is not lost when any of the downstream genes are mutated. Several explanations have been suggested for this observation (Desplan 1997). One possibility is that while they function endogenously in what appears to be a linear cascade, *ey*, *eya*, *so*, and *dac* are constantly cross-regulating each other as part of a larger network of genes. Overexpression of one gene results in concomitant upregulation of other members of the network. Alternatively, the feedback regulation of *ey* by *eya*, *so*, and *dac* might reflect a requirement for the same linear cascade during multiple stages of development. The induction of *ey* by its downstream targets would thus represent the progression from one iteration of the linear pathway to the next. Finally, the linear cascade suggested by epistasis analysis may be the entry point to a genetic regulatory loop which subsequently maintains optic development. Such a loop could be interrupted by mutations in any partner, and could be ectopically entered at any point by overexpression of one of the constituents in an appropriate cellular environment.

An understanding of this network is further complicated by the recent identification of a second *Pax6*-like gene in *Drosophila* designated *twin of eyeless (toy)* (Czerny et al. 1999). In fact, *toy* is more likely to be orthologous to vertebrate *Pax6* than *eyeless*, because although both Ey and Toy share high

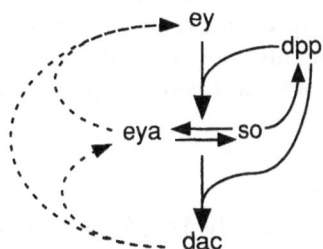

Fig. 2. A simple genetic model for early eye development. *eyeless (ey)* is required for the expression of both *eyes absent (eya)* and *sine oculis (so)*, which regulate each other and which also reside upstream of *dachshund (dac)*. *dpp* acts in conjunction with *ey* to regulate *eya*, *so*, and *dac*, and with *eya* and *so* to induce *dac*. Furthermore, *eya*, *so*, and *dac* are all capable of influencing *ey* and *dpp* expression through feedback regulation. Note that while the Eya protein interacts with those of So and Dac, this model depicts genetic interactions rather than interactions between gene products and that the arrows do not imply direct regulation. Evidence suggests that a similar epistasis may be present in the developing vertebrate lens

degrees (>90 %) of sequence identity with mammalian Pax6 in their paired and homeodomains, the PST domain of Toy is closer to the Pax6 PST domain than that of Ey. Ectopic *toy* expression induces ectopic expression of *ey*, hence, *toy* shares the same ability to induce ectopic eyes in *Drosophila* as *ey*. However, such ectopic *toy* expression is unable to induce ectopic eyes in *ey* mutant flies. Furthermore, several *toy* binding sites are present in the *ey* promoter which are essential for *ey* expression in the eye. Thus, *toy* appears to act upstream of *ey*.

A second *Pax6* gene has also been identified in zebra fish (Nornes et al. 1998), though careful attempts to find another *Pax6* in mouse have proved unsuccessful (M. Busslinger, pers. comm.). The question then arises as to why there should be two *Pax6*-like genes in fish and fly, but only one in mammals. *ey* and *toy* map close to each other on the *Drosophila* fourth chromosome (Halder et al. 1998), and may thus represent a duplication of the fly *Pax6* gene during evolution. The lack of a second mammalian *Pax6*-like gene suggests that this duplication may have occurred after divergence from a common ancestor. Given that the 5a exon is found in both of the zebra fish *Pax6* genes (Nornes et al. 1998), but is not found in *ey* in flies (W. Gehring, pers. comm.), it is likely that the second zebra fish *Pax6* gene is the result of a relatively recent duplication, independent of the duplication event in flies. In fact, evidence points to several genome-wide duplications in the fish during evolution, supporting this hypothesis (Sidow 1996).

A second possibility relates to the finding that the mammalian *Pax6* gene product may autoregulate its own expression. Thus, if *toy* acts to positively regulate *ey*, the functions of both *toy* and *ey* might be subsumed by the positive autoregulatory capacity of a single mammalian *Pax6* gene. If this were true, a second *Pax6* gene would no longer be necessary in mammals, and might be subsequently lost. This hypothesis is consistent with the finding that expression of *Pax6* mutant transcripts are lost in the prospective lens ectoderm of *Sey/Sey* homozygous embryos (Grindley et al. 1995), implying that *Pax6* gene expression requires the presence of its own gene product. Moreover, experiments show that the quail *Pax6* gene product can positively autoregulate its own expression in quail neuroretina (QNR) cells (Plaza et al. 1993).

6
From Fly to Mouse: Identifying *Pax6* Target Genes

With some important exceptions, the major targets for *Pax6* regulation in the developing vertebrate eye are presently unknown. One of these exceptions appears to be the crystallin genes (reviewed in Cvekl and Piatigorsky 1996). The α-, β-, and the taxon-specific ζ-*crystallin* genes all have *Pax6* PD recognition sites in their promoters (Cvekl et al. 1994, 1995a, b; Richardson et al. 1995; Gopal-Srivastava et al. 1996; Sharon-Frilling et al. 1998), and α-*crystallin* and the chick δ1-*crystallin* enhancers are positively regulated by *Pax6* in cell culture, while β1- and βA3/A1-*crystallins* appear to be repressed. *Pax6* has also been suggested to participate in a functional interaction with *Sox2* and 3, members of the HMG box family of architectural proteins, on the chick δ-crystallin enhancer (Kamachi et al. 1998; and H. Kondoh, pers. comm.). However, none of these presumptive targets appears to account for the role of *Pax6* in specifying the prospective lens placode, nor for its inferred function in retinal development.

Though the eyes of flies and vertebrates are quite different, the presence of conserved *Pax6* genes across metazoan phyla suggests conservation of the functional mechanisms controlling oculogenesis. This is underscored by the fact that *Drosophila* and mouse *Pax6* are equally capable of inducing ectopic eyes in fly imaginal discs (Halder et al. 1995). Given this apparent functional conservation between mammalian and insect *Pax6*, one approach to identifying *Pax6* target genes has been to take advantage of knowledge of the *Drosophila* eye-forming gene hierarchy. Based on this approach, mammalian homologues of *dac*, *eya*, and *so* have all been cloned, and examination of their roles in vertebrate development has begun.

Four *eya* homologues have been cloned in the mouse. All are expressed in subsets of the cranial placodes during placodal differentiation, with additional expression in craniofacial mesenchyme, dermamyotome, and the developing limb (Xu et al. 1997a; Zimmerman et al. 1997; Borsani et al. 1999). Analysis of the vertebrate *Eya* gene products reveals two distinct domains. A nonconserved proline-, serine-, threonine-rich (PST) region resides at the N-terminus of the protein (Xu et al. 1997a). This PST domain is capable of functioning as a transcriptional activator when fused to a GAL-4 DNA-binding domain (Xu et al. 1997b). The second domain, called the Eya domain, is a highly conserved 271-amino acid region at the C-terminus of the protein (Xu et al. 1997a). Neither the Eya nor the PST domains binds DNA (P. Xu, unpubl. results). However, in the fly, the Eya domain appears to be the site of protein-protein interactions between Eya and its partners (Chen et al. 1997; Pignoni et al. 1997), suggesting that the vertebrate *Eya* gene products may also interact with *so* and *dac* homologues to regulate transcription.

Eya 1 is expressed weakly in the lens placode and *Eya1*, 2, and 4 are expressed in the nasal placode, both structures which are affected in *Sey/Sey* mice (Xu et al. 1997a). Furthermore, expression of both *Eya1* and 2 is altered in the absence of functional *Pax6*, leading to loss of *Eya* expression in the cranial ectoderm. This suggests that, as in the fly, *Eya* might lie downstream of *Pax6* in a genetic pathway. Vertebrate *eya* homologues have subsequently been identified in human, where heterozygous mutations in *EYA1* are responsible for branchio-oto-renal (BOR) syndrome (Abdelhak et al. 1997a, b; Zimmerman et al. 1997). However, since both BOR syndrome and a mouse knockout of *Eya1* (Xu et al. 1999b) do not include major optic abnormalities, the precise role of *Eya* genes in vertebrate eye development remains unclear.

In the mouse, the *Six* gene family consists of six homologues of *Drosophila sine oculis* (Oliver et al. 1995a, b; Kawakami et al. 1996a, b; Jean et al. 1999; López-Rios et al. 1999; Toy and Sundin 1999). Like *so* in the fly, all of these genes contain a homeodomain and a 110-amino acid domain immediately 5' of the homeodomain called the *Six* domain. cDNAs for *Six2, 3, 5* are all expressed in the mouse retina, and *Six3* is expressed in the invaginating lens vesicle and optic cup, suggesting a role in the development of these structures (Kawakami et al. 1995b; Oliver et al. 1995b). Consistent with the epistasis in flies, *Six3* expression is absent from the presumptive lens ectoderm of *Sey/Sey* embryos (S. Wawersik and P. Purcell, unpubl. results).

A role for *Six3* in optic development can be inferred from experiments in the teleost fish medaka. Ectopic expression of mouse *Six3* in medaka leads to the development of ectopic lenses in the otic placode and forebrain (Oliver et al. 1996; Loosli et al. 1999). Interestingly, these ectopic lenses express *Pax6*, indicating that expression of *Six3* can activate *Pax6* expression, though this activation appears to be indirect.

Though expressed in the developing eye, *Six3* is not the most closely related *Six* family member to *Drosophila sine oculis*. Recently, *Optx2*, another member of the *sine oculis* gene family, was cloned in chick and mouse (Toy et al. 1998; Jean et al. 1999; López-Rios et al. 1999; Toy and Sundin 1999). However, while *Optx2* (also referred to as *Six6* and *Six9*) is expressed in the optic vesicle, it does not appear to be expressed in the developing lens placode. Subsequent to the discovery of *Optx2*, a second *sine oculis*-related gene has been identified in *Drosophila* that appears to be the true orthologue of both *Six3* and *Optx2* (Toy et al. 1998). This gene, named *optix*, is closely linked to *sine oculis*, and may represent a duplication of *so*. However, as no fly eye mutants appear to map near the *optix* locus, the function of this gene in insect eye formation is currently unknown.

In vertebrates, however, *Optx2* appears to be important in eye development. Several human chromosomal deletions of the *Optx2*-containing region demonstrate that *Optx2* haploinsufficiency leads to bilateral anophthalmia (Gallardo et al. 1999). In addition, misexpression of *Optx2* in primary explant cultures of retinal pigment epithelium (RPE) caused these cells to express neural retina-specific markers and overexpression of *Optx2* in *Xenopus* eyes increases the proliferation of retinal cells, suggesting a role for *Optx2* in neural retina development (Toy et al. 1998; Zuber et al. 1999).

Of the *Drosophila* eye regulatory genes that have been cloned in vertebrates, the most recent are two homologues of *dac*. The mouse and human *Dach* 1 and 2 genes both contain two domains of high conservation with *dac* (Hammond et al. 1998; Caubit et al. 1999; Davis et al. 1999; Heanue et al. 1999; Kozmik et al. 1999). As one of these domains resides near the N-terminus and the other near the C-terminus, they are referred to as Dachbox-N and -C respectively. Both domains share weak but significant homology with the proto-oncogenes Ski and Sno, which act as negative regulators of TGF-β signaling (Luo et al. 1999; Stroschein et al. 1999; Sun et al. 1999). Whether the vertebrate *Dach* genes function similarly remains to be determined. The role of the *Dach* genes in mammalian eye development is also unclear, though *Dach1* is expressed in perioptic mesenchyme and the developing neural retina (Hammond et al. 1998; Caubit et al. 1999; Davis et al. 1999; Kozmik et al. 1999). However, though not expressed in the eye, *Dach2* gene function during muscle development suggests that the regulatory hierarchy seen in *Drosophila* also functions in vertebrates (Heanue et al. 1999). In myogenesis, *Dach2* interacts with members of the vertebrate *Pax*, *Six*, and *Eya* families, suggesting that the fly eye regulatory cascade has been incorporated into the patterning of other vertebrate structures.

7
Conclusions

In recent years, understanding of the mechanisms controlling early vertebrate oculogenesis has progressed tremendously. Identification of the critical role of *Pax6* and definition of its functional properties have begun to elucidate the molecular events controlling eye development. These studies have further sparked the surprising realization that a number of regulators of eye development are conserved in mammals and *Drosophila*. This apparent similarity in the control of oculogenesis has led to the identification of vertebrate homologues of *eyes absent*, *sine oculis*, and *dachshund*. The high conservation of the functional domains of these genes suggests that, like *Pax6*, they may also have important functions in vertebrate eye formation. In addition, the expression patterns of members of the *Eya*, *Six*, and *Dach* gene families point to a role in vertebrate lens induction (Oliver et al. 1995b; Xu et al. 1997a; Hammond et al. 1998), and it has already been shown that *Pax6* is required for normal expression of the *Eya1* gene (Xu et al. 1997a). Though further investigation is warranted, we suggest that the epistatic relationships demonstrated in the fly between *ey*, *dpp* and their downstream targets may be preserved in vertebrates (Fig. 2). In support of this hypothesis is the recent finding that vertebrate BMPs participate in the process of lens induction in mouse (Furuta and Hogan 1998; Wawersik et al. 1999). Like *dpp* in the fly, BMPs 4 and 7 appear to cooperate with *Pax6* in regulating eye formation, apparently serving to maintain lens placode development after its initiation. Thus, rather than simply conserving the use of individual genes, the possibility arises that evolution has preserved the use of an entire genetic cassette in the regulation of optic development.

Even as the constituents of the *Pax6*-dependent genetic cascade are being identified, several critical questions remain. Among these is the question of why the fly homologues of *eya*, *so*, and *dac* are found as members of gene families in vertebrates. The eye is only one of a number of vertebrate sensory organs which develops through invagination of a placodal structure. In addition to their expression in the lens placode, members of the vertebrate *Eya*, *Six*, and *Dach* gene families are also expressed in otic and nasal placodes, suggesting a role in the development of these structures as well. Mutations in the human and mouse *Eya1* genes lead to defects in facial and otic development (Abdelhak et al. 1997a, b; Xu et al. 1999b). This, along with analysis of *Dach*, *Eya*, *Six*, and *Pax* family members in myogenesis (Heanue et al. 1999), suggests that members of each of these gene families have specialized to regulate the development of multiple organs.

The existence of multiple members of the *Eya* and *Six* gene families also raises questions as to the role of protein-protein interactions in their biological function. The recent demonstration that members of the *Six* gene family induce nuclear translocation of *Eya* genes in COS7 cells (Ohto et al. 1999) suggests that the synergistic action of *so* and *eya* seen in flies is preserved in vertebrates. Furthermore, it appears that different Eya proteins interact preferentially with subsets of Six proteins (Ohto et al. 1999), raising the possibility that each of these combinations has a distinct functional outcome. At present none of the *Eya* genes have been shown to be translocated by *Six3*, the only *Six* family member yet shown to be expressed in the lens placode (Ohto et al. 1999). Thus, even though a number of the vertebrate homologues of *Drosophila* eye regulatory genes have been identified, it has not yet been demonstrated whether their biological roles have also been conserved.

These questions aside, it is clear that the *Drosophila* eye, though structurally distinct from that of vertebrates, shows striking parallels at the molecular level. By exploiting these homologies, elements of the *Pax6*-dependent genetic pathway regulating vertebrate eye development have been identified. The challenge in future research will be to assign biological function to these newly identified factors through analysis of phenotypes from targeted mutation and misexpression experiments, biochemical assays, and gene regulation studies. Ultimately, an understanding of the collective functions of *Pax6*, *Dach*, and members of the *Eya* and *Six* families is likely to shed light not only on lens induction but on the inductive mechanisms involved in the development of multiple organ systems.

References

Abdelhak S, Kalatzis V, Heilig R, Compain S, Samson D, Vincent C, Weil D, Cruaud C, et al. (1997a) A human homologue of the *Drosophila eyes absent* gene underlies branchio-oto-renal (BOR) syndrome and identifies a novel gene family. Nat Genet 15:157–164

Abdelhak S, Kalatzis V, Heilig R, Compain S, Samson D, Vincent C, Levi-Acobas F, Cruaud C, et al. (1997b) Clustering of mutations responsible for branchio-oto-renal (BOR) syndrome in the *eyes absent* homologous region (eyaHR) of EYA1. Hum Mol Genet 6:2247–2255

Altmann CR, Chow RL, Lang RA, Hemmati-Brivanlou A (1997) Lens induction by *Pax-6* in *Xenopus laevis*. Dev Biol 185:119–123

Azuma N, Nishina S, Yanagisawa H, Okuyama T, Yamada M (1996) *PAX6* missense mutation in isolated foveal hypoplasia. Nat Genet 13:141-142

Berk M, Desai SY, Heyman HC, Colmenares C (1997) Mice lacking the *ski* proto-oncogene have defects in neurulation, craniofacial patterning and skeletal muscle development. Genes Dev 11:2029-2039

Bertuccioli C, Fasano L, Jun S, Wang S, Sheng G, Desplan C (1996) In vivo requirement for the paired domain and homeodomain of the *paired* segmentation gene product. Development 122:2673-2685

Bonini NM, Choi KW (1995) Early decisions in *Drosophila* eye morphogenesis. Curr Opin Genet Dev 5:507-515

Bonini NM, Leiserson WM, Benzer S (1993) The *eyes absent* gene: genetic control of cell survival and differentiation in the developing Drosophila eye. Cell 72:379-395.

Bonini NM, Bui QT, Gray-Board GL, Warrick JM (1997) The *Drosophila eyes absent* gene directs ectopic eye formation in a pathway conserved between flies and vertebrates. Development 124:4819-4826

Bopp D, Burri M, Baumgartner S, Frigerio G, Noll M (1986) Conservation of a large protein domain in the segmentation gene paired and in functionally related genes of Drosophila. Cell 47:1033-1040

Borsani G, DeGrandi A, Ballabio A, Bulfone A, Bernard L, Banfi S, Gattuso C, Mariani M, Dixon M, Donnai D, Metcalfe K, Winter R, Robertson M, Axton R, Brown A, van Heyningen V and Hanson I (1999) *EYA4*, a novel vertebrate gene related to *Drosophila eyes absent*. Hum Mol Genet 8:11-23

Burri M, Tromvoukis Y, Bopp D, Frigerio G, Noll M (1989) Conservation of the paired domain in metazoans and its structure in three isolated human genes. EMBO J 8:1183-1190

Callaerts P, Halder G, Gehring WJ (1997) *Pax-6* in development and evolution. Annu Rev Neurosci 20:483-532

Carrière C, Plaza S, Caboche J, Dozier C, Bailly M, Martin P, Saule S (1995) Nuclear localization signals, DNA binding, and transactivation properties of quail *Pax-6 (Pax-QNR)* isoforms. Cell Growth Differ 6:1531-1540

Caubit X, Thangarajah R, Theil T, Wirth J, Nothwang H-G, Rüther U, Krauss S (1999) Mouse Dac, a novel nuclear factor with homology to *Drosophila* dachshund shows a dynamic expression in the neural crest, the eye, the neocortex, and the limb bud. Dev Dyn 214:66-80

Chen R, Amoui M, Zhang Z, Mardon G (1997) *Dachshund* and *eyes absent* proteins form a complex and function synergistically to induce ectopic eye development in Drosophila. Cell 91:893-903

Chen R, Halder G, Zhang Z, Mardon G (1999) Signaling by the TGF-β homologue *decapentaplegic* functions reiteratively within the network of genes controlling retinal cell fate determination in *Drosophila*. Development 126:935-943

Cheyette BN, Green PJ, Martin K, Garren H, Hartenstein V, Zipursky SL (1994) The *Drosophila sine oculis* locus encodes a homeodomain-containing protein required for the development of the entire visual system. Neuron 12:977-996

Chisholm AD, Horvitz HR (1995) Patterning of the *Caenorhabditis elegans* head region of the *Pax-6* family member *vab-3*. Nature 377:52-55

Chow RL, Altmann CR, Lang RA, Hemmati-Brivanlou A (1999) *Pax6* induces ectopic eyes in a vertebrate. Development 126:4213-4222

Cvekl A, Piatigorsky J (1996) Lens development and crystallin gene expression: many roles for *Pax-6*. Bioessays 18, 621-630

Cvekl A, Sax CM, Bresnick EH, Piatigorsky J (1994) A complex array of positive and negative elements regulates the chicken *alpha A-crystallin* gene: involvement of Pax-6, USF, CREB and/or CREM, and AP-1 proteins. Mol Cell Biol 14:7363-7376

Cvekl A, Kashanchi F, Sax CM, Brady JN, Piatigorsky J (1995a) Transcriptional regulation of the mouse *alpha A-crystallin* gene: activation dependent on a cyclic AMP-responsive element (DE1/CRE) and a Pax-6-binding site. Mol Cell Biol 15:653-660

Cvekl A, Sax CM, Li X, McDermott JB, Piatigorsky J (1995b) Pax-6 and lens-specific transcription of the *chicken delta 1-crystallin* gene. Proc Natl Acad Sci USA 92:4681-4685

Czerny T, Schaffner G, Busslinger M (1993) DNA sequence recognition by Pax proteins: bipartite structure of the paired domain and its binding site. Genes Dev 7:2048-2061

Czerny T, Halder G, Kloter U, Souabni A, Gehring WJ, Busslinger M (1999) *Twin of eyeless*, a second *Pax-6* gene of *Drosophila*, acts upstream of *eyeless* in the control of eye development. Mol Cell 3:297–307

Davis RJ, Shen W, Heanue TA, Mardon G (1999) Mouse *Dach*, a homologue of *Drosophila dachshund*, is expressed in the developing retina, brain, and limbs. Dev Genes Evol 209:526–536

Desplan, C (1997) Eye development: governed by a dictator or a junta? Cell 91:861–864

Ekker SC, Ungar AR, Greenstein P, von Kessler DP, Porter JA, Moon RT, Beachy PA (1995) Patterning activities of vertebrate *hedgehog* proteins in the developing eye and brain. Curr Biol 5:944–955

Epstein J, Cai J, Glaser T, Jepeal L, Maas R (1994a) Identification of a Pax paired domain recognition sequence and evidence for DNA-dependent conformational changes. J Biol Chem 269:8355–8361

Epstein JA, Glaser T, Cai J, Jepeal L, Walton DS, Maas RL (1994b) Two independent and interactive DNA-binding subdomains of the Pax6 paired domain are regulated by alternative splicing. Genes Dev 8:2022–2034

Ericson J, Rasbass P, Schedl A, Brenner-Moroton S, Kawakami A, van Heyningen V, Jessell TM, Briscoe J (1997) *Pax6* controls progenitor cell identity and neuronal fate in response to graded Shh signaling. Cell 90:169–180

Fickenscher HR, Chalepakis G, Gruss P (1993) Murine Pax-2 protein is a sequence-specific *trans*-activator with expression in the genital system. DNA Cell Biol 12:381–391

Fortin AS, Underhill DA, Gros P (1997) Reciprocal effect of Waardenburg syndrome mutations on DNA binding by the *Pax-3* paired domain and homeodomain. Hum Mol Genet 6:1781–1790

Fortin AS, Underhill DA, Gros P (1998) Helix 2 of the paired domain plays a key role in the regulation of DNA-binding by the Pax-3 homeodomain. Nuc Acids Res 26:4574–4581

Fujioka M, Miskiewicz P, Raj L, Gullege AA, Weir M, Goto T (1996) *Drosophila paired* regulates late *even-skipped* expression through a composite binding site for the paired domain and the homeodomain. Development 122:2697–2707

Fujiwara M, Uchida T, Osumi-Yamashita N, Eto K (1994) Uchida rat (rSey): a new mutant rat with craniofacial abnormalities resembling those of the mouse Sey mutant. Differentiation 57:31–38

Furuta Y, Hogan BLM (1998) BMP4 is essential for lens induction in the mouse embryo. Genes Dev 12:3764–3775

Gajovic S, St-Onge L, Yokota Y, Gruss P (1997) Retinoic acid mediates *Pax6* expression during in vitro differentiation of embryonic stem cells. Differentiation 62:187–192

Gallardo ME, Lopaz-Rios J, Fernaud-Espinosa I, Granadino B, Sanz R, Ramos C, Ayuso C, Seller MJ, Brunner HG, Bovolenta P, Rodriguez de Cordoba S (1999) Genomic cloning and characterization of the human homeobox gene *SIX6* reveals a cluster of SIX genes in chromosome 14 and associates *SIX6* hemizygosity with bilateral anophthalmia and pituitary anomalies. Genomics 61:82–91

Glardon S, Callaerts P, Halder G, Gehring WJ (1997) Conservation of *Pax-6* in a lower chordate, the ascidian *Phallusia mammillata*. Development 124:817–825

Glaser T, Walton DS, Maas RL (1992) Genomic structure, evolutionary conservation and aniridia mutations in the human *PAX6* gene. Nat Genet 2:232–238

Glaser T, Jepeal L, Edwards JG, Young SR, Favor J, Maas RL (1994) *Pax6* gene dosage effect in a family with congenital cataracts, aniridia, anophthalmia and central nervous system defects. Nat Genet 7:463–471

Glaser T, Cai J, Epstein J, Walton DS, Jepeal L, Maas RL (1995) *PAX6* mutations in aniridia. Wiggs JR, ed. Molecular genetics of eye diseases. John Wiley, New York pp 51–82

Gopal-Srivastava R, Cvekl A, Piatigorsky J (1996) Pax-6 and alphaB-crystallin/small heat shock protein gene regulation in the murine lens. Interaction with the lens-specific regions, LSR1 and LSR2. J Biol Chem 271:23029–23036

Grindley JC, Davidson DR, Hill RE (1995) The role of *Pax-6* in eye and nasal development. Development 121:1433–1442

Halder G, Callaerts P, Gehring W (1995) Induction of ectopic eyes by targeted expression of the *eyeless* gene in *Drosophila*. Science 267:1788–1792

Halder G, Callaerts P, Flister S, Walldorf U, Kloter U, Gehring WJ (1998) *Eyeless* initiates the expression of both *sine oculis* and *eyes absent* during *Drosophila* compound eye development. *Development* 125: 2181-2191

Hammond KL, Hanson IM, Brown AG, Lettice LA, Hill RE (1998) Mammalian and *Drosophila dachshund* genes are related to the *Ski* proto-oncogene and are expressed in eye and limb. Mech Dev 74: 121-131

Hanson IM, Seqwright A, Hardman K, Hodgson S, Zaletayev D, Fekete G, van Heyningen V (1993) *Pax6* mutations in aniridia. Hum Mol Genet 2: 915-920

Hanson IM, Fletcher JM, Jordan T, Brown A, Taylor D, Adams RL, Punnet HH, van Heyningen V (1994) Mutations at the *PAX6* locus are found in heterogeneous anterior segment malformations including Peters' anomaly. Nat Genet 6: 168-173

Hazelett DJ, Bourouis M, Walldorf U, Treisman JE (1998) *Decapentaplegic* and *wingless* are regulated by *eyes absent* and *eyegone* and interact to direct the pattern of retinal differentiation in the eye disc. Development 125: 3741-3751

Heanue TA, Reshef R, Davis RJ, Mardon G, Oliver G, Tomarev S, Lassar AB, Tabin CJ (1999) Synergistic regulation of vertebrate muscle development by *Dach2*, *Eya2*, and *Six1*, homologues of genes required for *Drosophila* eye formation. Genes Dev 13: 3231-3243

Hill RE, Favor J, Hogan BL, Ton CC, Saunders GF, Hanson IM, Prosser J, Jordan T, et al. (1991) Mouse *Small eye* results from mutations in a paired-like homeobox-containing gene. Nature 354: 522-525

Hogan BL, Horsburgh G, Cohen J, Hetherington CM, Fisher G, Lyon MF (1986) *Small eyes (Sey)*: a homozygous lethal mutation on chromosome 2 which affects the differentiation of both lens and nasal placodes in the mouse. J Embryol Exp Morphol 97: 95-110

Hogan BL, Hirst EM, Horsburgh G, Hetherington CM (1988) *Small eye (Sey)*: a mouse model for the genetic analysis of craniofacial abnormalities. Development 103: 115-119

Hyatt GA, Schmitt EA, Marsh-Armstrong N, McCaffery P, Drager UC, Dowling JE (1996) Retinoic acid establishes ventral retinal characteristics. Development 122: 195-204

Jean D, Bernier G, Gruss P (1999) *Six6 (Optx2)* is a novel murine *Six3*-related homeobox gene that demarcates the presumptive pituitary/hypothalamic axis and the ventral optic stalk. Mech Dev 84: 31-40

Jun S, Desplan C (1996) Cooperative interactions between paired domain and homeodomain. Development 122: 2639-2650

Jun S, Wallen RV, Goriely A, Kalionis B, Desplan C (1998) Lune/eye gone, a Pax-like protein, uses a partial paired domain and a homeodomain for DNA recognition. Proc Natl Acad Sci USA 95: 13720-13725

Kamachi Y, Uchikawa M, Collignon J, Lovell-Badge R, Kondoh H (1998) Involvement of *Sox1, 2* and *3* in the early and subsequent molecular events of lens induction. Development 125: 2521-2532

Kammandel B, Chowdhury K, Stoykova A, Aparicio S, Brenner S, Gruss P (1999) Distinct *cis*-essential modules direct the time-space pattern of the *Pax6* gene activity. Dev Biol 205: 79-97

Kawakami K, Ohto H, Takizawa T, Saito T (1996a) Identification and expression of *six* family genes in mouse retina. FEBS Lett 393: 259-263

Kawakami K, Ohto H, Ikeda K, Roeder RG (1996b) Structure, function and expression of a murine homeobox protein AREC3, a homologue of *Drosophila sine oculis* gene product, and implication in development. Nucleic Acids Res 24: 303-310

Koroma BM, Yang JM, Sundin OH (1997) The *Pax-6* homeobox gene is expressed throughout the corneal and conjunctival epithelia. Invest Ophthalmol Visual Sci 38: 108-120

Kozmik Z, Czerny T, Busslinger M (1997) Alternatively spliced insertions in the paired domain restrict the DNA sequence specificity of *Pax6* and *Pax8*. EMBO J 16: 6793-6803

Kozmik Z, Pfeffer P, Kralova J, Paces J, Paces V, Kalousova A, Cvekl A (1999) Molecular cloning and expression of the human and mouse homologues of the *Drosophila dachshund* gene. Dev Genes Evol 209: 537-545

Krauss S, Johansen T, Korzh V, Fjose A (1991) Expression pattern of zebrafish *pax* genes suggests a role in early brain regionalisation. Nature 353: 267-270

Li HS, Yang JM, Jacobson RD, Pasko D, Sundin O (1994) *Pax-6* is first expressed in a region of ectoderm anterior to the early neural plate: implications for stepwise determination of the lens. Dev Biol 162:181-194

Loosli F, Winkler S, Wittbrodt J (1999) Ectopic retina in response to *Six3* overexpression. Genes Dev 13:649-654

López-Rios J, Gallardo ME, Rodriguez de Cordoba S, Bovolenta P (1999) *Six9 (Optx2)*, a new member of the six gene family of transcription factors, is expressed at early stages of vertebrate ocular and pituitary development. Mech Dev 83:155-159

Luo K, Stroschein SL, Wang W, Chen D, Martens E, Zhou S, Zhou Q (1999) The Ski oncoprotein interacts with the Smad proteins to repress TGFβ signaling. Genes Dev 13:2196-2206

MacDonald R, Barth KA, Xu Q, Holder N, Mikkola I, Wilson SW (1995) Midline signaling is required for Pax gene regulation and patterning of the eyes. Development 121:3267-3278

Mardon G, Solomon NM, Rubin GM (1994) *Dachshund* encodes a nuclear protein required for normal eye and leg development in Drosophila. Development 120:3473-3486

Matsuo T, Osumi-Yamashita N, Noji S, Ohuchi H, Koyama E, Myokai F, Matsuo N, Taniguchi S, Doi H, Iseki S, Ninomiya Y, Fujiwara M, Watanabe T, Eto K (1993) A mutation in the *Pax-6* gene in rat *Small eye* is associated with impaired migration of midbrain crest cells. Nat Genet 3:299-304

Mermod N, O'Neill EA, Kelly TJ, Tjian R (1989) The proline-rich transcriptional activator or CTF/NF-1 is distinct from the replication and DNA binding domain. Cell 58:741-753

Miskiewicz P, Morrissey D, Lan Y, Raj L, Kessler S, Fujioka M, Goto T, Weir M (1996) Both the paired domain and homeodomain are required for in vivo function of *Drosophila paired*. Development 122:2709-2718

Nelson LB, Spaeth GL, Nowinski TS, Margo CE, Jackson L (1984) Aniridia. A Review. Surv Ophthalmol 28:621-642

Niimi T, Seimiya M, Kloter U, Flister S, Gehring WJ (1999) Direct regulatory interaction of the eyeless protein with an eye-specific enhancer in the *sine oculis* gene during ey induction in *Drosophila*. Development 126:2253-2260

Nomura N, Sasamoto S, Ishii S, Date T, Ishizaka R (1989) Isolation of human cDNA clones of *ski* and the *ski*-related gene, *sno*. Nucleic Acids Res 25:5489-5500

Nornes S, Clarkson M, Mikkola I, Pedersen M, Bardsley A, Martinez JP, Krauss S, Johansen T (1998) Zebrafish contains two *Pax6* genes involved in eye development. Mech Dev 77:185-196

Ohto H, Kamada S, Tago K, Tominaga K, Ozaki H, Sato S, Kawakami K (1999) Cooperation of Six and Eya in activation of their target genes through nuclear translocation of Eya. Mol Cell Biol 19:6815-6824

Okladnova O, Syagailo YV, Mössner R, Riederer P, Lesch K-P (1998) Regulation of *PAX-6* gene transcription: alternate usage in human brain. Mol Brain Res 60:177-192

Oliver G, Wehr R, Jenkins NA, Copeland NG, Cheyette BNR, Hartenstein V, Zipursky SL, Gruss P (1995a) Homeobox genes and connective tissue patterning. Development 121:693-705

Oliver G, Mailhos A, Wehr R, Copeland NG, Jenkins NA, Gruss P (1995b) *Six3*, a murine homologue of the *sine oculis* gene, demarcates the most anterior border of the developing neural plate and is expressed during eye development. Development 121:4045-4055

Oliver G, Loosli F, Koster R, Wittbrodt J, Gruss P (1996) Ectopic lens induction in fish in response to the murine homeobox gene *Six3*. Mech Dev 60:233-239

Pei YF, Rhodin JAG (1970) The prenatal development of the mouse eye. Anat Rec 168:105-126

Pignoni F, Hu B, Zavitz KH, Xiao J, Garrity PA, Zipursky SL (1997) The eye-specification proteins *So* and *Eya* form a complex and regulate multiple steps in *Drosophila* eye development. *Cell* 91:881-891

Pituello F, Yamada G, Gruss P (1995) Activin A inhibits *Pax-6* expression and perturbs cell differentiation in the developing spinal cord in vitro. Proc Natl Acad Sci USA 92:6952-6956

Plaza S, Dozier C, Saule S (1993) Quail *Pax-6 (Pax-QNR)* encodes a transcription factor able to bind and trans-activate its own promoter. Cell Growth Differ 4:1041-1050

Plaza S, Dozier C, Turque N, Saule S (1995a) Quail *Pax-6 (Pax-QNR)* mRNAs are expressed from two promoters used differentially during retina development and neuronal differentiation. Mol Cell Biol 15:3344-3353

Plaza S, Dozier C, Langlois MC, Saule S (1995b) Identification and characterization of a neuro-retina-specific enhancer element in the quail Pax-6 *(Pax-QNR)* gene. Mol Cell Biol 15 : 892–903

Porter FD, Drago J, Xu Y, Cheema SS, Wassif C, Huang S-H, Lee E, Grinberg A, Massalas JS, Bodine D, Alt F, Westphal H (1997) *Lhx2*, a LIM homeodomain gene, is required for eye forebrain, and definitive erythrocyte development. Development 124 : 2935–2944

Prosser J, van Heyningen V (1998) *PAX6* mutations reviewed. Hum Mut 11 : 93–108

Quinn JC, West JD, Hill RE (1996) Multiple functions for *Pax6* in mouse eye and nasal development. Genes Dev 10 : 435–446

Quiring R, Walldorf U, Kloter U, Gehring W (1994) Homology of the *eyeless* gene of *Drosophila* to the *Small eye* gene in mice and aniridia in humans. Science 265 : 785–789

Ransom R (1979) The time of action of three mutations affecting Drosophila eye morphogenesis. J Embryol Exp Morphol 53 : 225–235

Richardson J, Cvekl A, Wistow G (1995) *Pax-6* is essential for lens-specific expression of *zeta-crystallin*. Proc Natl Acad Sci USA 92 : 4676–4680

Schedl A, Ross A, Lee M, Engelkamp D, Rashbass P, van Heyningen V, Hastie ND (1996) Influence of *Pax6* gene dosage on development: overexpression causes severe eye abnormalities. Cell 86 : 71–82

Sharon-Friling R, Richardson J, Sperbeck S, Lee D, Rauchman M, Maas R, Swaroop A, Wistow G (1998) Lens-specific gene recruitment of zeta-crystallin through Pax6, Nrl-Maf, and brain suppressor sites. Mol Cell Biol 18 : 2067–2076

Shen W, Mardon G (1997) Ectopic eye development in *Drosophila* induced by directed *dachshund* expression. Development 124 : 45–52

Sidow A (1996) Gen(om)e duplications in the evolution of early vertebrates. Curr Opin Genet Dev 6 : 715–722

Singh S, Tang HK, Lee J-Y, Saunders GF (1998) Truncation mutations in the transactivation region of PAX6 result in dominant-negative mutations. J Biol Chem 273 : 21531–21541

Stroschein SL, Wang W, Zhou S, Zhou Q, Luo K (1999) Negative feedback regulation of TGF-beta signaling by the SnoN oncoprotein. Science 286 : 771–774

Sun Y, Liu X, Eaton EN, Lane WS, Lodish WS, Weinberg RA (1999) Interaction of the Ski oncoprotein with Smad3 regulates TGF-beta signaling. Mol Cell 4 : 499–509

Tang HK, Chao L-Y, Saunders GF (1997) Functional analysis of paired box missense mutations in the *PAX6* gene. Hum Mol Genet 6 : 381–386

Tang HK, Singh S. Saunders GF (1998) Dissection of the transactivation function of the transcription factor encoded by the eye developmental gene *PAX6*. J Biol Chem 273 : 7210–7221

Theill LE, Castrillo JL, Wu D, Karin M (1989) Dissection of functional domains of the pituitary-specific transcription factor *GHF-1*. Nature 342 : 945–948

Tomarev SI, Callaerts P, Kos L, Zinovieva R, Halder G, Gehring W, Piatigorsky J (1997) Squid *Pax-6* and eye development. Proc Natl Acad Sci USA 94 : 2421–2426

Tomlinson A, Ready DF (1987) Neuronal differentiation in the *Drosophila* ommatidium. Dev Biol 120 : 366–376

Ton CCT, Hirvonen H, Miwa H, Weil MM, Monaghan P, Jordan T, van Heyningen V, Hastie ND, Meijers-Heijboer H, Dreschler M, Royer-Pokora B, Collins F, Swaroop A, Strong LC, Saunders GF (1991) Positional cloning and characterization of a paired box-and homeobox-containing gene from the Aniridia region. Cell 67 : 1059–1074

Toy J, Sundin OH (1999) Expression of the *optx2* homeobox gene during mouse development. Mech Dev 83 : 183–186

Toy J, Yang J-M, Leppert GS, Sundin OH (1998) The *Optx2* homeobox gene is expressed in the early precursors of the eye and activates retina-specific genes. Proc Natl Acad Sci USA 95 : 10643–10648

Treisman J, Harris E, Desplan C (1991) The paired box encodes a second DNA-binding domain in the paired homeodomain protein. Genes Dev 5 : 594–604

Underhill DA, Gros P (1997) The paired domain regulates DNA binding by the homeodomain within the intact Pax-3 protein. J Biol Chem 272 : 14175–145182

Underhill DA, Gros P (1995) Analysis of the mouse Splotch-delayed mutation indicates that the Pax-3 paired domain can influence homeodomain DNA-binding activity. Proc Natl Acad Sci USA 92 : 3692–3296

Walther C, Gruss P (1991) *Pax-6*, a murine paired box gene, is expressed in the developing CNS. Development 113 : 1435–1449

Wawersik S, Purcell P, Rauchman M, Dudley AT, Robertson EJ, Maas R (1999) BMP7 acts in murine lens placode development. Dev Biol 207 : 176–188

Williams SC, Altmann CR, Chow RL, Hemmati-Brivanlou A, Lang RA (1998) A highly conserved lens transcriptional control element from the *Pax-6* gene. Mech Dev 73, 225–229

Wilson D, Sheng G, Lecuit T, Dostatni N, Desplan C (1993) Cooperative dimerization of paired class homeo domains on DNA. Genes Dev 7 : 2120–2134

Wilson DS, Guenther B, Desplan C and Kuriyan J (1995) High resolution crystal structure of a paired (Pax) class cooperative homeodomain dimer on DNA. Cell 82 : 709–719

Wolff T, Ready DF (1993) Pattern formation in the *Drosophila* retina. The Development of *Drosophila melanogaster*. Cold Spring Harbor Laboratory Press, Cold Spring Harbor, New York, pp 1277–1325

Xu HE, Rould MA, Xu W, Epstein JA, Maas RL, Pabo CO (1999) Crystal structure of the human Pax6 paired domain-DNA complex reveals specific roles for the linker region and C-terminal subdomain in DNA binding. Genes Dev 13 : 1263–1275

Xu P-X, Woo I., Her H, Beier DR, Maas RL (1997a) Mouse *Eya* homologues of the Drosophila *eyes absent* gene require *Pax6* for expression in lens and nasal placode. Development 124 : 219–231

Xu P-X, Cheng J, Epstein JA, Maas RL (1997b) Mouse *Eya* genes are expressed during limb tendon development and encode a transcriptional activation function. Proc Natl Acad Sci USA 94 : 11974–11979

Xu P-X, Zhang X, Heaney S, Yoon A, Michelson AM, Maas RL (1999a) Regulation of *Pax6* expression is conserved between mice and flies. Development 126 : 383–395

Xu P-X, Adams J, Peters H, Brown MC, Heaney S, Maas R (1999b) *Eya1*-deficient mice lack ears and kidneys and show abnormal apoptosis of organ primordia. Nat Genet 23 : 113–117

Xu W, Rould MA, Jun S, Desplan C, Pabo CO (1995) Crystal structure of a paired domain-DNA complex at 2.5 A resolution reveals structural basis for Pax developmental mutations. Cell 80 : 639–650

Zhang Y, Emmons SW (1995) Specification of sense-organ identity by a *Caenorhabditis elegans Pax-6* homologue. Nature 377 : 55–59

Zimmerman JE, Bui QT, Steingrímsson E, Nagle DL, Fu W, Genin A, Spinner NB, Copeland NG, Jenkins NA, Bucan M, Bonini NM (1997) Cloning and characterization of two vertebrate homologues of the *Drosophila eyes absent* gene. Genome Res 7 : 128–141

Zipursky SL, Rubin GM (1994) Determination of neuronal cell fate: lessons from the R7 neuron of *Drosophila*. Annu Rev Neurosci 17 : 373–397

Zuber ME, Perron M, Philpott A, Bang A, Harris WA (1999) Giant eyes in *Xenopus laevis* by overexpression of *Xoptx2*. Cell 98 : 341–352

Zuker CS (1994) On the evolution of eyes: would you like it simple or compound? Science 265 : 742–743

Zwaan J, Pearce TL (1971) Cell population kinetics in the chicken lens primordium during and shortly after its contact with the optic cup. Dev Biol 25 : 96–118

Zwaan J, Bryan PR Jr, Pearce TL (1969) Interkinetic nuclear migration during the early stages of lens formation in the chicken embryo. J Embryol Exp Morph 21 : 71–83

Early Retinal Development in *Drosophila*

Ulrike Heberlein[1] and Jessica E. Treisman[2]

1
Introduction

Like other insects, the fruit fly *Drosophila melanogaster* has a compound eye, made up of approximately 800 individual eyes or ommatidia. The eye arises from the eye imaginal disk, which is part of the compound eye-antennal disk. The cells that will form this disk invaginate from the ectoderm during late embryogenesis (Jurgens and Hartenstein 1993); the disk primordium then grows inside the larva, where it consists of an epithelial bilayer. One columnar layer will give rise to all the cell types of the retina; it is covered by a squamous epithelial sheet called the peripodial membrane, which later contributes to the surface of the head (Haynie and Bryant 1986). Differentiation of the retina begins in the third of the three larval stages, known as instars, and continues during the first few days of pupal development (Ready et al. 1976; Fig. 1). In spite of the many morphological and developmental differences between fly and vertebrate eyes, many of the molecules involved appear to be conserved.

2
Specification of the Eye Primordium

Six genes have been implicated in the specification of the eye field because they are necessary for the formation of the eye and/or sufficient to cause ectopic eye development in other imaginal disks. These genes all encode nuclear proteins that are likely to function in a transcriptional regulatory hierarchy. The first of them to be expressed, in a broad domain in the an-

---

[1] Department of Anatomy, Programs in Neuroscience and Developmental Biology, University of California, San Francisco, 513 Parnassus Avenue, San Francisco, California 94143-0452, USA

[2] Skirball Institute for Biomolecular Medicine, Developmental Genetics Program, NYU Medical Center, 540 First Avenue, New York, New York 10016, USA

Results and Problems in Cell Differentiation, Vol. 31
M. E. Fini (Ed.): Vertebrate Eye Development
© Springer-Verlag Berlin Heidelberg 2000

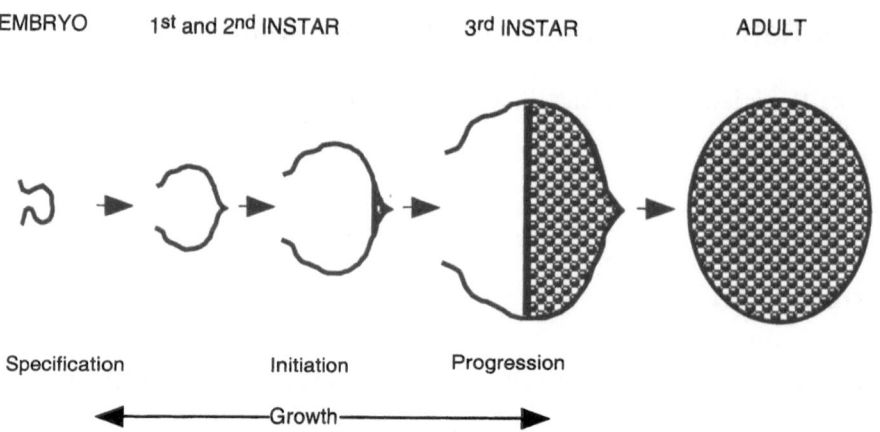

EMBRYO   1st and 2nd INSTAR   3rd INSTAR   ADULT

Specification   Initiation   Progression

◄─────────Growth─────────►

Fig. 1. Schematic representation of the retina at different stages of development

terior of the embryo, is *twin of eyeless* (*toy*), a homologue of the paired box/
homeobox gene *Pax-6* (Halder et al. 1998). Its expression is then refined to
the eye-antennal anlagen, where it is coexpressed with another *Pax-6* homologue, *eyeless* (*ey*) (Quiring et al. 1994; Halder et al. 1998). As its name
suggests, mutations in *ey* result in the absence of the eye, due to a failure in
photoreceptor differentiation in the eye disk (Halder et al. 1998). Both *ey* and
toy have a striking ability to induce the formation of eye tissue when either is
misexpressed in other imaginal disks (Halder et al. 1995; Czerny 1999). In
this case, *toy* activates the ectopic expression of *ey* but the reverse is not true,
suggesting that *toy* functions upstream of *ey* (Czerny 1999); however, the
phenotype of mutations in *toy* is not known. The placement of these genes
very early in the process of eye development is intriguing in light of the
phenotypes of mouse (*Small eye*) and human (*Aniridia*) mutations in *Pax-6*,
both of which cause defects in eye development (Hill et al. 1991; Ton et al.
1991). *Pax-6* expression has subsequently been found in photoreceptive organs in a variety of evolutionarily distant and developmentally diverse species
(Loosli et al. 1996; Glardon et al. 1997, 1998; Tomarev et al. 1997), suggesting
that it was associated with visual or at least anterior sensory structures in a
common ancestor.

    ey is required for the expression in the second instar eye disk of *sine oculis*
(*so*) (Halder et al. 1998), which encodes a homeodomain protein (Cheyette
et al. 1994; Serikaku and O'Tousa 1994), and *eyes absent* (*eya*) (Bonini et al.
1997; Halder et al. 1998), which encodes a novel nuclear protein (Bonini et al.
1993). Again, mutations in either *so* or *eya* result in the complete absence of
the eyes (Bonini et al. 1993; Cheyette et al. 1994). However, these genes are
less efficient in inducing ectopic eye development than *ey*; ectopic eyes are
only observed at a high frequency when *so* and *eya* are expressed in combination, and they are predominantly found in the antennal disk (Bonini
et al. 1997; Pignoni et al. 1997). This ectopic eye development is associated

with the activation of *ey* transcription, suggesting a positive feedback of *so* and *eya* on the earlier gene (Bonini et al. 1997; Pignoni et al. 1997). The reason for their combinatorial activity appears to be the association of the two proteins into a complex transcription factor with the DNA-binding domain of so and the transcriptional activation domain of eya (Pignoni et al. 1997). The proteins may be unstable outside this complex, as *eya* and *so* are each required for the normal level of expression of the other in the eye disk (Halder et al. 1998). However, *so* has an additional early role in the formation of the optic lobes of the brain that appears to be independent of *eya* (Cheyette et al. 1994; Serikaku and O'Tousa 1994).

Both *eya* and *so* are required for the expression of *dachshund* (*dac*) (Chen et al. 1997; Pignoni et al. 1997), which encodes a novel nuclear protein (Mardon et al. 1994), in the early third instar eye disk. *dac* mutants lack eyes (Mardon et al. 1994), and *dac* expression can induce, although weakly, ectopic eye development (Shen and Mardon 1997). Coexpression of *eya* with *dac* enhances the production of ectopic eyes, perhaps because the two proteins can physically interact (Chen et al. 1997). However, as neither protein contains a known DNA-binding domain and both contain domains capable of transcriptional activation (Chen et al. 1997; Pignoni et al. 1997), the basis of this synergy is less clear than for *so* and *eya*. Unlike *ey*, *so*, and *eya*, *dac* is not required for normal growth of the early eye disk and thus seems to act at a slightly later stage of development (Mardon et al. 1994). Vertebrate homologues of *so*, *eya* and *dac* are all expressed in eye tissues (Oliver et al. 1995; Xu et al. 1997; Zimmerman et al. 1997; Hammond et al. 1998; Seo et al. 1998), suggesting the possibility that these three genes define a conserved regulatory network downstream of *Pax-6* (Treisman 1999). However, these genes are not limited to the eye in either flies or vertebrates; it is therefore not entirely clear how they determine the eye fate.

An additional gene required for formation of the eye is *eyegone* (*eyg*), which encodes a protein containing a Pax-6-like homeodomain but only the C-terminal half of the paired domain (Jang et al. 1999). *eyg* is expressed at the same stage as *ey* in the embryonic eye primordium, but neither gene requires the other for its expression; both might be targets of *toy* (Jang et al. 1999). *eyg* is not sufficient to induce eye development outside the eye disk, though it enhances the ability of *ey* to do so (Jang et al. 1999). A final gene that can induce eye development when misexpressed in the antennal disk, although it is not required for normal eye development, is *teashirt* (*tsh*), which encodes a zinc finger transcription factor (Pan and Rubin 1998). It is possible that ectopic *tsh* results in the maintenance of *ey* expression present in the embryonic antennal primordium. A diagram outlining the known regulatory interactions between the eye specification genes *ey*, *eya*, *so*, and *dac* is shown in Fig. 2.

Finally, two homeobox-containing genes, *homothorax* (*hth*) and *extradenticle* (*exd*) act to inhibit eye formation (Pai et al. 1998). Both genes are coexpressed along the anterior-lateral margins of the eye disk and the

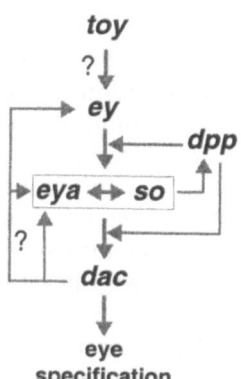

Fig. 2. A cartoon depicting known and possible interactions in the genetic hierarchy that controls retinal specification in *Drosophila*. After Chen et al. (1999)

peripodial membrane, regions that will give rise to head cuticle. Clones of cells lacking either *hth* or *exd* cause ectopic eye formation near the ventral margin of the eye; conversely, ectopic expression of *hth* blocks retinal development (Pai et al. 1998). It is therefore likely that these genes delimit the eye field. Hth protein acts by regulating nuclear localization of exd. How *hth* and *exd* interact with genes that promote (*ey, eya, so, eyg,* and *dac*) or inhibit (*wingless,* see below) eye development remains to be determined.

3
Dorsal-Ventral Patterning and Growth of the Eye Disk

Growth of the eye primordium and its further development depends on asymmetry in the dorsal-ventral (DV) axis; signaling between the dorsal and ventral fields specifies the midline of the disk, which defines the site where differentiation will initiate. The outcome of these signals appears to be the activation of the Notch (N) cell-surface receptor at the DV midline. This activation is accomplished by two ligands, Delta (Dl) on the dorsal side of the midline, and Serrate (Ser) on the ventral side (Cho and Choi 1998; Dominguez and de Celis 1998; Papayannopoulos et al. 1998). The asymmetric distribution of these ligands is critically dependent on the restriction of the signaling molecule fringe (fng) to the ventral half of the disk (Cho and Choi 1998; Dominguez and de Celis 1998; Papayannopoulos et al. 1998). During both wing and eye development, fng has been shown to potentiate the response of N to Dl, leading to the activation of *Ser* expression; fng also inhibits the response of N to Ser, preventing it from activating *Dl* in ventral cells (Panin et al. 1997; Cho and Choi 1998; Dominguez and de Celis 1998; Papayannopoulos et al. 1998). A boundary of *fng* expression is thus essential for N activation; either loss of *fng* activity or misexpression of *fng* throughout the disk causes a failure of the eye disk to grow and differentiate (Cho and Choi

1998; Dominguez and de Celis 1998; Papayannopoulos et al. 1998). This failure can be rescued by a constitutively activated form of N, which leads to overgrowth of the eye field (Dominguez and de Celis 1998).

The ventral restriction of *fng* expression may be controlled by three related homeobox genes in the *Iroquois* complex, *araucan* (*ara*), *caupolican* (*caup*), and *mirror* (*mrr*) (Gomez-Skarmeta et al. 1996; McNeill et al. 1997). All three are expressed specifically in the dorsal half of the eye (McNeill et al. 1997; Dominguez and de Celis 1998; Netter et al. 1998), and ectopic expression of either *caup* or *mrr* has been shown to repress *fng* expression (Cho and Choi 1998; Dominguez and de Celis 1998). The basis for their own dorsal restriction is less clear. Expression of *mrr* requires the function of the Wnt protein encoded by *wingless* (*wg*), and overexpression of *wg* can expand *mrr* expression, shifting the DV midline ventrally (Heberlein et al. 1998). Maintenance of *mrr* repression ventrally requires the *Polycomb* group of genes (Netter et al. 1998), implicating a repressive chromatin structure, though there is no evidence that these genes provide the spatial localization necessary to establish its initial repression. In addition to promoting growth of the eye disk and contributing to the selection of the site at which differentiation initiates, the DV boundary emits a signal defining ommatidial polarity (reviewed by Jarman 1996).

4
Initiation of Differentiation

In the third larval instar, photoreceptor clusters begin to differentiate in the eye disk, starting at the posterior margin and spreading anteriorly. Initiation of neuronal differentiation requires the function of the *hedgehog* (*hh*) gene, which encodes a secreted protein (Dominguez and Hafen 1997; Royet and Finkelstein 1997; Borod and Heberlein 1998). *hh* is present at the central posterior margin of the eye disk prior to the start of differentiation and may thus define the site of initiation (Dominguez and Hafen 1997; Royet and Finkelstein 1997; Borod and Heberlein 1998). *decapentaplegic* (*dpp*), a BMP family member, is also required for initiation (Burke and Basler 1996; Wiersdorff et al. 1996; Chanut and Heberlein 1997; Pignoni and Zipursky 1997). *dpp* is expressed around the posterior and lateral margins of the early eye disk (Masucci et al. 1990); its posterior expression is dependent on *hh* (Dominguez and Hafen 1997; Royet and Finkelstein 1997; Borod and Heberlein 1998). Clones of cells mutant for downstream, cell-autonomous components of the *dpp* pathway, such as the receptors encoded by *thick veins* (*tkv*) and *punt* or the transcription factor encoded by *Mothers against dpp* (*Mad*), fail to initiate photoreceptor differentiation (Burke and Basler 1996; Wiersdorff et al. 1996). Ectopic expression of *dpp* in the eye disk induces ectopic differentiation, but only initiating at the anterior margin of the disk

(Chanut and Heberlein 1997; Pignoni and Zipursky 1997), suggesting that the margins express a factor required in conjunction with *dpp*. This ectopic initiation requires *hh*, the expression of which is activated at the anterior margin by *dpp*. This may reflect a normal feedback regulation of *hh* by *dpp*, which must be limited to the margins of the eye disk (Borod and Heberlein 1998).

Localization of the site of initiation also requires negative regulation by *wg*, which is expressed at the dorsal and ventral margins of the eye disk. Its removal using a temperature-sensitive allele results in ectopic initiation, particularly from the dorsal margin (Ma and Moses 1995; Treisman and Rubin 1995), that is *hh*-dependent (Borod and Heberlein 1998). Conversely, ectopic expression of *wg* at the posterior margin can inhibit initiation (Treisman and Rubin 1995). *wg* is expressed throughout the second instar eye disk and is subsequently restricted to the anterior dorsal and ventral margins by *dpp* (Royet and Finkelstein 1997). *wg* plays a role in promoting the differentiation of head cuticle by these regions of the eye disk, mediated in part by its activation of the expression of the homeobox gene *orthodenticle* (*otd*) (Royet and Finkelstein 1997). Although ectopic *wg* is present in posterior margin cells that are unable to receive the dpp signal (Burke and Basler 1996; Wiersdorff et al. 1996), it is not the only reason for their failure to differentiate as photoreceptors, as clones of cells doubly mutant for *Mad* and for *wg* still form no photoreceptors (Hazelett et al. 1998). wg does not interfere with the expression or function of the dpp protein; it may act further downstream, preventing the epidermal growth factor (EGF) receptor pathway (see below) from triggering photoreceptor differentiation (Hazelett et al. 1998). Neuronal differentiation is severely impaired in eye disks lacking EGFR function (Kumar et al. 1998).

The initiation of neuronal differentiation in the eye disk also requires the function of the proneural gene *atonal* (*ato*). Flies carrying a loss-of-function *ato* mutation are nearly eyeless and the remaining narrow eyes contain no photoreceptor neurons (Jarman et al. 1994). The failure of photoreceptor differentiation in *ato* mutants is due to a defect in the specification of photoreceptor R8, the so-called ommatidial founder cell, which is the first photoreceptor cell to differentiate in each cluster. It is believed that in the absence of an R8 cell the remaining photoreceptors fail to be recruited and induced to differentiate as neurons. The expression of *hh*, *dpp*, and *wg* is normal in young third instar *ato* mutant eye disks. However, neuronal differentiation is absent and *dpp* expression decays (Jarman et al. 1994, 1995; Borod and Heberlein 1998). Ato protein is expressed along the posterior margin just prior to the start of differentiation, and this expression requires *hh* function for its initiation and/or maintenance. This observation, together with the fact that *hh* expression at the margin is *ato*-independent, suggests that *ato* functions genetically downstream of *hh* in the pathway that leads to the initiation of differentiation in the eye disk (Dominguez and Hafen 1997; Borod and Heberlein 1998).

ato encodes a basic helix-loop-helix (bHLH) protein and is thus a potential transcriptional regulator. In vitro, ato protein forms heteromultimers with another bHLH protein encoded by the *daughterless (da)* gene; these ato-da multimers act as sequence-specific DNA-binding complexes (Jarman et al. 1993). Loss of *da* function in the retina leads to a failure of neuronal differentiation similar to that observed for *ato* (Brown et al. 1996). It is therefore likely that *ato* and *da* act as partners during proneural specification in the eye. A cartoon diagram summarizing our current understanding of the genetic hierarchy that controls initiation is shown in Fig. 3.

5
How are Specification and Initiation Related?

The connection between the early specification of the eye disk and its subsequent differentiation is not well understood. One important area of investigation is the regulatory relationship between the eye specification genes *ey*, *eya*, *so*, *dac*, and *eyg* on one hand, and the patterning genes *hh*, *dpp*, and *wg* on the other. *eya* and *so* are necessary for *dpp* expression, which is only weakly initiated in eye disks mutant for either of these genes and is completely absent in *eya* or *so* mutant cells at the third instar (Pignoni et al. 1997; Hazelett et al. 1998). *eya* and *eyg* both play a role in the repression of *wg*, which is misexpressed at the posterior margin of *eyg* mutant disks and in third instar *eya* mutant clones (Hazelett et al. 1998; Jang et al. 1999). The lack of photoreceptor differentiation in *eyg* mutant disks can be rescued by inhibiting wg signaling at the posterior margin, indicating that repressing *wg* posteriorly is a critical function of *eyg* (Hazelett et al. 1998). This is not true for *eya* (Hazelett et al. 1998), consistent with a requirement for *eya* even at late stages of photoreceptor differentiation (Pignoni et al. 1997). *dpp* expression is normal in *dac* mutant eye disks (Mardon et al. 1994), though *wg* is ectopically expressed at the posterior of the disk (Treisman and Rubin 1995); the role this ectopic *wg* plays in the *dac* mutant phenotype has not been

Fig. 3. A cartoon showing a complex series of genetic interactions that directly or indirectly regulate the initiation of photoreceptor differentiation in *Drosophila*. *Arrows* indicate positive interactions while intersecting lines depict negative interactions

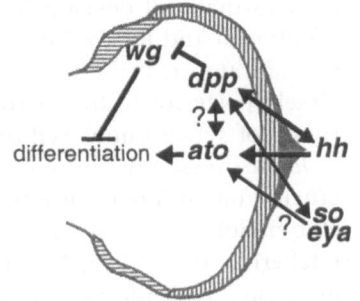

tested. Potential regulatory interactions between the eye specification genes and *hh* have not yet been reported. However, both *eya* and *so* are required for normal *ato* expression (Jarman et al. 1995); whether *hh* acts as an intermediary in this process is unknown.

There is also evidence that *dpp* and *wg* may act upstream of or in conjunction with the specification transcription factors. Although *ey* is still present in *dpp* mutant eye disks, *eya* and *so* are only weakly expressed (Chen et al. 1999). Also, widespread misexpression of *ey* in other imaginal disks results in photoreceptor formation only in regions of the disks where *dpp* and *hh* are present and *wg* is absent (Halder et al. 1998). Coexpression of *dpp* with *ey* expands the domain in which it can induce eye development considerably, though it is still restricted to the posterior of the disks, where *hh* is expressed (Chen et al. 1999). *dpp* also enhances the efficiency of ectopic eye induction by *ey*, *eya*, or *so* (Chen et al. 1999). *hh* and *dpp* also act to pattern other imaginal disks and the embryo, so in themselves they have no specificity for eye development. Presumably the late requirement observed for *eya* and *so* reflects their function in interpreting *hh/dpp* as a signal for photoreceptor differentiation.

6
Progression of Differentiation

Differentiation of the retina is asynchronous; once initiated at the posterior tip of the eye disk, it progresses across the epithelium one row at a time, reaching the anterior margin approximately 2 days later. The anterior edge of this differentiation wave is marked by an indentation in the apical surface of the epithelium, the morphogenetic furrow (MF), which spans the disk along its dorsoventral axis (Ready et al. 1976). The MF is a consequence of localized and transient changes in cell shape (Wolff and Ready 1991). The MF coincides temporally and spatially with several important events in retinal morphogenesis. Ahead of the furrow, cells are unpatterned and undifferentiated, and they divide asynchronously. In the MF, cells become synchronized in the G1 phase of their cell cycle as they begin to associate into evenly spaced clusters that will develop into the individual ommatidia (Ready et al. 1976; Tomlinson and Ready 1987; Wolff and Ready 1991). The first five cells of the ommatidium exit the cell cycle and begin differentiating as neurons immediately posterior to the furrow, while the remaining cells undergo one more round of synchronous cell division prior to differentiation.

hh, expressed in and secreted by cells located posterior to the MF, plays a crucial role in furrow progression (Heberlein and Moses 1995). Reduction or elimination of *hh* function in the eye disk leads to a halt in MF progression (Heberlein et al. 1993; Ma et al. 1993). Conversely, ectopic expression of *hh* or ectopic activation of the *hh* signal transduction pathway in undifferentiated cells located ahead of the MF causes their precocious differentiation (Heberlein et al. 1995; Ma and Moses 1995; Pan and Rubin 1995; Wehrli and Tomlinson 1995). However, clones of cells mutant for the *smoothened* (*smo*) gene, which is required for the transduction of the *hh* signal, still allow neuronal differentiation and furrow progression, albeit at a reduced rate (Strutt and Mlodzik 1997). This suggests that other signals, in addition to *hh*, may contribute to proper MF progression (see below).

How does *hh* regulate furrow progression? *hh* induces the expression of several genes in and around the MF, including *dpp* and *ato*. *dpp* is expressed in the furrow (Blackman et al. 1991), but its function in MF progression is not clear. Clones of cells mutant for the autonomous components of the *dpp* signal transduction pathway, such as *tkv*, *punt*, and *Mad*, display only minor irregularities when located in the interior regions of the eye disk (Burke and Basler 1996; Wiersdorff et al. 1996; Penton et al. 1997). However, global reduction of *dpp* function, by means of a temperature-sensitive allelic combination, significantly slows the progression of differentiation (Chanut and Heberlein 1997). It is possible that *dpp* plays a permissive role in furrow progression by ensuring proper cell cycle control; indeed, it has been shown recently that *dpp* is required for G1 arrest in the furrow (Horsfield et al. 1998).

A more direct mechanism by which *hh* regulates furrow progression is through its induction of *ato* expression. Ato protein is expressed in a strip of cells immediately anterior to the MF. In the furrow, expression is restricted to evenly spaced groups of cells and shortly thereafter to single R8 cells (Jarman et al. 1994). Removal of *hh* function, by means of a temperature-sensitive mutation, leads to a loss of ato in the MF (Borod and Heberlein 1998); ectopic expression of *hh* induces ectopic expression of ato (Heberlein et al. 1995). During the process of MF progression, cells that receive the *hh* signal must in turn become cells that send the *hh* signal to their more anteriorly located neighbors. This transition from recipient to sender of the *hh* signal requires the receiving cell to differentiate as a photoreceptor neuron, a process to which *ato* and *da* function are central (for a review see Treisman and Heberlein 1998). The mechanisms by which *hh*, *ato*, and *da* regulate each other's expression is thus an important and unresolved issue.

Two additional genes of the HLH family negatively regulate furrow progression: *hairy* (*h*) and *extramacrochaete* (*emc*). *h* is expressed in a dorsoventral strip located immediately anterior to the domain of *ato* expression (Brown et al. 1991); *emc*, while fairly ubiquitous, is highest near the anterior margin (Brown et al. 1995). Clones of cells mutant for *h* or *emc* function in the eye have little effect on normal development. However, the MF and the front of differentiation are strikingly accelerated when crossing clones of cells doubly mutant for *h* and *emc* (Brown et al. 1995). It is likely that these two genes act normally by inhibiting the proneural functions of *ato*. Curiously, ectopic expression of *hh* or ectopic activation of the *hh* pathway ahead of the furrow leads to ectopic expression of *h* in nearby cells (Heberlein et al. 1995; Pan and Rubin 1995). Thus, *hh* appears to induce not only an activator of neural differentiation (*ato*), but also an inhibitor (*h*). The correct balance of these two molecules may ensure the orderly patterning of the retina.

Another signaling pathway that has recently been shown to modulate *ato* expression in the MF is that activated by the EGFR: *ato* expression is reduced in clones of cells lacking the EGFR and induced in cells in which the EGFR pathway has been ectopically activated (Dominguez et al. 1998; Spencer et al. 1998). However, *ato* expression appears relatively normal in eye disks in which the EGFR has been inactivated by a temperature-sensitive mutation, perhaps reflecting some residual function (Kumar et al. 1998). Thus, the EGFR pathway may not only induce the normal differentiation of the R1-6 photoreceptors (reviewed in Freeman 1996), but also play a role in modulating furrow progression through the regulation of *ato* expression. This effect could, however, be mediated indirectly by EGFR induction of *hh*. Interestingly, ectopic activation of the EGFR pathway can induce neuronal differentiation in the eye disk even in the absence of *ato* function (Dominguez et al. 1998), showing that the normally obligate order of retinal differentiation can be bypassed by artificial activation of the EGFR. However, the resulting ommatidia are disorganized, underscoring the importance of a gradual and asynchronous differentiation process in proper patterning and, ultimately, proper vision.

A third signaling system that regulates *ato* expression and photoreceptor differentiation is the *Notch* (*N*) pathway. Clones of cells lacking the *N* receptor or its ligand Delta show reduced *ato* expression and impaired neuronal differentiation (Baker and Yu 1997). It is not known whether *N* is required for cells to respond to *hh* or if it plays a more direct role in neuronal differentiation.

Finally, the steroid hormone ecdysone has recently been shown to control furrow progression (Brennan et al. 1998). Mutants that fail to produce or are impaired in the response to ecdysone show a disruption in furrow progression. Expression of *hh* and *ato* is lost in these mutants, providing a likely mechanism for their effect on the furrow. Curiously, loss of function of Ultraspiracle (Usp), the *Drosophila* RXR, in clones of cells leads to a slight acceleration of the furrow (Zelhof et al. 1997). Usp forms heterodimers with the Ecdysone receptor (EcR), and would therefore have been predicted to act as a positive modulator of furrow progression. It is possible that both USP and EcR have other dimerization partners that determine their specific roles in furrow progression.

In summary, the progression of the wave of neuronal differentiation is a highly regulated and complex process that involves the coordinate function of multiple signaling systems. An understanding of the interactions between the hedgehog, EGFR, Notch, and ecdysone signaling pathways will be crucial to our understanding of how the furrow moves across the eye disk leaving in its wake a beautifully precise developmental pattern. Vertebrate retinae also differentiate asynchronously; thus lessons learned from flies are likely to provide insights into their developmental mechanism.

Acknowledgements. The eye development projects in our laboratories are funded by NIH/ EY11410 and NSF/IBN-9604216 (to U. H.), and NIH/GM56131 and NSF/IBN-9728140 (to J.E.T.). We thank members of our laboratories for their insights on eye development and comments on the manuscript.

References

Baker NE, Yu S-Y (1997) Proneural function of neurogenic genes in the developing *Drosophila* eye. Curr Biol 7 : 122–132

Banerjee U, Zipursky SL (1990) The role of cell–cell interaction in the development of the *Drosophila* visual system. Neuron 4 : 177–187

Blackman RK, Sanicola M, Raftery LA, Gillevet T, Gelbart WM (1991) An extensive 3' *cis*-regulatory region directs the imaginal disk expression of *decapentaplegic*, a member of the TGF-b family in *Drosophila*. Development 111 : 657–666

Bonini NM, Leiserson WM, Benzer S (1993) The *eyes absent* gene: genetic control of cell survival and differentiation in the developing *Drosophila* eye. Cell 72 : 379–395

Bonini NM, Bui QT, Gray-Board GL, Warrick JM (1997) The *Drosophila eyes absent* gene directs ectopic eye formation in a pathway conserved between flies and vertebrates. Development 124 : 4819–4826

Borod ER, Heberlein U (1998) Mutual regulation of *decapentaplegic* and *hedgehog* during the initiation of differentiation in the *Drosophila* retina. Dev. Biol. 197 : 187–197

Brennan CA, Ashburner M, Moses K (1998) Ecdysone pathway is required for furrow progression in the developing *Drosophila* eye. Development 125 : 2653–2664

Brown NL, Sattler CA, Markey DR, Carroll SB (1991) *hairy* gene function in the *Drosophila* eye: normal expression is dispensable but ectopic expression alters cell fates. Development 113 : 1245–1256

Brown NL, Sattler CA, Paddock SW, Carroll SB (1995) Hairy and emc negatively regulate morphogenetic furrow progression in the *Drosophila* eye. Cell 80 : 879–887

Brown NL, Paddock SW, Sattler CA, Cronmiller C, Thomas BJ, Carroll SB (1996) Daughterless is required for Drosophila photoreceptor cell determination, eye morphogenesis, and cell cycle progression. Developmental Biology 179 : 65–78

Burke R, Basler K (1996) Hedgehog-dependent patterning in the *Drosophila* eye can occur in the absence of dpp signaling. Dev Biol 179 : 360–368

Chanut F, Heberlein U (1997) Role of decapentaplegic in initiation and progression of the morphogenetic furrow in the developing *Drosophila* retina. Development 124 : 559–567

Chen R, Amoui M, Zhang Z, Mardon G (1997) Dachshund and eyes absent proteins form a complex and function synergistically to induce ectopic eye development in *Drosophila*. Cell 91 : 893–904

Chen R, Halder G, Zhang Z, Mardon G (1999) Signaling by the TGF-b homolog *decapentaplegic* functions reiteratively within the network of genes controlling retinal cell fate determination in *Drosophila*. Development 126 : 935–943

Cheyette BNR, Green PJ, Martin K, Garren H, Hartenstein V, Zipursky SL (1994) The *Drosophila sine oculis* locus encodes a homeodomain-containing protein required for the development of the entire visual system. Neuron 12 : 977–996

Cho K-O, Choi K-W (1998) Fringe is essential for mirror symmetry and morphogenesis in the *Drosophila* eye. Nature 396 : 272–276

Czerny T, Halder G, Kloter U, Souabni A, Gehring WJ, Busslinger M (1999) Twin of eyeless, a second Pax-6 gene of Drosophila, acts upstream of eyeless in the control of eye development. Molecular Cell 3 : 297–307

Dominguez M, de Celis JF (1998) A dorsal/ventral boundary established by Notch controls growth and polarity in the *Drosophila* eye. Nature 396:276-278

Dominguez M, Hafen E (1997) Hedgehog directly controls initiation and propagation of retinal differentiation in the *Drosophila* eye. Genes Dev 11:3254-3264

Dominguez M, Wasserman JD, Freeman M (1998) Multiple functions of the EGF receptor in *Drosophila* eye development. Curr Biol 8:1039-1048

Freeman M (1997) Cell determination strategies in the *Drosophila* eye. Development 124:261-270

Glardon S, Callaerts P, Halder G, Gehring WJ (1997) Conservation of Pax-6 in a lower chordate, the ascidian *Phallusia mammillata*. Development 124:817-825

Glardon S, Holland LZ, Gehring WJ, Holland ND (1998) Isolation and developmental expression of the amphioxus *Pax-6* gene (*AmphiPax-6*): insights into eye and photoreceptor evolution. Development 125:2701-2710

Gomez-Skarmeta JL, Glavic A, de la Calle-Mustienes E, Modolell J, Mayor R (1998) Xiro, a Xenopus homolog of the *Drosophila* Iroquois complex genes, controls development at the neutral plate. Embo Journal 17:181-190

Halder G, Callaerts P, Gehring WJ (1995) Induction of ectopic eyes by targeted expression of the *eyeless* gene in *Drosophila*. Science 267:1788-1792

Halder G, Callaerts P, Flister S, Walldorf U, Kloter U, Gehring WJ (1998) Eyeless initiates the expression of both *sine oculis* and *eyes absent* during *Drosophila* compound eye development. Development 125:2181-2191

Hammond KL, Hanson IM, Brown AG, Lettice LA, Hill RE (1998) Mammalian and *Drosophila* *dachshund* genes are related to the *Ski* proto-oncogene and are expressed in eye and limb. Mech Dev 74:121-131

Haynie JL, Bryant PJ (1986) Development of the eye-antenna imaginal disk and morphogenesis of the adult head in *Drosophila melanogaster*. J Exp Zool 237:293-308

Hazelett DJ, Bourouis M, Walldorf U, Treisman JE (1998) *Decapentaplegic* and *wingless* are regulated by *eyes absent* and *eyegone* and interact to direct the pattern of retinal differentiation in the eye disk. Development 125:3741-3751

Heberlein U, Moses K (1995) Mechanisms of *Drosophila* retinal morphogenesis: the virtues of being progressive. Cell 81:987-990

Heberlein U, Wolff T, Rubin GM (1993) The TGFb homolog *dpp* and the segment polarity gene *hedgehog* are required for propagation of a morphogenetic wave in the *Drosophila* retina. Cell 75:913-926

Heberlein U, Singh CM, Luk AY, Donohoe TJ (1995) Growth and differentiation in the *Drosophila* eye coordinated by *hedgehog*. Nature 373:709-711

Heberlein U, Borod ER, Chanut FA (1998) Dorsoventral patterning in the *Drosophila* retina by *wingless*. Development 125:567-577

Hill RE, Favor J, Hogan BL, Ton CC, Saunders GF, Hanson IM, Prosser J, Jordan T, Hastie ND, van Heyningen V (1991) Mouse *small eye* results from mutations in a paired-like homeobox-containing gene. Nature 354:522-525

Horsfield J, Penton A, Secombe J, Hoffman FM, Richardson H (1998) Decapentaplegic is required for arrest in G1 phase during *Drosophila* eye development. Development 125:5069-5078

Jang C-C, Jones N, Chao J-L, Bessarab DA, Kuo YM, Jun S, Desplan C, Beckendorf S, Sun YH (1999) Two Pax genes, *eye gone* and *eyeless*, act in parallel in determining *Drosophila* eye development. Development (submitted)

Jarman AP (1996) Epithelial polarity in the *Drosophila* compound eye: eyes left or right? Trends Genet 12:121-123

Jarman AP, Grau Y, Jan LY, Jan YN (1993) *Atonal* is a proneural gene that directs chordotonal organ formation in the *Drosophila* peripheral nervous system. Cell 73:1307-1321

Jarman AP, Grell EH, Ackerman L, Jan LY, Jan YN (1994) *atonal* is the proneural gene for *Drosophila* photoreceptors. Nature 369:398-400

Jarman AP, Sun Y, Jan LY, Jan YN (1995) Role of the proneural gene, *atonal*, in formation of *Drosophila* chordotonal organs and photoreceptors. Development 121:2019-2030

Jurgens G, Hartenstein V (1993) The terminal regions of the body pattern. In: The development of *Drosophila Melanogaster*. Bate M, Martinez-Arias A (ed) Cold Spring Harbor Laboratory Press; Cold Spring Harbor, New York, pp 687-746

Kumar JP, Tio M, Hsiung F, Akopyan S, Gabay L, Seger R, Shilo B-Z, Moses K (1998) Dissecting the roles of the *Drosophila* EGF receptor in eye development and MAP kinase activation. Development 125:3875-3885

Loosli F, Kmita-Cunisse M, Gehring WJ (1996) Isolation of a *Pax-6* homolog from the ribbonworm *Lineus sanguineus*. Proc Natl Acad Sci USA 93:2658-2663

Ma C, Moses K (1995) *Wingless* and *patched* are negative regulators of the morphogenetic furrow and can affect tissue polarity in the developing *Drosophila* compound eye. Development 121:2279-2289

Ma C, Zhou Y, Beachy PA, Moses K (1993) The segment polarity gene *hedgehog* is required for progression of the morphogenetic furrow in the developing *Drosophila* eye. Cell 75:927-938

Mardon G, Solomon NM, Rubin GM (1994) *Dachshund* encodes a nuclear protein required for normal eye and leg development in *Drosophila*. Development 120:3473-3486

Masucci JD, Miltenberger RJ, Hoffmann FM (1990) Pattern-specific expression of the *Drosophila decapentaplegic* gene in imaginal disks is regulated by 3' cis-regulatory elements. Genes Dev 4:2011-2023

McNeill H, Yang CH, Brodsky M, Ungos J, Simon MA (1997) *Mirror* encodes a novel PBX-class homeoprotein that functions in the definition of the dorsal-ventral border in the *Drosophila* eye. Genes Dev 11:1073-1082

Netter S, Fauvarque MO, Diez del Corral R, Dura JM, Coen D (1998) *White*[+] transgene insertions presenting a dorsal/ventral pattern define a single cluster of homeobox genes that is silenced by the polycomb-group proteins in *Drosophila melanogaster*. Genetics 149:257-275

Oliver G, Mailhos A, Wehr R, Copeland NG, Jenkins NA, Gruss P (1995) *Six3*, a murine homologue of the *sine oculis* gene, demarcates the most anterior border of the developing neural plate and is expressed during eye development. Development 121:4045-4055

Pai C-Y, Kuo T-S, Jaw JJ, Kurant E, Chen C-T, Bessarab DA, Salzberg A, Sun YH (1998) The homothorax homeoprotein activates the nuclear localization of another homeoprotein, Extradenticle, and suppresses eye development in *Drosophila*. Genes Dev 12:435-446

Pan D, Rubin GM (1995) cAMP-dependent protein kinase and *hedgehog* act antagonistically in regulating *decapentaplegic* transcription in *Drosophila* imaginal disks. Cell 80:543-552

Pan D, Rubin GM (1998) Targeted expression of *teashirt* induces ectopic eyes in *Drosophila*. Proc Natl Acad Sci USA 95:15508-15512

Panin VM, Papayannopoulos V, Wilson R, Irvine KD (1997) Fringe modulates Notch-ligand interactions. Nature 387:908-912

Papayannopoulos V, Tomlinson A, Panin VM, Rauskolb C, Irvine KD (1998) Dorsal-ventral signaling in the *Drosophila* eye. Science 281:2031-2034

Penton A, Selleck SB, Hoffmann FM (1997) Regulation of cell cycle synchronization by *decapentaplegic* during *Drosophila* eye development. Science 275:203-206

Pignoni F, Zipursky SL (1997) Induction of *Drosophila* eye development by Decapentaplegic. Development 124:271-278

Pignoni F, Hu B, Zavitz KH, Xiao J, Garrity PA, Zipursky SL (1997) The eye specification proteins so and eya form a complex and regulate multiple steps in *Drosophila* eye development. Cell 91:881-892

Quiring R, Walldorf U, Kloter U, Gehring WJ (1994) Homology of the *eyeless* gene of *Drosophila* to the *Small eye* gene in mice and *Aniridia* in humans. Science 265:785-789

Ready DF, Hanson TE, Benzer S (1976) Development of the *Drosophila* retina, a neurocrystalline lattice. Dev Biol 53:217-240

Royet J, Finkelstein R (1997) Establishing primordia in the *Drosophila* eye-antennal imaginal disk: the roles of *decapentaplegic*, *wingless* and *hedgehog*. Development 124:4793-4800

Seo H-C, Drivenes O, Ellingsen S, Fjose A (1998) Expression of two zebrafish homologs of the murine *Six3* gene demarcates the initial eye primordia. Mech Dev 73:45–57

Serikaku MA, O'Tousa JE (1994) *Sine oculis* is a homeobox gene required for *Drosophila* visual system development. Genetics 138:1137–1150

Shen W, Mardon G (1997) Ectopic eye development in *Drosophila* induced by directed *dachshund* expression. Development 124:45–52

Spencer SA, Powell PA, Miller DT, Cagan RL (1998) Regulation of EGF receptor signaling establishes pattern across the developing *Drosophila* retina. Development 125:4777–4790

Strutt DI, Mlodzik M (1997) Hedgehog is an indirect regulator of morphogenetic furrow progression in the *Drosophila* eye disk. Development 124:3233–3240

Tomarev SI, Callaerts P, Kos L, Zinovieva R, Halder G, Gehring W, Piatigorsky J (1997) Squid Pax-6 and eye development. Proc Natl Acad Sci USA 94:2421–2426

Tomlinson A, Ready DF (1987) Neuronal differentiation in the *Drosophila* ommatidium. Dev Biol 120:366–376

Ton CCT, Hirvonen H, Miwa H, Weil MM, Monaghan P, Jordan T, van Heyningen V, Hastie ND, Meijers-Heijboer H, Drechsler M, et al. (1991) Positional cloning and characterization of a paired box- and homeobox-containing gene from the aniridia region. Cell 67:1059–1074

Treisman JE (1999) A conserved blueprint for the eye? Bioessays 21:843–850

Treisman JE, Heberlein U (1998) Eye development in *Drosophila*: Formation of the eye field and control of differentiation. Current Topics in Dev Biol 39:119–158

Treisman JE, Rubin GM (1995) *Wingless* inhibits morphogenetic furrow movement in the *Drosophila* eye disk. Development 121:3519–3527

Wehrli M, Tomlinson A (1995) Epithelial planar polarity in the developing *Drosophila* eye. Development 121:2451–2459

Wiersdorff V, Lecuit T, Cohen SM, Mlodzik M (1996) *Mad* acts downstream of dpp receptors, revealing a differential requirement for *dpp* signaling in initiation and propagation of morphogenesis in the *Drosophila* eye. Development 122:2153–2162

Wolff T, Ready DF (1991) The beginning of pattern formation in the *Drosophila* compound eye: the morphogenetic furrow and the second mitotic wave. Development 113:841–850

Xu P-X, Woo I, Her H, Beier DR, Maas RL (1997) Mouse *Eya* homologues of the *Drosophila eyes absent* gene require *Pax6* for expression in lens and nasal placode. Development 124:219–231

Zelhof AC, Ghbeish N, Tsai C, Evans RM, McKeown M (1997) A role for Ultrapiracle, the *Drosophila* RXR, in morphogenetic furrow movement and photoreceptor cluster formation. Development 124:2499–2506

Zimmerman JE, Bui QT, Steingrimsson E, Nagle DL, Fu W, Genin A, Spinner NB, Copeland NG, Jenkins NA, Bucan M, Bonini NM (1997) Cloning and characterization of two vertebrate homologs of the *Drosophila eyes absent* gene. Genome Res 7:128–141

Induction of the Lens

Nicolas Hirsch and Robert M. Grainger[1]

1
Introduction

The eye may be a window to the soul, but for developmental biologists it has provided a window of another kind: one into the processes of embryonic induction. Since the turn of the century, when embryologists first noted that the single lens of a cyclopic embryo is in perfect register with the other parts of its single, median eye, the induction of the vertebrate lens has been a classical model for the study of embryonic induction. Although a comprehensive model exists describing the timing and tissue interactions involved in lens induction (Grainger et al. 1992; Grainger 1996), only recently have data become available which identify some of the molecular biological components that may underlie this process. In this chapter, we will briefly discuss early experiments in lens induction and the flaws inherent in their methodology as well as review model systems commonly used in the study of lens induction. We will also discuss a number of genes thought to play critical roles in lens induction and examine how they fit into the current model.

2
A Historical Overview of the Study of Lens Induction

The early observations of cyclopic embryos alluded to above suggested that parts of the eye are integrated into a whole unit even when the location of the eye is disturbed. During normal development the retina and lens are derived from different sites at the anterior end of the embryo; at the neural plate stage the presumptive retinal regions can be mapped to anterior neural tissue and the lens to adjacent "sensory" ectoderm (Fig. 1A). At the time of neural tube closure the presumptive retina comprises the protruding tip of the newly formed optic vesicle, which contacts the presumptive lens ectoderm (PLE) for the first time. Lens differentiation begins many hours later in development.

---

[1] Department of Biology, University of Virginia, Charlottesville, Virginia 22903, USA

Results and Problems in Cell Differentiation, Vol. 31
M. E. Fini (Ed.): Vertebrate Eye Development
© Springer-Verlag Berlin Heidelberg 2000

GASTRULA   NEURAL PLATE   NEURAL TUBE   TADPOLE

A. Normal lens development

Lens field
Retinal field

Lens ectoderm
Optic vesicle

Lens
Optic cup

B. Classical model of lens induction

Necessity of optic vesicle

Ablation of retinal field

No lens forms

Sufficiency of optic vesicle

Transplant of ectopic optic vesicle

Ectopic lens

C. Current model of lens induction

COMPETENCE

BIAS

Area of inhibition
Area of specification

INHIBITION AND SPECIFICATION

DIFFERENTIATION

The first experiments demonstrating the relationship of presumptive retina and lens were performed by Spemann (1901), who showed that the developing retina is necessary to induce the lens; in cases where the presumptive retina was ablated at the neural plate stage, a lens failed to develop. Additional work by Lewis (1904) suggested that the optic vesicle may also be sufficient to induce a lens. When he transplanted the optic vesicle of a late-stage neurula donor beneath nonlens regions of ectoderm, he saw lens-like structures develop. Much early experimental work tended to reinforce this view of the necessity and sufficiency of the optic vesicle for lens induction (Fig. 1B). In this way, the induction of the lens came to be viewed as a model for other embryonic inductions: a single tissue, the optic vesicle, being responsible for a single response, induction of the lens. Some early evidence questioned the simplicity of this model, however, particularly Mencl's study of the formation of "free" lenses; i.e. lenses formed without the presence of an optic vesicle (Mencl 1903). Spemann also found, in *Rana esculenta*, that lenses formed in anterior ectoderm even after the ablation of the entire neural plate with a hot needle (Spemann 1907), implying that the optic vesicle was not absolutely required for lens induction.

In addition, a rigorous reexamination of the literature by Saha and colleagues (1989) argued that many early studies were flawed because of an inability of the experimenter to distinguish two critical factors. First, whether the actual source of an induced lens was donor ectoderm or contaminating host PLE and, second, whether the induced structure was truly a differentiated lens or simply an artifactual thickening of the ectoderm. A large body of more recent work has taken these objections into consideration. Early studies of contamination showed that lenses formed after the removal of the PLE arose almost exclusively from host lens cells which remained adherent to the optic vesicle during the surgery (Stone and Dinnean 1940), casting doubt on Lewis' argument that the optic vesicle is sufficient for lens induction. The use of molecular markers, in particular antibodies to crystallin, the predominant protein expressed in differentiated lens cells, allowed researchers to unam-

---

Fig. 1A–C. Models of lens development in *Xenopus*. All views of embryos are dorsal. Neural plate and neural tube are outlined with *solid black lines*. Lens and retinal/optic areas are marked at each stage. Nieuwkoop and Faber (N&F) stages are: *Gastrula* (N&F stage 11), *Neural plate* (N&F stage 14), *Neural tube* (N&F stage 19), and *Tadpole* (N&F stage 37/8). **A.** Normal lens development in a wildtype embryo. **B.** Classical model of lens induction. *First line* summarizes Spemann's experiments (Spemann 1901) showing the necessity of the optic vesicle for lens formation by ablating the early retinal field on one side of the embryo. *Second line* summarizes Lewis' experiments (Lewis 1904) showing the sufficiency of the optic vesicle for lens induction by transplanting an ectopic vesicle under the flank ectoderm. **C.** Current model of lens induction. *Shaded areas* show tissues involved in the different phases of the lens induction model: competence (*light gray*), bias (*medium gray*), inhibition (*light gray*), specification (*dark gray*) and differentiation (*black*)

biguously distinguish lens tissue from a nonlens artifact (Fedestova and Barabanov 1978; Karkinen-Jaaskelainen 1978).

A synthesis of the data from classic experiments, modern transplantation studies, gene knockouts, and molecular cloning experiments has led to a model for lens induction considerably different from the single-inducer model of 30 years ago. We divide the current model (Fig. 1C) into five different phases: competence, bias, inhibition, specification, and differentiation. Each phase occurs at a progressively more advanced stage of embryonic development and each phase is associated with the expression of a number of gene products in the lens ectoderm (Table 1). The earliest step in this process occurs at mid/late gastrula stages, when a region of ectoderm becomes competent to respond to lens-inducing signals. Competence, defined by Waddington (1936) as the ability of cells to respond to a signal, is the earliest phase of lens development; the ectoderm of a *Xenopus* embryo is competent to receive lens-inducing signals for only a few hours (Servetnick and Grainger 1991). Between mid-gastrula and neural plate stages the lens forming potential, or bias, is enhanced in anterior non-neural ectoderm by planar interactions with the developing neural plate (Henry and Grainger 1987). Lens bias is correlated with the expression of a number of putative transcription factors including *Six3, Pax-6,* and *Otx2.* The biased region, however, is sig-

Table 1. Timing of gene expression in the presumptive lens ectoderm (PLE) in *Xenopus.* The timing of gene expression and the timing of development start with early embryonic stages (*top*) and progress to later ones (*bottom*)

Gene	Stage of embryonic development	Phase of lens induction	Reference
Sox-3	Early gastrula	Competence ↓	Zygar et al. (1998)
Optx2	Late gastrula	Bias	Toy et al. (1998) (chick)
Pax-6			Zygar et al. (1998)
Otx2	Early neural plate		Pannese et al. (1995)
Six3			Zygar et al. (1998)
Eya 3		↓	Xu et al. (1997) (mouse)
	Neural ridge		
Lens-1			Kenyon et al. (1999)
	Neural fold	Inhibition and Specification	
	Neural tube	↓	
Sox-3			Zygar et al. (1998)
L-Maf		Differentiation	Ogino and Yasuda, (1998) (chick)
Maf-B			Kawauchi et al. (1999) (mouse)
Prox1			Del-Rio Tsonis et al. (1999)
Pitx-3			Semina et al. (1997) (mouse)
Crystallins	Early tadpole	↓	Smolich et al. (1993)

nificantly larger than the actual lens ectoderm, and an inhibition of bias in nonlens head ectoderm is hypothesized to take place at early neural tube stages, restricting lens-forming ability to ectoderm directly overlying the optic vesicle. This ectoderm, if it is removed immediately after making contact with the optic vesicle and cultured in isolation, will go on to form a rudimentary lens and is said to be specified. The differentiation of the lens occurs after optic vesicle contact with the PLE and is characterized by synthesis of crystallins and the formation of fiber cells. Recent experiments suggest that the timing mechanism governing the onset of differentiation is plastic and can be reset by environmental influences (M. Offield, N. Hirsch and R. Grainger, unpubl.).

While this model encompasses many of the observations made about the location and timing of the different steps of lens induction, much less is known about the molecular interactions which underlie these steps. For example, a number of transcription factors are known to be expressed in the PLE at the stage when lens-forming bias is acquired (Table 1), but it is not presently known whether any, or all of these factors, contribute to bias. In addition, the nature of the inductive signal which travels between the presumptive retina and the PLE has not yet been fully characterized, nor has the nature of the later interaction between the optic vesicle and the lens epithelium which results in the onset of crystallin expression and changes in lens cell morphology.

A number of model systems have been used in the study of lens induction, each with its own particular strengths and weaknesses. Much of the classical work has been done on the embryos of urodele or anuran amphibians, particularly *Xenopus laevis*, since tissue transplantation and ablation is easily done in these species. Although *X. laevis* has a number of drawbacks as a genetic system, recent advances in transgenic technology (Amaya and Kroll 1999; Kroll and Amaya 1996) and in the use of the diploid *Xenopus tropicalis* (Amaya et al. 1998) have made amphibians a more attractive system in which to pursue questions about genetic interactions in the process of lens induction. By far the greatest amount of genetic analysis of lens development, however, has been done in the mouse. The use of both naturally occurring mutations in eye development such as *Small eye* and *Aphakia*, as well as knockouts of genes such as *BMP-4* (Wawersik et al. 1999) and *Otx2* (Ang et al. 1996), have contributed a great deal to the understanding of lens development. A number of zebra fish mutants have been identified with abnormal or ectopic lens development (Easter and Nicola 1996; Karlstrom et al. 1999). A great deal of embryology has been done in the chick to address the necessity and sufficiency of the optic vesicle for lens induction; moreover chick and mouse cell culture lines have been instrumental in defining the regulatory regions for a number of the crystallin genes (Cvekl and Piatigorsky 1996; Duncan et al. 1996, 1998). In the next sections we will consider the potential roles of specific gene products in the biological processes which define lens induction.

3
Competence

Lens competence is one of a series of competences through which animal cap ectoderm progresses in early embryogenesis. Prior to becoming lens-competent, this tissue is first competent to respond to mesoderm-inducing signals from the late morula stage, Nieuwkoop and Faber stage 6 (N&F 6) (Nieuwkoop and Faber 1956) through the early gastrula stage (N&F 10.5) and then to neural-inducing signals from about N&F stage 9 to 11.5. One hypothesis explaining the control of these competences invokes a tissue-autonomous developmental timer, since ectoderm explanted from early gastrulae and cultured in vitro for a period of time gains and loses competence in synchrony with tissue left in vivo (Servetnick and Grainger 1991). The autonomy of this timer is thought to be strong enough that even single, nondividing cells isolated from animal caps will lose mesodermal competence in synchrony with cells in whole embryos (Grainger and Gurdon 1989). While there is evidence that the timing mechanism governing the loss of mesodermal competence may be controlled by the relative levels of maternal to zygotic histone proteins (Steinbach et al. 1997), very little is currently known about the component parts of a cell-autonomous timer. Recent evidence also suggests that this process may be controlled not by the appearance or disappearance of a receptor for an inducing signal, as is often suggested, but by a signal transduction component downstream of the receptor (K. Curran and R. Grainger, unpubl.). While there is support for an autonomous element in the mesoderm competence process, there is also evidence that some aspects of competence within the animal cap are maintained through cell–cell signaling pathways involving FGF (Cornell et al. 1995). An intercellular signaling component within the animal cap is suggested by experiments in which disruption of cell contact by dissociation of the animal cap can prolong the period of mesodermal competence (K. Curran and R. Grainger, unpubl.).

The molecular basis of lens competence has not yet been investigated, though there are some genes, expressed during the lens-competent period in *Xenopus* animal cap ectoderm, that are thereby candidates for regulators of competence. One gene expressed in the ectoderm during this period is the HMG-box family member *Sox-3* (Zygar et al. 1998). While its role is unclear, there is evidence that the related gene *Sox-2* increases the responsiveness of neurally competent gastrula ectoderm to FGF signaling. It has been shown that bFGF alone cannot fully neuralize this ectoderm, but it can cause ectoderm that is overexpressing *Sox-2* to make neural markers such as N-CAM and N-tubulin (Mizuseki et al. 1998). *Sox-3* may act in a similar fashion in lens-competent ectoderm, possibly by making certain regions of genomic DNA accessible for other transcriptional regulatory complexes to bind. Another family of potential regulators of competence are the bone morphogenetic proteins (BMPs). The dynamic expression pattern of these genes in animal cap tissue corresponds to periods when competences are changing. Both *BMP-7* and *BMP-4* expression in *Xenopus* animal cap cells is increasing during the period when mesodermal competence is normally lost (Nishimatsu and Thomsen 1998). In the mouse, *BMP-7* expression in the presumptive lens ectoderm also occurs at the right developmental time to suggest a role in competence (Wawersik et al. 1999). Perhaps not coincidentally, *BMP-4* misexpression in animal caps virtually eliminates *Sox-2* expression (Mizuseki et al. 1998), and may be a mechanism for reducing the aforementioned responsiveness of animal cap cells to inducing molecules such as FGF.

Although lens development may perhaps best be viewed as a single, unified process, competence and bias at present appear to be distinguishable stages in this process with discrete, identifiable characteristics. The competence process is governed by an autonomous timer within responsive tissue; in contrast, the acquisition of bias by the head ectoderm requires signaling from the anterior neural plate. Thus, to characterize the molecular basis of lens bias, one must look at signaling molecules both within and outside the anterior neural plate and at genes within the presumptive lens ectoderm which may be responding to the inducers and maintaining lens bias in the ectoderm.

4
Bias

The findings on lens competence have important implications in defining when lens-inducing signals act: they must begin acting within the short competent period during gastrulation. This finding is consistent with experiments suggesting that the later-stage optic vesicle is not sufficient to induce lens formation and that inducing signals must act at earlier stages of development. Work in *Xenopus* has shown that planar signaling from the anterior neural plate at gastrula and early neurula stages is critical for the development of a lens (Henry and Grainger 1990). Tissues such as dorsolateral mesoderm, which migrate beneath the presumptive lens ectoderm during the course of normal development, are unable to induce a lens by themselves but have been shown to enhance the inductive effect of the neural plate almost twofold (Henry and Grainger 1990). These inductive signals from the neural plate and other tissues confer a lens-forming bias on a large region of head ectoderm.

The identity of the signals responsible for lens bias has not been established. There is evidence, however, that the BMPs and retinoic acid (RA) may play some role in carrying a biasing signal to the PLE region. Immunolocalization of *BMP-7* in the mouse shows that it is in the head ectoderm during early lens induction and that inhibition of *BMP-7* signaling at that time

inhibits lens formation (Wawersik et al. 1999). In addition, this study shows
that expression of *Pax-6*, a paired-type homeobox gene critical in eye development, is lost prior to lens placode formation in *BMP-7* knockouts while
BMP-7 expression is unaffected in *Pax-6* mutant animals, suggesting that
BMP-7 acts upstream of *Pax-6* in the hierarchy of lens biasing factors. Retinoid signaling in the lens ectoderm has been assayed using transgenic mice
that express a retinoic acid response element fused to *lac-Z*, and thereby have
β-galactosidase activity in cells where retinoic acid receptors (RARs) are
activated (J. Enwright and R. Grainger, unpubl.). These mice demonstrate
that RARs are active in the developing retina and lens ectoderm before neural
tube closure, at a time when a lens bias is thought to be established. Further
evidence shows that when this transgenic strain is crossed with *Small eye*
(*Sey*) mice carrying a loss-of-function mutation of *Pax-6*, retinoid signaling is
dramatically decreased in the retina and lens.

The lens-forming bias of the anterior ectoderm is gradually acquired over
time between neural plate (N&F 14) and neural tube (N&F 19) stages.
Transplantation experiments using ectoderm posterior, anterior, and ventral
to the PLE shows increased lens-forming ability when transplanted from a
N&F stage 19 donor to a N&F stage 19 host, relative to the equivalent ectoderm taken from a N&F stage 14 donor (Grainger et al. 1997). The acquisition
of bias is correlated temporally and spatially with the expression of a number
of genes expressed in the PLE that may play a role in regulating the response
to a biasing signal.

One of the earliest genes expressed in biased ectoderm is the homeobox
gene *Otx2*, a member of the *orthodenticle/Otx* gene family. In *Xenopus*, *Otx2*
is expressed in anterior neural and placodal ectoderm at N&F stage 14
(Pannese et al. 1995) and is one of the earliest genes expressed in the presumptive lens ectoderm (Zygar et al. 1998). Although *Otx2* is required for the
formation of head structures, it is not known whether it is required for lens
development per se. In the *Sey* mouse, *Otx2* expression is unaffected (M.
Fisher and R. Grainger, unpubl.) suggesting that this gene may function
either upstream of *Pax-6* or in a different pathway. Recent work in *Xenopus*
has shown that *Otx2* expression is strongly and rapidly upregulated in lens-
competent ectoderm transplanted to the PLE region of a N&F stage 14 host.
This expression is far weaker if the ectoderm is transplanted to a more
posterior region of the same stage host (Zygar et al. 1998). This result suggests that *Otx2* is part of an early response to inductive signals from the
neural plate.

The necessity of the paired-box gene *Pax-6* in the formation of the retina
and lens during embryonic development has been demonstrated in a number
of different species. In the *Sey* mouse the retina and lens are significantly
reduced or fail to form entirely (Hill et al. 1992). The same phenotype
characterizes the human disease Aniridia (AN), which has a similar genetic
basis (Ton et al. 1991; Glaser et al. 1992). That it is also sufficient to induce
eye and lens formation has been shown in *Drosophila*, where misexpression

of *Pax-6/eyeless* in leg and wing imaginal disc tissue gives rise to complete, ectopic eyes (Halder et al. 1995). Recent work has shown that ectopic eye formation can also be induced in vertebrates by misexpression of *Pax-6* (Chow et al. 1999). The expression pattern of *Pax-6* in early *Xenopus* embryos is certainly consistent with its playing a role in bias. It is first expressed shortly after *Otx2*, and in a similar region consisting of anterior neuroectoderm and sensory ectoderm containing the presumptive lens and olfactory regions (Hirsch and Harris 1997; Zygar et al. 1998). *Pax-6* mRNA is present continuously in the PLE from N&F stage 14 through N&F stage 42, well past the time of lens cell differentiation. Culture experiments which take advantage of the *Small eye* mutation in rat and mouse suggest that *Pax-6* is primarily required in the ectoderm to enable it to respond to lens-inducing signals, although it may play a minor role in the transmission of that signal as well. Tissue recombination experiments with a *Small eye* rat (*rSey*) show that *rSey/rSey* optic vesicles can induce a lens response from wildtype (WT) or *rSey/+* lens ectoderm, but *rSey/rSey* ectoderm cannot be induced to form lens even by WT optic vesicle (Fujiwara et al. 1994). Similar experiments with the *Small eye* mouse show poor lens responsiveness in *Sey/+* ectoderm, and additionally show that *Sey/Sey* optic vesicles cannot induce lenses even in WT ectoderm (J. Enwright and R. Grainger, unpubl.). Strong *Pax-6* ectodermal expression is quite clearly needed in the response to inductive signals, since even heterozygotic ectoderm is deficient in its response to induction, while its requirement for the production of signals by the optic vesicle is less clear.

A group of genes that are expressed in a close spatial and temporal parallel with *Pax-6* are the members of the *eyes absent/Eya* gene family. In *Drosophila*, *eyes absent* is known to interact with other genes, including *eyeless/Pax-6*, to initiate ectopic eye development (Bonini et al. 1997; Chen et al. 1997; Pignoni et al. 1997). The murine genes *Eya1* and *Eya2* are differentially expressed in the sensorial ectoderm before neural tube closure with *Eya1* in both the nasal and lens placodes, while *Eya2* is only in the nasal placode (Xu et al. 1997). This suggests that these genes may have a role in determining placodal identity using a combinatorial code to distinguish nasal (*Eya1* and *2*) versus lens (*Eya1* only). Furthermore, this expression in the nasal and lens placodes is coincident with *Pax-6*, and placodal *Eya1* expression is significantly reduced in *Sey* homozygous mutants (Xu et al. 1997), suggesting a regulatory role for *Pax-6* in *Eya 1* expression.

Two related genes expressed in anterior ectoderm, *Six3* and *Optx2*, are members of the *sine oculis/Six* homeobox-containing gene family known to be required for eye formation in *Drosophila* (Cheyette et al. 1994). In *Xenopus*, *Optx2* is first expressed early in gastrulation in dorsal ectoderm that is becoming lens-competent (N&F 10.5). It remains on in anterior ectoderm throughout neurulation, but then becomes much reduced in the PLE after the optic vesicle contacts the overlying ectoderm (N. Hirsch and R. Grainger, unpubl.). Another *Optx2* homologue in *Xenopus* has been found to regulate eye size during development, as overexpression of the cDNA or an *Optx2/*

Engrailed repressor construct increases the size of the eye (Zuber et al. 1999). The *Xenopus* expression pattern is different from that in the chick, where *Optx2* expression remains strong in the PLE until a significantly later stage of lens development (Toy et al. 1998). This suggests that a role for *Optx2* in *Xenopus* may be during the competence or bias phases rather than the specification or differentiation phase. *Six3* is also expressed in anterior ectoderm during the acquisition of lens bias (Zygar et al. 1998). There is evidence that *Six3* is able to induce ectopic lenses in the killifish medaka, but this occurs at a very low frequency (2.5 %) and only if the ectopic expression occurs in or near the otic placode (Oliver et al. 1996). This implies that *Six3* requires an additional influence, such as a predisposition to become a sensory structure like the ear, before it can change cell fates and induce ectopic lens formation. Another potential bias gene, *Lens-1*, is a member of the forkhead family of transcription factors. It is expressed during open neural plate stages in both the PLE and presumptive nasal epithelium (PNE), but it becomes restricted to the PLE after neural tube closure (Kenyon et al. 1999).

Because a lens bias in *Xenopus* is acquired in large regions of head ectoderm (Grainger et al. 1997), one might propose that genes involved in bias would be expressed in areas extending beyond the PLE. In fact, a number of transcriptional regulators are expressed in large areas of sensorial ectoderm (*Pax-6*, *Eya*, *Lens-1*) during neurula stages. However, none of these genes is expressed in all of the areas that acquire bias, implying that, at the very least, bias must result from expression of more than one of these genes, although other as yet unidentified genes may play a part in the process. The fact that bias is already present in neural plate stage ectoderm (Henry and Grainger 1987) suggests that if there is a single gene which controls bias, it has yet to be identified. Furthermore, bias in the PLE is most likely established too early to involve genes such as *Pax-6* or *Lens-1*, which have an onset of expression at or later than N&F stage 14.

5
Inhibition

Many of the genes implicated in establishing a lens bias are expressed in regions of anterior ectoderm significantly larger than the presumptive lens ectoderm. The extent of this large region of lens bias has been demonstrated by transplanting ear or cement gland ectoderm to the PLE at N&F stage 19 and showing well-developed lens formation in 25% of cases (Grainger et al. 1997). In order for the embryo to ensure that only the biased region overlying the eye goes on to form lens, a lateral inhibitory mechanism is clearly required.

Evidence leading to the hypothesis that there might be a lens-induction inhibitor goes back to experiments of King (1905), who was the first to show that lenses still form in embryos from which the neural plate had been removed. This was at first a confounding result since it suggested that the retinal rudiment was not required, at least after the neural plate stage, for lens formation, in contrast to what was first shown by Spemann. These "free lenses" are a phenomenon which has now been observed across numerous anuran and urodele species, but not by every investigator. A study by Von Woellwarth (1961), following up a report by Mangold (1954), suggested an explanation. When Von Woellwarth ablated only anterior neural tissue he saw no free lenses, but when he also removed posterior neural plate free lenses were observed. His experiments pointed toward an inhibitory factor in the posterior neural folds: he suggests that neural crest derived from this region may be suppressing free lens formation. The inhibition is presumed to occur during the migration of neural crest cells around the optic vesicle following neural tube closure. Recent work in our laboratory has shown that head mesenchyme, much of which is neural-crest-derived, inhibits a lens bias in the PLE in mouse embryos (J. Enwright and R. Grainger, unpubl.). *Eya* genes are expressed in the periooptic mesenchyme at this time and may play a role in inhibiting a lens fate in the ectoderm (Xu et al. 1997).

6
Specification

At the time of neural tube closure the lens ectoderm, if explanted into culture, will express crystallin without any further inductive influence (Henry and Grainger 1990) and is said to be specified. While the developmental commitment process is one of the most important issues in developmental biology, little is known about the genes that control this process in general, and in formation of the lens in particular. A number of the genes mentioned earlier in the bias section, including *Pax-6*, are expressed in the PLE during the later part of the bias period and remain on in the lens after specification. Since genes like *Pax-6* are expressed before specification, their expression may possibly be more consistent with a role in specification. Genes of this group, especially *Pax-6*, are also likely to act during later stages of the lens-forming program since they are still expressed at these stages. There is strong evidence based on the ability of *Pax-6* to activate the crystallin promoter (see below) that it certainly plays later roles as well.

   Pax-6 is one of a group of genes that have a central role in eye determination in both *Drosophila* and in vertebrates. Misexpression of these genes can cause ectopic eye and lens formation, as discussed previously. Experiments showing that *Pax-6* misexpression can cause a low frequency of ectopic lens-like structures and eyes to form in *Xenopus* suggest that at least this gene has the potential to activate a significant part of the eye development program. Whether this gene is sufficient to elicit determination in vivo remains to be established.

7
Differentiation

Although specified ectoderm does not require an inductive signal to subsequently activate the differentiation program, it has not previously been established whether the time when differentiation will commence is also established when lens tissue is specified. Experiments by Gurdon (Gurdon et al. 1985) suggest that the differentiation program for muscle is always activated at a particular developmental age. Recent work from our laboratory, however, shows that the interval between specification and differentiation (as measured by the onset of crystallin expression) is not fixed and can be influenced by the host environment. In these experiments embryos from *X. tropicalis* lines bearing a γ-crystallin promoter-GFP transgene were utilized to follow the onset of crystallin expression in real time in vivo. Specified ectoderm (N&F 19), when transplanted to a younger host (N&F 14), shows a significant delay (4 h) in the onset of crystallin expression when compared to unoperated controls (M. Offield, N. Hirsch and R. Grainger, unpubl.). This suggests that the timing mechanism is plastic and that the progressive buildup of "biasing" or "specifying" gene products in this ectoderm does not predetermine the beginning of differentiation.

After the lens has been specified, several genes are expressed in the lens ectoderm which are thought to interact with the crystallin promoter directly and which may have a direct role in controlling the onset of differentiation. *Xenopus Sox-3* is expressed in lens-competent ectoderm at gastrulation, but *Sox-3* mRNA cannot be detected in the PLE at bias stages. However, after neural tube closure, just following the time of lens specification, *Sox-3* expression is activated strongly in the PLE and continues in this region throughout the remaining period of lens formation. This biphasic pattern of expression suggests that *Sox-3* may play a dual role in lens development: early, during competence, and then later, during differentiation. *Sox-3* expression in biased lens ectoderm has been shown to require signaling from retinal tissue (Zygar et al. 1998) in a similar fashion to *Sox-2* and *Sox-3* in chick (Kamachi et al. 1998). Experimental evidence further suggests that *Sox-2* and *Sox-3* are required for the activation of crystallin genes (Kamachi et al. 1995, 1998).

Another gene expressed after specification and known to regulate crystallin expression is *L-Maf* (Ogino and Yasuda 1998), a member of the *maf* proto-oncogene family of bZIP transcription factors. This gene was cloned in a screen to identify gene products that bound to the consensus DNA sequence for the chicken αA-crystallin promoter αCE2. The presence of *Maf* mRNA has also been shown in the lens of the rat (Yoshida et al. 1997). During lens differentiation, different maf family members are expressed in different cell types, with *MafB* found in epithelial cells (Kawauchi et al. 1999) while *L-maf* is found in the fiber cells (Ogino and Yasuda 1998), suggesting that different *Maf* genes may act to specify different cell types in the developing lens.

Two other transcription factors are implicated in the lens differentiation process. The homeobox gene *Pitx3* had been found to be expressed in the mouse lens starting at day 11 of embryonic development, at the time of lens fiber cell formation, and continues to be expressed in lens epithelial cells at day 15. Furthermore, *Pitx3* maps close to the *Aphakia* locus on mouse chromosome 19, making it a strong candidate gene for the *Aphakia* phenotype, which is characterized by small eyes lacking lenses (Semina et al. 1997). Mutations in human *Pitx3* have been associated with a familial history of cataracts (Semina et al. 1998). Another homeobox gene, *Prox1*, homologous to the *Drosophila* gene *prospero*, is also expressed in the lens during early stages of lens differentiation (Del Rio-Tsonis et al. 1999). Knockouts of this gene in mice show deficiencies in the elongation and survival of lens fiber cells (Wigle et al. 1999).

In addition to the genes noted above, there is strong evidence that *Pax-6* plays a direct role in the regulation of crystallin synthesis. In mouse and chick fibroblasts that contain a reporter gene fused to the crystallin promoter, the coexpression of *Pax-6* is necessary and sufficient for reporter gene expression (Cvekl et al. 1995a, 1995b). As noted earlier, evidence suggests that ectopic *Pax-6* expression in *Xenopus* is sufficient to form ectopic lenses (Altmann et al. 1997). However, recent work by the authors suggests that *Pax-6* is insufficient to activate many early markers of lens development and that it may function by ectopically activating only the β-crystallin promoter (N. Hirsch and R. Grainger, unpubl.). There is a great deal of evidence that *Pax-6* directly regulates crystallin promoter activity in cell culture (Cvekl et al. 1995a, 1995b; Richardson et al. 1995). In addition to factors previously mentioned (*Sox-2/3, L-Maf,* and *Pax-6*), other transcriptional regulators, including AP-1, CREB or CREM, and USF, are also implicated in crystallin gene activation (Cvekl et al. 1995a).

Although such a large number of transcription factors suggests a complex scheme of regulation by the crystallin promoter, the proper regulation of the onset of crystallin synthesis at the appropriate time undoubtedly requires such complexity. *Pax-6*, for example, is present in lens ectoderm from neural plate stages onward, and yet crystallin genes are not activated in the lens until tadpole stages (Smolich et al. 1994). This implies that either there are additional factors that act early in lens development to block the activation of the crystallin promoter, or that other positive factors are required later. Although many factors, like the *Sox* and *Maf* genes, do act in important ways at later stages, they cannot alone direct crystallin expression since both gene families are also activated in the lens ectoderm hours before crystallin synthesis begins. There are many unresolved issues that need to be clarified to gain a better picture of the genetic control of crystallin synthesis.

8
Conclusions

The model of lens induction set forth in this chapter fits well with current experimental results. Additionally, the availability of a number of molecular markers for each phase of lens induction makes it possible to ask more detailed questions concerning the molecular mechanisms which underlie competence, bias, inhibition, specification, and differentiation. The earliest phase of lens induction, lens competence, is part of a series of competences through which the animal cap ectoderm passes; first is mesodermal competence, then neural, and then lens (Servetnick and Grainger 1991). Although the mechanism by which the ectoderm gains and loses lens competence is poorly understood, the use of the HMG-box gene *Sox-3* as a marker for competent ectoderm should prove a useful tool in answering this question. The lens bias phase, where lens-forming potential is strengthened and restricted to the anterior non-neural ectoderm, is marked by changes in the expression of a number of genes, including *Pax-6*, *Six3*, *Optx2* (*Six9*), and *Otx2*. While misexpression studies have shown that both *Pax-6* and *Six3* can activate crystallin expression in nonlens ectoderm, the question of whether these genes are required to establish a true lens bias or specification in ectoderm is not yet clear. The next phase in lens development requires that the lens be aligned with the eye, a step that requires the elimination of lens bias in nonlens ectoderm and the strengthening of lens bias in ectoderm directly overlying the optic vesicle. Thus, the two phases of inhibition and specification are closely interrelated. Evidence that neural crest mesenchyme is involved in inhibition comes from studies in both frog and mouse, although the question of whether the mesenchyme inhibits lens formation through the secretion of an inhibiting factor, by physical contact, or by some other mechanism, is still unclear. The mechanism underlying specification is also unsettled, although it most likely involves a complex interaction of transcriptional regulators, including *Pax-6* and *Six3*. *Pax-6*, *Sox-3* and the *maf*-family members *Maf-B* and *L-Maf* are all implicated in controlling crystallin gene regulation, a major component of the last step in the lens formation process, lens differentiation, but how these transcription factors act in concert to effect this process has not been resolved.

Owing to the large body of molecular genetic studies on lens development, biologists can now see a great deal more of the complexities of embryonic induction through the window of the eye. What is necessary for the future is an organization of the many genes expressed during lens development into some sort of functional hierarchy. An increasing reliance on genetic systems will be required for this kind of understanding. Studies in mice and fish are therefore likely to be very informative, as may be the use of the new model amphibian system *X. tropicalis* that combines traditional embryological strengths with the potential for genetic manipulation, including the ability to make transgenic lines and perform mutagenesis.

References

Altmann CR, Chow RL, Lang RA, Hemmati-Brivanlou A (1997) Lens induction by *Pax-6* in Xenopus laevis. Dev Biol 185(1):119–123

Amaya E, Kroll KL (1999) A method for generating transgenic frog embryos. In: Sharpe P, Mason I. (eds) Methods in molecular biology, molecular embryology: methods and protocols. Humana Press, Totowa, New Jersey, pp 393–414

Amaya E, Offield MF, Grainger RM (1998) Frog genetics: *Xenopus tropicalis* jumps into the future. Trends Genet 14(7):253–255

Ang SL, Jin O, Rhinn M, Daigle N, Stevenson L, Rossant J (1996) A targeted mouse *Otx2* mutation leads to severe defects in gastrulation and formation of axial mesoderm and to deletion of rostral brain. Development 124(23):4819–4826

Bonini NM, Bui QT, Gray-Board GL, Warrick JM (1997) The *Drosophila eyes absent* gene directs ectopic eye formation in a pathway conserved between flies and vertebrates. Development 124(23):4819–4826

Chen R, Amoui M, Zhang Z, Mardon G (1997) *Dachshund* and *eyes absent* proteins form a complex and function synergistically to induce ectopic eye development in *Drosophila*. Cell 91(7):893–903

Cheyette BN, Green PJ, Martin K, Garren H, Hartenstein V, Zipursky SL (1994) The *Drosophila sine oculis* locus encodes a homeodomain-containing protein required for the development of the entire visual system. Neuron 12(5):977–996

Chow RL, Altmann CR, Lang RA, Hemmati-Brivanlou A (1999) *Pax-6* induces ectopic eyes in a vertebrate. Development 126:4213–4222

Cornell RA, Musci TJ, Kimelman D (1995) FGF is a prospective competence factor for early activin-type signals in *Xenopus* mesoderm induction. Development 121(8):2429–2437

Cvekl A, Piatigorsky J (1996) Lens development and crystallin gene expression: many roles for *Pax-6*. Bioessays 18(8):621–630

Cvekl A, Kashanchi F, Sax CM, Brady JN, Piatigorsky J (1995a) Transcriptional regulation of the mouse alpha A-crystallin gene: activation dependent on a cyclic AMP-responsive element (DE1/CRE) and a *Pax-6*-binding site. Mol Cell Biol 15(2):653–660

Cvekl A, Sax CM, Li X, McDermott JB, Piatigorsky J (1995b) *Pax-6* and lens-specific transcription of the chicken delta 1-crystallin gene. Proc Natl Acad Sci USA 92(10):4681–4685

Del Rio-Tsonis K, Tomarev SI, Tsonis PA (1999) Regulation of *Prox 1* during lens regeneration. Invest Ophthalmol Visual Sci 40(9):2039–2045

Duncan MK, Banerjee-Basu S, McDermott JB, Piatigorsky J (1996) Sequence and expression of chicken beta A2- and beta B3-crystallins. Exp Eye Res 62(1):111–119

Duncan MK, Haynes JI, 2nd, Cvekl A, Piatigorsky J (1998) Dual roles for *Pax-6*: a transcriptional repressor of lens fiber cell-specific beta-crystallin genes. Mol Cell Biol 18(9):5579–5586

Easter SS, Jr., Nicola GN (1996) The development of vision in the zebrafish (*Danio rerio*). Dev Biol 180(2):646–663

Fedestova NG, Barabanov VM (1978) Lens and adenohypophesis differentiation in the ectoderm of the oral region in the chick embryos under in vitro cultivation. Ontogenez 9:609–615

Fujiwara M, Uchida T, Osumi-Yamashita N, Eto K (1994) Uchida rat (*rSey*): a new mutant rat with craniofacial abnormalities resembling those of the mouse *Sey* mutant. Differentiation 57(1):31–38

Glaser T, Walton DS, Maas RL (1992) Genomic structure, evolutionary conservation and aniridia mutations in the human PAX6 gene. Nat Genet 2(3):232–239

Grainger RM (1996) New perspectives on embryonic lens induction. Sem in Cell Dev Biol 7:149–155

Grainger RM, Gurdon JB (1989) Loss of competence in amphibian induction can take place in single nondividing cells. Proc Natl Acad Sci USA 86(6):1900–1904

Grainger RM, Henry JJ, Saha MS, Servetnick M (1992) Recent progress on the mechanisms of embryonic lens formation. Eye 6(2):117–122

Grainger RM, Mannion JE, Cook TL, Zygar CA (1997) Defining intermediate stages in cell determination – acquisition of a lens-forming bias in head ectoderm during lens determination. Dev Gen 20(3):246–257

Gurdon JB, Fairman S, Mohun TJ, Brennan S (1985) Activation of muscle-specific actin genes in *Xenopus* development by an induction between animal and vegetal cells of a blastula. Cell 41:913–922

Halder G, Callaerts P, Gehring WJ (1995) Induction of ectopic eyes by targeted expression of the *eyeless* gene in *Drosophila*. Science 267(5205):1788–1792

Henry JJ, Grainger RM (1987) Inductive interactions in the spatial and temporal restriction of lens-forming potential in embryonic ectoderm of *Xenopus laevis*. Dev Biol 124(1):200–214

Henry JJ, Grainger RM (1990) Early tissue interactions leading to embryonic lens formation in *Xenopus laevis*. Dev Biol 141:149–163

Hill RE, Favor J, Hogan BL, Ton CC, Saunders GF, Hanson IM, Prosser J, Jordan T, Hastie ND, van Heyningen V (1992) Mouse *Small eye* results from mutations in a paired-like homeobox-containing gene. Nature 355(6362):522–525

Hirsch N, Harris WA (1997) *Xenopus Pax-6* and Retinal Development. J Neurobiol 32(1):45–61

Kamachi Y, Sockanathan S, Liu Q, Breitman M, Lovell-Badge R, Kondoh H (1995) Involvement of *SOX* proteins in lens-specific activation of crystallin genes. EMBO J 14(14):3510–3519

Kamachi Y, Uchikawa M, Collignon J, Lovell-Badge R, Kondoh H (1998) Involvement of *Sox1, 2* and *3* in the early and subsequent molecular events of lens induction. Development 125(13):2521–2532

Karkinen-Jaaskelainen M (1978) Permissive and direct interaction in lens induction. J Embryol Exp Morphol 44:167–179

Karlstrom RO, Talbot WS, Schier AF (1999) Comparative synteny cloning of zebrafish *you-too*: mutations in the *Hedgehog* target *gli2* affect ventral forebrain patterning. Genes Dev 13(4):388–393

Kawauchi S, Takahashi S, Nakajima O, Ogino H, Morita M, Nishizawa M, Yasuda K, Yamamoto M (1999) Regulation of lens fiber cell differentiation by transcription factor c-Maf. J Biol Chem 274(27):19254–19260

Kenyon K, Moody S, Jamrich M (1999) A novel *forkhead* gene mediates early steps during *Xenopus* lens formation. Development 126(22):5107–5116

King HD (1905) Experimental studies on the eye of the frog embryo. Arch Entwicklungs mech 19:85–107

Kroll KL, Amaya E (1996) Transgenic *Xenopus* embryos from sperm nuclear transplantations reveal FGF signaling requirements during gastrulation. Development 122(10):3173–3183

Lewis WH (1904) Experimental studies on the development of the eye in Amphibia. I. On the origin of the lens, *Rana palustris*. Am J Anat 3:505–536

Mangold O (1954) Entwicklung und Differenzierung der prasumptiven Epidermis und ihres unterlagernden Entomesodermus aus der Neurula von Triton alpestris als Isolat. Acta Morphol Acad Sci Hung 10:153–176

Mencl EM (1903) Ein Fall von beiderseitiger Augenlinsenausbildung der Abwesenheit von Augenblasen. Arch Entwicklungs mech 16:328–339

Mizuseki K, Kishi M, Matsui M, Nakanishi S, Sasai Y (1998) *Xenopus Zic-related-1* and *Sox-2*, two factors induced by chordin, have distinct activities in the initiation of neural induction. Development 125(4):579–587

Nieuwkoop PD, Faber J (1956) Normal table of *Xenopus laevis*. North Holland Publishing, Amsterdam

Nishimatsu S, Thomsen GH (1998) Ventral mesoderm induction and patterning by bone morphogenetic protein heterodimers in *Xenopus* embryos. Mech Dev 74(1–2):75–88

Ogino H, Yasuda K (1998) Induction of lens differentiation by activation of a bZIP transcription factor, *L-Maf*. Science 280(5360):115–118

Oliver G, Loosli F, Koster R, Wittbrodt J, Gruss P (1996) Ectopic lens induction in fish in response to the murine homeobox gene *Six3*. Mech Dev 60(2):233–239

Pannese M, Polo C, Andreazzoli M, Vignali R, Kablar B, Barsacchi G, Boncinelli E (1995) The *Xenopus* homologue of *Otx2* is a maternal homeobox gene that demarcates and specifies anterior body regions. Development 121(3):707–720

Pignoni F, Hu B, Zavitz KH, Xiao J, Garrity PA, Zipursky SL (1997) The eye-specification proteins *So* and *Eya* form a complex and regulate multiple steps in *Drosophila* eye development. Cell 91(7):881–891

Richardson J, Cvekl A, Wistow G (1995) *Pax-6* is essential for lens-specific expression of zeta-crystallin. Proc Natl Acad Sci USA 92(10):4676–4680

Saha MS, Spann CL, Grainger RM (1989) Embryonic lens induction: more than meets the optic vesicle. Cell Differ Dev 28(3):153–171

Semina EV, Reiter RS, Murray JC (1997) Isolation of a new homeobox gene belonging to the *Pitx/Rieg* family: expression during lens development and mapping to the *aphakia* region on mouse chromosome 19. Hum Mol Genet 6(12):2109–2116

Semina EV, Ferrell RE, Mintz-Hittner HA, Bitoun P, Alward WL, Reiter RS, Funkhauser C, Daack-Hirsch S, Murray JC (1998) A novel homeobox gene PITX3 is mutated in families with autosomal-dominant cataracts and ASMD. Nat Genet 19(2):167–170

Servetnick M, Grainger RM (1991) Changes in neural and lens competence in *Xenopus* ectoderm: evidence for an autonomous developmental timer. Development 112(1):177–188

Smolich BD, Tarkington SK, Saha MS, Stathakis DG, Grainger RM (1993) Characterization of *Xenopus laevis* gamma-crystallin-encoding genes. Gene 128(2):189–195

Smolich BD, Tarkington SK, Saha MS, Grainger RM (1994) *Xenopus* gamma-crystallin gene expression: evidence that the gamma-crystallin gene family is transcribed in lens and nonlens tissues. Mol Cell Biol 14(2):1355–1363

Spemann HH (1901) Uber Korrelationen in der Entwicklung des Auges. Verh Anat Ges 15:61–79

Spemann HH (1907) Neue Tatsachen zum Linsenproblem. Zool Anz 31:379–386

Steinbach OC, Wolffe AP, Rupp RA (1997) Somatic linker histones cause loss of mesodermal competence in *Xenopus*. Nature 389(6649):395–399

Stone LS, Dinnean FL (1940) Experimental studies on the relation of the optic vesicle and cup to lens formation in *Ambystoma punctuatum*. J Exp Zool 83:95–126

Ton CC, Hirvonen H, Miwa H, Weil MM, Monaghan P, Jordan T, van Heyningen V, Hastie ND, Meijers-Heijboer H, Drechsler M, Royer-Pokora B, Collins F, Swaroop A, Strong LC, Saunders GF (1991) Positional cloning and characterization of a paired box- and homeobox-containing gene from the *aniridia* region. Cell 67(6):1059–1074

Toy J, Yang JM, Leppert GS, Sundin OH (1998) The *optx2* homeobox gene is expressed in early precursors of the eye and activates retina-specific genes. Proc Natl Acad Sci USA 95(18):10643–10648

Von Woellwarth C (1961) Die Rolle des Neuralleistenmaterials und der Temperatur bei der Determination der Augenlinse. Embryologia 6:219–242

Waddington CH (1936) The origin of competence for lens formation in the amphibia. J Exp Biol 13:86–91

Wawersik S, Purcell P, Rauchman M, Dudley AT, Robertson EJ, Maas R (1999) *BMP7* acts in murine lens placode development. Dev Biol 207(1):176–188

Wigle JT, Chowdhury K, Gruss P, Oliver G (1999) *Prox1* function is crucial for mouse lens-fibre elongation. Nat Genet 21(3):318–322

Xu PX, Woo I, Her H, Beier DR, Maas RL (1997) Mouse *Eya* homologues of the *Drosophila eyes absent* gene require *Pax6* for expression in lens and nasal placode. Development 124(1):219–231

Yoshida K, Imaki J, Koyama Y, Harada T, Shinmei Y, Oishi C, Matsushima-Hibiya Y, Matsuda A, Nishi S, Matsuda H, Sakai M (1997) Differential expression of *maf-1* and *maf-2* genes in the developing rat lens. Invest Ophthalmol Visual Sci 38(12):2679–2683

Zuber ME, Perron M, Philpott A, Bang A, Harris WA (1999) Giant Eyes in *Xenopus laevis* by Overexpression of *XOptx2*. Cell 98:341–352

Zygar CA, Cook TL, Grainger RM (1998) Gene activation during early stages of lens induction in *Xenopus*. Development 125(17):3509–3519

Molecular Control of Cell Diversification in the Vertebrate Retina

Sabine Fuhrmann, Lely Chow, and Thomas A. Reh[1]

1
Introduction

Any discussion of the molecular mechanisms that are responsible for the generation of cell diversity in the retina requires some consideration of what constitutes neuronal diversity. Traditionally, retinal neurobiologists have classified retinal cell types on the basis of their morphology, in terms both of their laminar position and their dendritic structure, as well as by their electrophysiological properties. While it is generally accepted that there are five basic classes of retinal neurons, photoreceptors, horizontal cells, bipolar cells, amacrine cells, and ganglion cells, there is a considerable amount of debate as to how these basic classes are subdivided. While there are critical differences in gene expression between the classes of cone photoreceptors that determine the wavelength of light to which they respond, there is little difference in their morphology; by contrast, the various classes of retinal ganglion cells have critical differences in their morphology and connectivity, but as yet few molecules have been identified that are differentially expressed in subsets of ganglion cells, and it is not at all clear whether these differences correspond to the morphological subclasses of ganglion cells. Nevertheless, most retinal neurobiologists place the number of retinal cell types somewhere between "70 and the low hundreds (Cook 1996)." Although the attributes of neurons, from cell morphology to specific gene expression, are present in a continuum across these cell classes, "we can think of (these attributes) as switches, partitioning cell fates in discontinuous ways to create the entities we call cell types" (Cook 1996). Schemes for classifying the different retinal neurons into functional or anatomical classes have been of enormous use in the understanding of retinal structure-function relations. Rodieck championed the idea of "natural classes" of retinal cell subtypes, where the various attributes of a particular type of retinal cell were classified by a factor analysis, and this allowed for clear distinctions between many of the subclasses of retinal ganglion cells. More recently, Cook has proposed that the subtypes of

[1] Department of Biological Structure, Box 357420, Health Sciences Center, University of Washington, Seattle, Washington 98195, USA

Results and Problems in Cell Differentiation, Vol. 31
M. E. Fini (Ed.): Vertebrate Eye Development
© Springer-Verlag Berlin Heidelberg 2000

wide field retinal cells can be more easily identified by their independent mosaics of distribution. However, these schemes have been of less use in defining the nature of the problem of neuronal diversity for those of us interested in the developmental mechanisms that govern its formation. For example, these studies cannot tell us what aspects of the mature cell types in the retina are controlled by intrinsic programs operating within the cells, and which aspects of cell type are a property of the surrounding cells in the retina. It is also not clear from a developmental point of view which properties of a cell type are the primary ones and which properties come about as a result of some type of refinement of the cell phenotype. Reliance on the classification of mature cells is not necessarily appropriate. For example, a classification scheme of amacrine cells might put all amacrine cells in one group and dopamine-containing amacrine cells in a subgroup. However, evidence from *Xenopus* blastomere injections indicates that the specification of the neuro-transmitter phenotype takes place long before the cell is committed to the amacrine cell identity (Huang and Moody 1997).

The aim of this chapter is to address the question of retinal cell diversity in molecular terms. In addition to providing an understanding of how cells with diverse patterns of cellular gene expression arise from multipotent progenitor cells, there may be additional benefits to defining retinal cell types by their gene expression. Retinal cell types are frequently defined by the morpho-logical attributes of the cells. While consideration of dendritic arborizations and connectivity are critical for the understanding of the function of a par-ticular cell class, the expression of transmitters, receptors, and channels is clearly as important for the function of the retinal circuitry.

Before embarking on the discussion of the molecular mechanisms in-volved in generating neuronal diversity, it is helpful to review a few of the basic assumptions in the field. It is widely assumed that retinal neurons, once generated, have relatively stable fates or phenotypes. For example, once a rod photoreceptor has been generated and differentiates its mature morphology, it cannot change its phenotype into an amacrine cell or a ganglion cell. This appears to be true even for cells that have been removed from their normal environment and cultured in isolation. Thus, once a stable pattern of gene expression is attained in a retinal neuron, most of these genes must remain activated in a cell-autonomous manner. While this appears to be the case for the five main retinal cell classes, the situation is less clear for the subtypes of retinal cells; for example, while rod photoreceptor cells have not been ob-served to express cone opsins, there is some evidence that cone cells might express more than one opsin at some stage of development. Studies of retinas in which particular cell classes have been eliminated by lesions or degener-ation have provided insight into the degree of plasticity of retinal phenotypes. Elimination of dopaminergic amacrine cells in the frog retina, for example, does not result in the conversion of some other type of amacrine cell into this cell class (Negishi et al. 1982; Reh and Tully 1986), although the amacrine cells of this class at the retinal margin undergo major morphological changes, growing dendrites hundreds of microns to cover the retinal territory denuded by the neurotoxic lesion. These results highlight the principle that some fundamental pattern of gene expression within retinal neurons is very stable, even though the morphology of the cells can be quite variable, regulated to a large degree by the environment in which they find themselves. An interesting example of this property of retinal neurons is found in studies where retinal neurons were isolated and cultured on different cell adhesion molecules; the ganglion cells had dramatically different morphologies when isolated from other retinal cells and cultured on NCAM, N-cadherin, or L1 substrates; yet they continued to express the proteins characteristic of these cells, such as thy1.1 antigen and neurofilament protein (Kljavin et al. 1994).

A second common assumption that many of the researchers in this field share is that the basic five retinal cell classes represent a fundamental distinction in cell identity that reflects differences in critical transcription factors that control different patterns of gene expression. Frequently, studies (see below) interpret their results as demonstrating that a particular transcription factor directs the multipotent retinal progenitor to a specific retinal cell fate. Combinatorial codes of transcription factors may define the retinal identities; alternatively, unique transcription factors may direct cells to particular fates. At this point, it is not clear exactly when during development these basic five retinal cell identities are established, or how soon they become solidified. Currently, there is no clear picture of what types of transcription factors specify the distinct retinal identities, if we use as the gold standard that misexpression of the putative gene is both necessary and sufficient for a specific retinal cell type.

Despite the lack of an identified code of critical transcription factors that defines the basic classes of retinal neurons, there must be differences in the transcriptional regulators of the rod photoreceptor and the retinal ganglion cell, for example, because they are such clearly different cell types with distinct patterns of gene expression, even when isolated from their normal environment. Therefore, we can safely conclude that even if only small differences in gene expression incrementally contribute to distinct, stable cell identities, these differences should be discoverable.

2
Does a Code of Transcription Factors Define Retinal Cell Identities?

2.1
Homeodomain Transcription Factors

Several investigators have proposed that a combinatorial code of homeodomain transcription factors might pattern the different retinal cell classes

(Belecky-Adams et al. 1997; Passini et al. 1997), in a manner analogous to that observed for the *pax* gene patterning of other areas of the nervous system. In the spinal cord, for example, the dorsal ventral axis displays a complex pattern of paired-homeodomain transcription factor gene expression. *Nkx 2.1* is expressed in the most ventral cells of the neural tube, and adjacent to this domain the motoneurons develop. Just dorsal to the zone of ventral motoneuron differentiation is a domain of *pax6* expression. Dorsal to the cells that express *pax6*, another paired-class homeodomain transcription factor, *chx10*, is expressed in a domain of neural tube cells. The most dorsal cells of the neural tube express *pax3*. The expression of these transcription factors serves more than just to mark these different zones of the spinal cord, since deletion of any of them causes abnormalities in the differentiation of the specific classes of neurons that would normally develop from that domain.

Do the homeodomain gene products act to specify the development of the specific retinal cell identities? Several of the genes that pattern the developing spinal cord are also expressed in the developing and mature retina. One of the most important eye-specific homeodomain transcription factors is *pax6*, and the homologous gene in *Drosophila*, *eyeless*. These genes are critical for eye development in diverse species; mutations in the *Drosophila eyeless* gene result in the partial or complete absence of the compound eyes (Quiring et al. 1994). *Pax6* mutations cause the small eye (*Sey*) phenotype in mice in the heterozygous condition, while homozygous loss of this gene produces mice with no eyes at all (Hogan et al. 1986). Initially in retinal development, *pax6* is expressed in retinal progenitor cells, but it is later confined to differentiating amacrine cells, ganglion cells and horizontal cells (Krauss et al. 1991; Walther and Gruss 1991; Puschel et al. 1992; Li et al. 1994; Hitchcock et al. 1996; Belecky-Adams et al. 1997; Hirsch and Harris 1997; see Fig. 1).

Another paired-class homeodomain transcription factor is *chx10*. Like *pax6*, mutations in *chx10* also cause defects in ocular development. The ocular retardation (*or*) mutant, first characterized by Truslove (1962), has recently been shown to result from a mutation in the *chx10* gene (Burmeister et al. 1996). Like *pax6*, *chx10* is initially expressed throughout the retinal epithelium, including the progenitor zone; but becomes confined to the bipolar cells during late embryonic and postnatal mouse development (Liu et al. 1994; Fig. 1). Burmeister et al. (1996) have found that bipolar cells fail to develop in the *or* mutant retina, consistent with the idea that *chx10* plays a critical role in bipolar cell differentiation.

Crx is another paired-homeodomain transcription factor which has been shown to be highly expressed in the developing and mature retina. In this case, however, *crx* is not expressed in the progenitor cells, but rather is confined to the developing and mature photoreceptor cells and appears to be involved in regulating the expression of genes necessary for photoreceptor differentiation (Chen et al. 1997; Furukawa et al. 1997). Retinal progenitor cells which have been retrovirally transfected with a dominant-negative *crx* gene fail to form photoreceptor outer segments (Furukawa et al. 1997). In

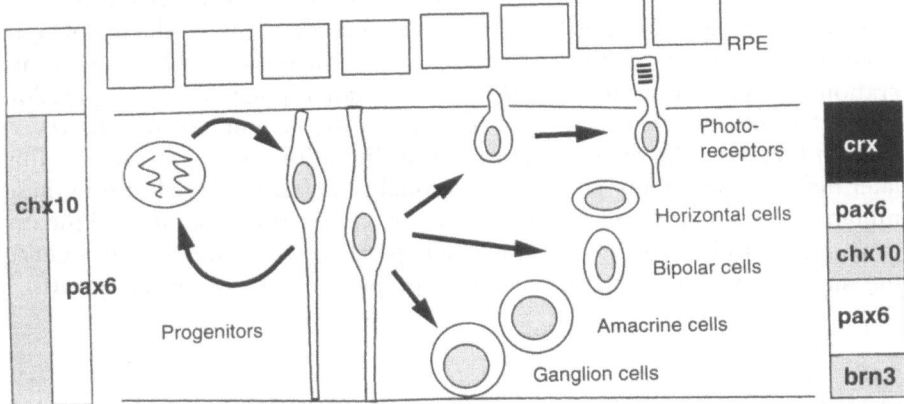

Fig. 1. Progenitor cells and differentiated retinal neurons express specific homeodomain transcription factors. *Chx10* and *pax6* are examples of homeodomain transcription factors that are expressed throughout the retinal epithelium in all the progenitor cells at early stages of development (left). The homeodomain "code" or pattern of expression in mature retinal neurons is shown schematically (right). In addition to *pax6* and *chx10*, which have now become restricted to particular subsets of retinal neurons, *crx* and *Brn 3* transcription factors are also expressed in retinal neurons

addition, *crx* appears to be essential for photoreceptor survival (Freund et al. 1996). *Crx* transactivates a number of genes characteristic of the differentiated photoreceptor phenotype, including interphotoreceptor retinoid-binding protein (IRBP), β-phospho-diesterase, and arrestin (Chen et al. 1997; Furukawa et al. 1997). Thus, while not critical for the initial formation of photoreceptors, the *crx* gene is important in photoreceptor-specific gene expression.

Another class of transcription factors that have important roles in cell-type specific gene expression in the retina are members of the POU class of homeodomain transcription factors. *Brn 3.0*, *Brn 3.1*, and *Brn 3.2* are all expressed in overlapping subsets of postmitotic retinal ganglion cells (Xiang et al. 1993; 1995; Turner et al. 1994; Erkman et al. 1996). Interestingly, targeted disruption of *Brn 3.2* results in a loss of approximately 70% of the retinal ganglion cells (Erkman et al. 1996; Gan et al. 1996). However, the loss of ganglion cells is not due to their failure to develop, but rather their loss comes after they have migrated to the ganglion cell layer and sent axons into the optic nerve. The death of these cells may be due to an inability to respond to trophic factors (Erkman et al. 1996). Targeted disruptions of *Brn 3.0* and *3.1* have no effects on ganglion cells (Erkman et al. 1996; McEvilly et al. 1996; Xiang et al. 1996), and this is thought to be due to redundancy in the expression and function of this class of genes.

In sum, homeodomain transcription factors are expressed in specific cell types in the retina, analogous to their regional patterns of expression in other areas of the developing and mature nervous system. However, the evidence that these genes are specifically required or sufficient to direct retinal progenitor cells to specific fates has not been forthcoming, due to the pleiotropic effect that mutations in these genes can cause on retinal progenitor proliferation. The best example is *chx10*, where the late bipolar specific expression correlates with the absence of this cell type in the *or* mutant mice; however, even in this case, the early defect in cell proliferation may contribute to this later defect, since all late-generated neuronal cell classes are reduced in these mutants. Nevertheless, the *Brn* POU domain transcription factors and the *crx* gene appear to be critical for retinal cell specific gene expression and hence the maintenance of the mature cell phenotype and survival of these cells.

2.2
bHLH Transcription Factors

The bHLH class of transcription factors is another group of proteins that have been implicated in the control of retinal cell identities. These proteins are critical for the development of sensory organs in *Drosophila* as well as for cell type-specific gene expression in a variety of vertebrate tissues. The achaete scute complex in *Drosophila* encodes four separate genes each of which is critical for some part of the fly nervous system. *Achaete* and *scute* are bHLH transcription factors that are required for the external sensory organs in the bristles that cover the mature fly, while *lethal of scute* and *asense* are necessary for the formation of the central nervous system. Another related gene, *atonal*, is required for the development of the internal sensory organs and, most importantly for this chapter, the *Drosophila* eye.

Several homologues to these genes have been identified in vertebrates; *MASH-1* (or mammalian achaete scute homologue 1) was the first one identified. *MASH-1* (or *Xash1* in *Xenopus* and *Cash1* in chick) is expressed in the developing neural retina in the progenitor cells. In the rat retina, where this gene has been most extensively analyzed, *MASH-1* is expressed in only a subset of retinal progenitor cells (Jasoni and Reh 1996). Its expression begins relatively late in embryonic development, first appearing at E15, and continuing into postnatal ages (up to P10). No expression is seen in the adult retina. Since the peak of *MASH-1* expression coincides temporally with the generation of late-appearing retinal cell types, such as rods, bipolar cells and Müller glia, Jasoni and Reh (1996) proposed that *MASH-1* is involved in the generation of these cell types. Consistent with this hypothesis, when Brown et al. (1998) overexpressed *MASH-1* in the developing *Xenopus* retinas, they found an increase in the differentiation of bipolar cells (Brown et al. 1998). Moreover, targeted deletion of *MASH-1* delays the differentiation of rod photoreceptors and horizontal cells, and also results in a reduction in the number of bipolar cells and Müller glia (Tomita et al. 1996a).

Several other related bHLH transcription factors have also been examined for a role in the generation of diversity of retinal cell types (Yan and Wang

1998) and in the rodent (Roztocil et al. 1997; Sommer et al. 1996). The bHLH protein *NeuroD*, was initially found to promote premature neuronal differentiation in *Xenopus* when overexpressed early in development. In the chick retina, this gene is expressed in the developing photoreceptors (Roztocil et al. 1997), while in the mouse and rat, *NeuroD* is expressed in the ganglion cell layer as well as in photoreceptors (Brown et al. 1998; Morrow et al. 1999). The results from overexpression of this gene in the chick are nicely consistent with the idea that a particular bHLH transcription factor can direct cells to a specific cell fate. When *NeuroD* is transfected into developing chick retinas using a replication-incompetent retrovirus, the infected cells develop into photoreceptors at a much higher rate than the uninfected cells, both in vivo and in dissociated cultures (Yan and Wang 1998). Moreover, misexpression of *NeuroD* in dissociated cultures of retinal pigmented epithelium can cause pigmented epithelial cells to transdifferentiate into photoreceptor-like cells (Yan and Wang 1998). Based on this evidence, *NeuroD* appears to be a transcription factor which specifically promotes the photoreceptor cell fate.

Although the evidence from the chick supports the idea that particular bHLH proteins direct retinal progenitor cells to specific fates, the results from similar experiments in mammals do not support this conclusion. Using a replication-defective retrovirus to overexpress *NeuroD*, Morrow et al. (1999) found that the clones derived from single infected progenitor cells have a greater number of amacrine cells and photoreceptors, but fewer bipolar cells and Müller glia. In addition, targeted deletion of *NeuroD* results in greater numbers of Müller glia, but no reported deficit in the number of photoreceptors (Morrow et al. 1999). Thus, in the mouse, *NeuroD* appears to be more important in promoting neurogenesis of multiple retinal cell classes and suppressing gliogenesis.

The data from studies of other bHLH genes are no more conclusive on the issue of whether specific members of this class of transcription factors direct cells to particular fates in the developing retina. *Xath5*, a *Xenopus* atonal homologue, (Kanekar et al. 1997) (or *Math5a* in the mouse) is expressed early in retinal development both in proliferating retinal progenitors and in early differentiating retinal neurons. As for *MASH-1*, this gene is not expressed in all progenitor cells, but appears to be expressed in only a subset. When *Xath5* is transfected into the optic vesicles of developing *Xenopus* embryos, there is an increase in the number of cells that assume the ganglion cell fate (Kanekar et al. 1997), consistent with a role in the development of the first born retinal cell class, the retinal ganglion cells. However, when *Math5* is overexpressed in the same assay, it promotes the differentiation of retinal bipolar cells, again complicating the simple notion of one bHLH gene for each retinal cell type (Brown et al. 1998).

NeuroM is another bHLH transcription factor expressed in the developing retina and other areas of the neural tube. This gene is expressed transiently in cells lining the ventricular zone of the CNS that have exited the cell cycle but have not yet begun to migrate into the inner layers of the neural tube

(Roztocil et al. 1997). Thus *NeuroM* defines a nonproliferating, premigratory population of cells. In the chick retina, *NeuroM* first appears at E2 in the central retina. As development proceeds, this expression expands peripherally. *NeuroM*-expressing cells are found throughout the retina, interspersed with nonexpressing cells. *NeuroM* is also expressed in the mature retina, in the horizontal cells, and the bipolar neurons. While this restricted pattern of expression in the mature retina suggests a role in the maintenance of the phenotype of these cells, no functional data have yet been reported for *NeuroM*.

Consistent with the idea that bHLH transcription factors control retinal cell differentiation are the results from studies of an inhibitor of this class of proteins known as *HES-1*. *HES-1* is a bHLH transcription factor which is a mammalian homologue of the *Drosophila* genes *hairy* and *enhancer-of-split* (Akazawa et al. 1992; Sasai et al. 1992; Feder et al. 1993; Takebayashi et al. 1994). The hairy and enhancer-of-split proteins act as dominant suppressors of the achaete scute proteins in *Drosophila*, titrating out their activity. The vertebrate homologues include *HES-1*, and appear to act as inhibitors of vertebrate achaete scute homologues (Ishibashi et al. 1994, 1995). A study by Tomita et al. (1996b) showed that *HES-1* is highly expressed in retinal progenitor cells. Using retroviral mediated gene transfer, they also demonstrated that retinal progenitor cells overexpressing *HES-1* failed to express markers characteristic of differentiated retinal cells and remained localized to the outer retina. Moreover, mice with a targeted deletion of *HES-1* had small eyes, and exhibited premature differentiation of rod photoreceptors in the outer retina. Interestingly, these mice also had fewer bipolar neurons, similar to the ocular retardation (*or*) phenotype.

In sum, there are several different bHLH transcription factors expressed in the developing and mature retina, in cell type-specific patterns that are consistent with this class of proteins playing a key role in the generation of retinal cell diversity. However, experimental manipulations of expression of these factors has not provided a clear answer as to whether they play an essential function in specifying retinal cell type. An alternative possibility for the function of these transcription factors is that they are critical for controlling the exit of the progenitor cell from the cell cycle. In this model, *MASH-1* maintains the progenitor cell in an undifferentiated state, and when the cell switches on *Math5* or one of the other "neuro-determination" bHLH genes, this may cause it to exit from the cell cycle and terminally differentiate.

3
Do Extracellular Signals Define Retinal Cell Identities?

Progenitor cells in the vertebrate retina are multipotent (Turner and Cepko 1987; Holt et al. 1988; Wetts and Fraser 1988; Turner et al. 1990; Fekete et al.

1994) but the potential to generate different cell types becomes restricted during retinal development. For example, clones of progenitors that are labeled postnatally in the rat retina give rise only to late-developing cell types like rod photoreceptors, bipolar cells, and Müller glia, but never to early-generated, like ganglion cells and horizontal cells (Turner and Cepko 1987; Turner et al. 1990). Similarly, in vitro studies indicate that if progenitor cells are isolated from early embryonic retina, they differentiate primarily into retinal ganglion cells, while when progenitor cells are isolated from postnatal retina, they do not differentiate into ganglion cells in vitro, but a significant number of them will express aspects of the rod photoreceptor cell fate (Reh and Kljavin 1989).

Further studies support the concept that progenitor cells change over the time course of development. Embryonic retinal progenitor cells are not induced to differentiate in response to cAMP, while progenitor cells from postnatal retina are (Taylor and Reh 1990). Prior to embryonic day 17 in the rat, retinal progenitor cells do not respond to TGF-α, while after this developmental age this factor is a potent mitogen (Lillien and Cepko 1992). As described above, retinal progenitor cells also express different complements of genes at different stages of development; as noted above, Jasoni et al. (1996) demonstrated that the bHLH transcription factors, MASH-1 and Cash1 (in the rat and chicken respectively) are expressed only in retinal progenitors in the later stages of retinal development, after embryonic day 16.

In addition to these studies that have shown intrinsic differences among the progenitor population in the retina, several studies have also demonstrated that developmental changes in the microenvironment influence the potential of progenitor cells to generate different cell types (Negishi et al. 1985; Reh and Tully 1986; Reh 1987, 1992; Watanabe and Raff 1990; Harris and Messersmith 1992). For example, in heterochronic cocultures consisting of reaggregated embryonic and postnatal rat retinal cells, following dissociation embryonic progenitors generate more rods in the presence of postnatal cells whereas less rods are generated by postnatal cells (Watanabe and Raff 1990; Reh 1992). Interestingly, the production of additional rods was not accelerated in these cultures, suggesting a fixed intrinsic schedule of cell type generation. In addition, early in vitro studies showed that cell–cell contact and diffusible activities produced by retinal cells influence retinal cell fate (Altshuler and Cepko 1992; Harris and Messersmith 1992; Reh 1992; Watanabe and Raff 1992) and since then a high number of possible extrinsic signal molecules has been identified (see below). These data support the involvement of epigenetic mechanisms by production of extrinsic signals in the surrounding microenvironment, regulating proliferation of progenitors and development of different retinal phenotypes.

Extrinsic signals that have been identified so far represent cell–cell interactions or diffusible molecules like neurotrophic proteins, or hormonal and other soluble factors. It has been generally difficult to distinguish effects on cell determination from those on early differentiation. In the following sections, those factors that have been shown to influence the development of each cell type will be described.

3.1
Ganglion Cells

Ganglion cells are one of the first cell types to be produced and few ganglion cells are generated late in development (Snow and Robson 1994). Isolated progenitor cells differentiate as ganglion cells in vitro (Reh and Kljavin 1989; Austin et al. 1995) within minutes following terminal mitosis (Waid and McLoon 1995). Although it is still not clear which mechanism initiates the generation of ganglion cells, several findings suggest that direct cell–cell interactions like lateral inhibition mediated by Notch-Delta signaling prevent other cells from differentiating into ganglion cells (Austin et al. 1995; Dorsky et al. 1995, 1997; Henrique et al. 1997). Absence of the surface cell receptor Notch leads to an increase in the number of developing ganglion cells, whereas the constitutive activation of Notch gives opposite results (Austin et al. 1995; Dorsky et al. 1995). Therefore, activation of Notch by certain cell–cell contacts blocks cells from assuming the ganglion cell phenotype. However, this mechanism is possibly only active during early development because older retinal cells produce a diffusible, not yet identified activity that blocks ganglion cell differentiation independently of Notch expression (Waid and McLoon 1998).

One family of factors that have been shown to be important for ganglion cell differentiation are the fibroblast growth factors (FGF). One or more FGFs appear to be important for the initiation of differentiation of the first ganglion cells that form in the retina. In organ cultures of chick optic vesicles, blocking FGF signaling with either antibodies or antisense oligonucleotides inhibits ganglion cell differentiation in the neural retina (Pittack et al. 1997; Desire et al. 1998). In addition, exogenous application of FGF1 or FGF2 stimulates ganglion cell differentiation in explant cultures of eye cups or optic vesicles in both birds and mammals (Guillemot and Cepko 1992; Pittack et al. 1997). Members of the FGF family and their receptors are expressed during retinal development (Mascarelli et al. 1987; Cirillo et al. 1990; Heuer et al. 1990; Wanaka et al. 1991; Conolly et al. 1992; de Iongh and McAvoy 1992; 1997; Consigli et al. 1993; Bugra et al. 1993; Jaquemin et al. 1993; Patstone et al. 1993; Ohuchi et al. 1994; Tcheng et al. 1994a; Gao and Hollyfield 1995; McFarlane et al. 1995, 1998; Fayein et al. 1996; Riou et al. 1998; McWhirter et al. 1997; Wilke et al. 1997). However, while several studies have shown that FGF is important for ganglion cell differentiation in chicks and mammals, FGF may not be critical for this process in all species; dominant negative FGF receptor expression in *Xenopus* does not interfere with ganglion cell development in this species (McFarlane et al. 1998).

3.2
Horizontal Cells

There is very little known about the factors that are important for horizontal cell differentiation. In the chick, the neurotrophin receptor trkA is present in early developing horizontal cells (Karlsson et al. 1998), suggesting a role for NGF in cell survival and/or neurite outgrowth, while CNTF receptor expression in horizontal cells is found during late development (Fuhrmann et al. 1998a). However, to date, no studies have found any effects of NGF or CNTF on horizontal cell development.

3.3
Amacrine Cells

Amacrine cells are by far the most diverse cell class in the vertebrate retina. There is an enormous variety of morphologically and functionally distinguishable subpopulations located in the inner part of the inner nuclear layer and displaced in the ganglion cell layer. While amacrine cells are well characterized with regard to expression of cell-specific markers, and morphological properties, much less is known about the regulation of their development.

Retinoic acid treatment of embryonic rat retinal cultures (Kelley et al. 1994) and FGF receptor blockade in frog (McFarlane et al. 1998) inhibit the development of amacrine cells, while TGF-β treatment of rat retinal cultures promotes the production of amacrine cells (Anchan and Reh 1995). These factors are normally expressed in the developing retina, and also have effects on the differentiation of other retinal cell types. For example, Kelley et al. (1994; see below) proposed that retinoic acid diverts retinal progenitor cells from an amacrine cell fate to the rod photoreceptor cell fate. However, it is also possible that the effects on amacrine cell development and rod photoreceptor differentiation are not linked, since other factors can influence their development independently (see below).

At least some classes of presumptive amacrine cells in the chick retina also express the high-affinity NGF receptor trkA very early in their development (Karlsson et al. 1998). Many of these cells appear to degenerate during the period of programmed cell death in the inner nuclear layer (Cook et al. 1998). Although no effects of NGF on amacrine cells have been reported, these observations suggest that the availability of NGF regulates survival of some amacrine cell subpopulations. BDNF and NT-3, as well as their receptors, are also localized to cells in the inner nuclear layer (Cohen-Cory and Fraser 1994; Rickman and Brecha 1995; Garner et al. 1996; Hallbook et al. 1996; Das et al. 1997) and both molecules support the survival of amacrine cells in vitro (de La Rosa et al. 1994). In rat retina in vivo, antisense oligonucleotides targeted to trkB mRNA specifically decreased the labeling of the AII-subclass of parvalbumin-positive amacrine cells (Rickman and Rickman 1996) and BDNF supports the development of the dopaminergic network (Cellerino et al. 1998). In addition to neurotrophins, the cytokine CNTF has also been shown to be important for the development of subtypes of amacrine cells. A subpopulation of amacrine cells, including a particular type of cholinergic neurons, expresses CNTF receptor (Heller et al. 1995; Kirsch et al. 1997; Fuhrmann et al. 1998a). Moreover, exogeneous application of CNTF to retinal cell cultures causes an upregulation of cholinergic differentiation in amacrine cells (Hofmann 1988a, b; Kirsch et al. 1997; Fuhrmann et al. 1998a).

3.4
Photoreceptor Cells

Results from in vitro studies have provided convincing evidence for the importance of extrinsic signals during photoreceptor development. In aggregate cultures, interactions via cell–cell contacts with appropriate neighbors promoted rod development (Harris and Messersmith 1992; Reh 1992; Morrow et al. 1998). By coculturing retinal cells of different age Watanabe and Raff (1990, 1992), Reh (1992), and Altshuler and Cepko (1992) demonstrated that expression of the photoreceptor marker opsin was stimulated by a developmentally regulated factor. In contrast to rodents, cone differentiation in monolayer cultures from chick retina was reported to occur in the absence of extrinsic factors as a developmental default pathway (Adler and Hatlee 1989).

Since these initial studies, an increasing number of extrinsic factors have been identified that influence generation, differentiation, and survival of rod photoreceptor cells in vivo and in vitro. Three factors have been shown to have these effects on rod differentiation in at least two species: retinoic acid, sonic hedgehog, and CNTF. Sonic hedgehog (shh) is a member of a gene family encoding secreted signaling molecules related to the hedgehog gene of *Drosophila* and is involved in patterning in different embryonic structures. Hedgehog proteins and their receptors (patched) are expressed very early in the developing eye (Jensen and Wallace 1997; Levine et al. 1997; Takabatake et al. 1997). Application of recombinant shh to rat retinal cultures has a transient mitogenic effects on progenitor cells and causes a specific increase in differentiating photoreceptors (Levine et al. 1997). In the zebrafish, antisense oligonucleotides against shh inhibit retinal growth and, specifically, the development of the rod photoreceptors, consistent with these results (D.L. Stenkamp, personal communication). However, in cultures of neonatal mouse retina, the number of amacrine cells and Müller glia was increased in shh-treated cultures (Jensen and Wallace 1997). Overall it is clear that shh has effects on the retinal progenitor cells, and promotes their development into late-generated retinal cell types, like rod photoreceptors and Müller glia.

Several studies have shown that ligands for members of the steroid/thyroid superfamily of receptors, retinoic acid (RA) and thyroid hormone, are important for photoreceptor development. In embryonic and neonatal cultures of mammalian retinas, RA causes an increase in the number of rod photoreceptors that differentiate (Kelley et al. 1994, 1995a) and a concomitant decrease in the number of amacrine cells. In zebrafish, RA is necessary for the generation of the appropriate rod/cone ratio; RA treatment causes a precocious differentiation of rod photoreceptors analogous to that observed in the rat (Hyatt et al. 1996). In chick retinal cultures, retinoids support survival of photoreceptors rather than their differentiation (Stenkamp et al. 1993; Fuhrmann et al. 1998b); it is interesting that this species difference appears to exist for other signaling systems involved in photoreceptor development, such as CNTF (see below). Retinoid receptors, retinoids, thyroid hormone, and its receptors are present in the developing retina (Dolle et al. 1990, 1994; Forrest et al. 1990; Rossant et al. 1991; Ruberte et al. 1991; Mangelsdorf et al. 1992; McCaffery et al. 1992; Prati et al. 1992; Sjoberg et al. 1992; McCaffery and Drager 1993; Nicotra et al. 1994; Seleiro et al. 1994; Mey et al. 1997; Hoover et al. 1998), and depletion of RA or its receptors causes severe defects in eye development (Drysdale and Crawford 1994; Grondona et al. 1996; Dickman et al. 1997; Kochhaar et al. 1998; for review see Kastner et al. 1995; Smith et al. 1998), whereas combinations of both RA and thyroid hormone cause progenitor cells to differentiate as either rods or cones, depending on the relative concentrations of these ligands (Kelley et al. 1995b).

Opposite effects of CNTF on photoreceptor differentiation are obtained in developing chick and rat retina in vitro. In chick, CNTF promotes photoreceptor differentiation (Fuhrmann et al. 1995, 1998b; Kirsch et al. 1996), while in rat retina CNTF suppresses rod differentiation (Kirsch et al. 1996, 1997, 1998; Ezzedine et al. 1997; Neophytou et al. 1997). In explant cultures of CNTF receptor-deficient mice, more cells express opsin than in wild-type animals (Ezzedine et al. 1997). In the chick, CNTF receptor is expressed in the developing retina prior to expression of opsin (Heller et al. 1995; Fuhrmann et al. 1998a, b) in cells of the presumptive photoreceptor layer. Although CNTF receptor is not expressed in the outer nuclear layer of the rat retina at the appropriate time during development in vivo, it is possible that progenitor cells or newly developing rods in the inner nuclear layer express the receptor. Alternatively, the effects of CNTF on opsin expression may not be direct, but rather may cause the expression of a different factor that supresses opsin in these cells. Müller glia has been identified as a source of CNTF in developing rat and chick retina (Hofmann 1988a, b; Kirsch et al. 1997). Other members of the cytokine family, LIF and oncostatin M but not Il-6, exert similar effects on photoreceptor differentiation in rodents in vitro (Ezzedine et al. 1997; Neophytou et al. 1997; Kirsch et al. 1998) and LIF could be identified as another Müller glia-derived factor (Neophytou et al. 1997).

In addition to the three factors described above, other molecules have been shown to promote rod photoreceptor development in retinal cultures. Treatment of cell cultures with laminin β2-containing extracellular matrices promotes the expression of opsin in dissociated retinal cells (Hunter et al.

1992; Libby et al. 1996, 1997). Furthermore, the trophic amino acid taurine has been shown to influence photoreceptor differentiation in rat retinal cultures (Altshuler et al. 1993) and its expression pattern implies a role in the development and maturation of the mammalian retina (Lake 1994; Nag et al. 1998). FGF promotes differentiation of opsin-immunoreactive cells in post-natal rat and embryonic chick retinal monolayer cultures and in fish retinal slice cultures (Hicks and Courtois 1988, 1992; Mack and Fernald 1993; Tcheng et al. 1994b). Inhibition of FGF receptor signaling by introducing a dominant negative form in *Xenopus* embryos causes a 50 % loss of photo-receptors (McFarlane et al. 1998). Similarly, injections of FGF antisense oligonucleotides into embryonic chick eyes result in a decrease of cell number in the outer nuclear layer (Desire et al. 1998). While these effects may be due to FGF action on retinal progenitors, recent results support the im-portance of FGF for the survival of photoreceptors (for review see Hicks 1996).

The mitogens TGF-a and EGF can act as inhibitors of photoreceptor dif-ferentiation (Anchan et al. 1991; Lillien and Cepko 1992; Reh 1992) and expression of both EGF receptors and ligands was found in the developing retina (Anchan et al. 1991; Lillien and Cepko 1992; Meyer and Birchmeier 1995; Lillien and Wancio 1998).

3.5
Bipolar Cells

One potential candidate for a factor involved in the production of bipolar cells is CNTF. CNTF promotes the differentiation or survival of retinal bi-polar cells in vitro (Ezzedine et al. 1997; Fuhrmann et al. 1998a) and a small subpopulation of bipolar cells express CNTF receptor transiently, but pre-cisely, during the appropriate developmental period (Fuhrmann et al. 1998a). CNTF itself is localized in Müller glia (Hofmann 1988a, b; Kirsch et al. 1997) and, therefore, could be available to these bipolar cells. FGF-2 and BDNF also have effects on rat bipolar cells in vitro (Wexler et al. 1998), but these appear to be survival effects.

3.6
Müller Glia

Two signaling systems are known to affect Müller glia cell differentiation in the retina: EGF/TGF-α/EGF receptor and sonic hedgehog. Activation of the EGF receptor, either by application of exogenous ligand or by overexpres-sion of the receptor using retroviral infection (Lillien 1995) causes retinal progenitor cells to differentiate into Müller glia. As noted above, this treat-ment also prevents photoreceptor differentiation (Lillien and Cepko 1992; Reh 1992). However, overexpression of EGF receptor at early developmental stages is not sufficient to cause progenitor cells to the Müller glia cell fate, suggesting that additional mechanisms regulate competence to generate Müller glia (Lillien and Wancio 1998). Jensen and Wallace (1997) reported that sonic hedgehog promotes Müller glia differentiation in embryonic rat retinal cultures, though these effects have not been observed by Levine et al. (1997) using a similar culture system.

4
Conclusion

The retina has served as a model system for developmental studies directed at understanding the molecular basis for cell identity in the nervous system. The ability to identify the various cell classes, both morphologically and chemically, in vitro and in vivo, has been of great advantage in these studies. However, despite a considerable effort, the molecules that generate retinal cell diversity are not well understood. Clearly, there are several important factors, both intracellular and extracellular, that contribute to cell-specific gene expression in the retina. To date, the approaches of localization, misexpression, and targeted disruption have failed to give a clear view of how, or

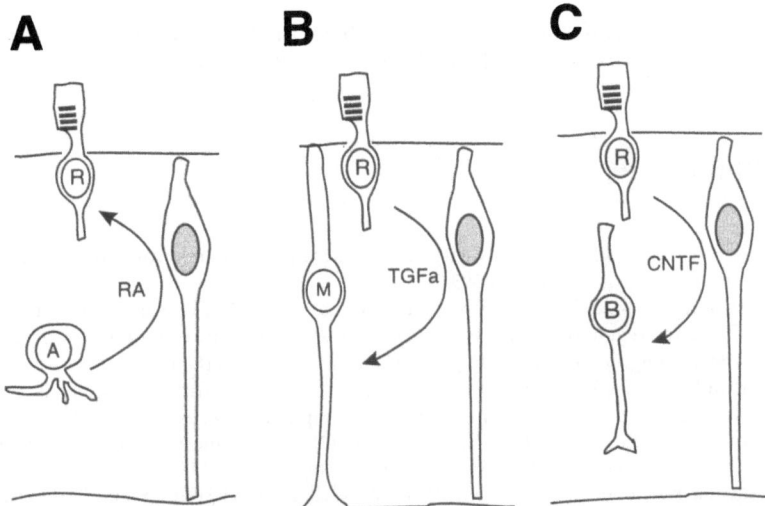

Fig. 2A–C. Soluble factors that affect retinal neurogenesis often promote one retinal cell fate while concomitantly inhibiting the development of a different cell fate. In *A* retinoic acid (RA) is shown to inhibit amacrine cell (A) differentiation while promoting the development of the rod photoreceptor (R). In *B* TGF-α suppresses rod photoreceptor differentiation, but promotes the progenitor cell to adopt a Müller glial cell fate (M). In *C* CNTF (in the mammalian retina) is shown to also suppress rod photoreceptor differentiation, but when this factor is in excess, the progenitor cells may develop as bipolar cells depending on the culture system

if, any of the known transcription factors are necessary and sufficient to define any retinal cell type. Nevertheless, these studies have primarily been limited by the lack of control of the timing and level of misexpression of these transcription factors – we know that many transcription factors can affect the process by which cell diversity arises in the retina, it is just not clear what exactly those effects are.

The experimental manipulations of soluble factors and signaling systems have led to the generation of several models of diversity, but here, too, the models are a good working start, but are not nearly complete in their details. One feature that they share is shown in Fig. 2. Several factors cause an increase in one cell class and a concomitant decrease in another. Retinoic acid causes cells to develop as rod photoreceptors and inhibits the acquisition of at least some type of amacrine cell fate. Activation of the EGF receptor blocks progenitor cells from acquiring the rod photoreceptor fate, but promotes Müller glial differentiation. CNTF, at least in the rat, also inhibits rod photoreceptor differentiation, but could promote bipolar development. These examples suggest that the processes that direct cells to specific fates are in some way linked. The challenge for the future will be to define how these signaling systems interact under normal physiological conditions, and how these control the function of the many transcription factors necessary to produce the diversity in retinal cell structure and function.

Acknowledgements. The authors wish to thank Drs. Olivia Bermingham-McDonogh, Andrew Davis, Andrew Fischer, and Edward Levine for critical comments.

References

Adler R, Hatlee M (1989) Plasticity and differentiation of embryonic retinal cells after terminal mitosis. Science 243:391–393

Akazawa C, Sasai Y, Nakanishi S, Kageyama R (1992) Molecular characterization of a rat negative regulator with a basic helix-loop-helix structure predominantly expressed in the developing nervous system. J Biol Chem 267:21879–21885

Altshuler D, Cepko C (1992) A temporally regulated, diffusible activity is required for rod photoreceptor development in vitro. Development 114:947–957

Altshuler D, Lo Turco JJ, Rush J, Cepko C (1993) Taurine promotes the differentiation of a vertebrate retinal cell type in vitro. Development 119:1317–1328

Anchan RM, Reh TA, Angello J, Balliet A, Walker M (1991) EGF and TGF-α stimulate retinal neuroepithelial cell proliferation in vitro. Neuron 6:923–936

Anchan RM, Reh TA (1995) Transforming growth factor-β-3 is mitogenic for rat retinal progenitor cells in vitro. J Neurobiol 28:133–145

Austin CP, Feldman DE, Ida JA, Jr., Cepko CL (1995) Vertebrate retinal ganglion cells are selected from competent progenitors by the action of Notch. Development 121:3637–3650

Belecky-Adams T, Tomarev S, Li HS, Ploder L, McInnes RR, Sundin O, Adler R (1997) *Pax-6*, *Prox 1*, and *Chx10* homeobox gene expression correlates with phenotypic fate of retinal precursor cells. Invest Ophthalmol Visual Sci 38:1293–1303

Brown NL, Kanekar S, Vetter ML, Tucker PK, Gemza DL, Glaser T (1998) Math5 encodes a murine basic helix-loop-helix transcription factor expressed during early stages of retinal neurogenesis. Development 125 : 4821–4833

Bugra K, Oliver L, Jacquemin E, Laurent M, Courtois Y, Hicks D (1993) Acidic fibroblast growth factor is expressed abundantly by photoreceptors within the developing and mature rat retina. Eur J Neurosci 5 : 1586–1595

Burmeister M, Novak J, Liang MY, Basu S, Ploder L, Hawes NL, Vidgen D, Hoover F, Goldman D, Kalnins VI, Roderick TH, Taylor BA, Hankin MH, McInnes RR (1996) Ocular retardation mouse caused by Chx10 homeobox null allele: impaired retinal progenitor proliferation and bipolar cell differentiation. Nat Genet 12 : 376–384

Cellerino A, Pinzon-Duarte G, Carroll P, Kohler K (1998) Brain-derived neurotrophic factor modulates the development of the dopaminergic network in the rodent retina. J Neurosci 18 : 3351–3362

Chen R, Amoui M, Zhang Z, Mardon G (1997) Dachshund and eyes absent proteins form a complex and function synergistically to induce ectopic eye development in Drosophila. Cell 91 : 893–903

Cirillo A, Arruti C, Courtois Y, Jeanny JC (1990) Localization of basic fibroblast growth factor binding sites in the chick embryonic neural retina. Differentiation 45 : 161–167

Cohen-Cory S, Fraser SE (1994) BDNF in the development of the visual system of Xenopus. Neuron 12 : 747–761

Connolly SE, Hjelmeland LM, LaVail MM (1992) Immunohistochemical localization of basic fibroblast growth factor in mature and developing retinas of normal and RCS rats. Curr Eye Res 11 : 1005–1017

Consigli SA, Lyser KM, Joseph-Silverstein J (1993) The temporal and spatial expression of basic fibroblast growth factor during ocular development in the chicken. Invest Ophthalmol Visual Sci 34 : 559–566

Cook JE (1996) Spatial properties of retinal mosaics: an empirical evaluation of some existing measures. Visual Neurosci 13 : 15–30

Cook B, Portera-Cailliau C, Adler R (1998) Developmental neuronal death is not a universal phenomenon among cell types in the chick embryo retina. J Comp Neurol 396 : 12–19

Das I, Hempstead BL, MacLeish PR, Sparrow JR (1997) Immunohistochemical analysis of the neurotrophins BDNF and NT-3 and their receptors trk B, trk C, and p75 in the developing chick retina. Visual Neurosci 14 : 835–842

de Iongh R, McAvoy JW (1992) Distribution of acidic and basic fibroblast growth factors (FGF) in the foetal rat eye: implications for lens development. Growth Factors 6 : 159–177

de Iongh RU, Lovicu FJ, Chamberlain CG, McAvoy JW (1997) Differential expression of fibroblast growth factor receptors during rat lens morphogenesis and growth. Invest Ophthalmol Visual Sci 38 : 1688–1699

de la Rosa EJ, Arribas A, Frade JM, Rodriguez-Tebar A (1994) Role of neurotrophins in the control of neural development: neurotrophin-3 promotes both neuron differentiation and survival of cultured chick retinal cells. Neuroscience 58 : 347–352

Desire L, Courtois Y, Jeanny JC (1998) Suppression of fibroblast growth factors 1 and 2 by antisense oligonucleotides in embryonic chick retinal cells in vitro inhibits neuronal differentiation and survival. Exp Cell Res 241 : 210–221

Dickman ED, Thaller C, Smith SM (1997) Temporally-regulated retinoic acid depletion produces specific neural crest, ocular and nervous system defects. Development 124 : 3111–3121

Dolle P, Ruberte E, Leroy P, Morriss-Kay G, Chambon P (1990) Retinoic acid receptors and cellular retinoid binding proteins. I. A systematic study of their differential pattern of transcription during mouse organogenesis. Development 110 : 1133–1151

Dolle P, Fraulob V, Kastner P, Chambon P (1994) Developmental expression of murine retinoid X receptor (RXR) genes. Mech Dev 45 : 91–104

Dorsky RI, Rapaport DH, Harris WA (1995) Xotch inhibits cell differentiation in the Xenopus retina. Neuron 14 : 487–496

Dorsky RI, Chang WS, Rapaport DH, Harris WA (1997) Regulation of neuronal diversity in the *Xenopus* retina by Delta signalling. Nature 385:67-70

Drysdale TA, Crawford MJ (1994) Effects of localized application of retinoic acid on *Xenopus laevis* development. Dev Biol 162:394-401

Erkman L, McEvilly RJ, Luo L, Ryan AK, Hooshmand F, O'Connell SM, Keithley EM, Rapaport DH, Ryan AF, Rosenfeld MG (1996) Role of transcription factors *Brn-3.1* and *Brn-3.2* in auditory and visual system development. Nature 381:603-606

Ezzeddine ZD, Yang X, DeChiara T, Yancopoulos G, Cepko CL (1997) Postmitotic cells fated to become rod photoreceptors can be respecified by CNTF treatment of the retina. Development 124:1055-1067

Fayein NA, Head MW, Jeanny JC, Courtois Y, Fuhrmann G (1996) Expression of the chicken cysteine-rich fibroblast growth factor receptor (CFR) during embryogenesis and retina development. J Neurosci Res 43:602-612

Feder JN, Jan LY, Jan YN (1993) A rat gene with sequence homology to the *Drosophila* gene *hairy* is rapidly induced by growth factors known to influence neuronal differentiation. Mol Cell Biol 13:105-113

Fekete DM, Perez-Miguelsanz J, Ryder EF, Cepko CL (1994) Clonal analysis in the chicken retina reveals tangential dispersion of clonally related cells. Dev Biol 166:666-682

Forrest D, Sjoberg M, Vennstrom B (1990) Contrasting developmental and tissue-specific expression of α and β thyroid hormone receptor genes. Embo J 9:1519-1528

Freund C, Horsford DJ, McInnes RR (1996) Transcription factor genes and the developing eye: a genetic perspective. Hum Mol Genet 5:1471-1488

Fuhrmann S, Kirsch M, Hofmann HD (1995) Ciliary neurotrophic factor promotes chick photoreceptor development in vitro. Development 121:2695-2706

Fuhrmann S, Kirsch M, Heller S, Rohrer H, Hofmann HD (1998a) Differential regulation of ciliary neurotrophic factor receptor-α expression in all major neuronal cell classes during development of the chick retina. J Comp Neurol 400:244-254

Fuhrmann S, Heller S, Rohrer H, Hofmann HD (1998b) A transient role for ciliary neurotrophic factor in chick photoreceptor development. J Neurobiol 37:672-683

Furukawa T, Kozak CA, Cepko CL (1997) *Rax*, a novel paired-type homeobox gene, shows expression in the anterior neural fold and developing retina. Proc Natl Acad Sci USA 94:3088-3093

Gan L, Xiang M, Zhou L, Wagner DS, Klein WH, Nathans J (1996) POU domain factor *Brn-3b* is required for the development of a large set of retinal ganglion cells. Proc Natl Acad Sci USA 93:3920-3925

Gao H, Hollyfield JG (1995) Basic fibroblast growth factor in retinal development: differential levels of bFGF expression and content in normal and retinal degeneration (rd) mutant mice. Dev Biol 169:168-184

Garner AS, Menegay HJ, Boeshore KL, Xie XY, Voci JM, Johnson JE, Large TH (1996) Expression of TrkB receptor isoforms in the developing avian visual system. J Neurosci 16:1740-1752

Grondona JM, Kastner P, Gansmuller A, Decimo D, Chambon P, Mark M (1996) Retinal dysplasia and degeneration in RARβ2/RARγ2 compound mutant mice. Development 122:2173-2188

Guillemot F, Cepko CL (1992) Retinal fate and ganglion cell differentiation are potentiated by acidic FGF in an in vitro assay of early retinal development. Development 114:743-754

Hallbook F, Backstrom A, Kullander K, Ebendal T, Carri NG (1996) Expression of neurotrophins and trk receptors in the avian retina. J Comp Neurol 364:664-676

Harris WA, Messersmith SL (1992) Two cellular inductions involved in photoreceptor determination in the *Xenopus* retina. Neuron 9:357-372

Heller S, Finn TP, Huber J, Nishi R, Geissen M, Puschel AW, Rohrer H (1995) Analysis of function and expression of the chick GPA receptor (GPAR α) suggests multiple roles in neuronal development. Development 121:2681-2693

Henrique D, Hirsinger E, Adam J, Le Roux I, Pourquie O, Ish-Horowicz D, Lewis J (1997) Maintenance of neuroepithelial progenitor cells by Delta-Notch signalling in the embryonic chick retina. Curr Biol 7:661-670

Heuer JG, von Bartheld CS, Kinoshita Y, Evers PC, Bothwell M (1990) Alternating phases of FGF receptor and NGF receptor expression in the developing chicken nervous system. Neuron 5:283-296

Hicks D (1996) Characterization and possible roles of fibroblast growth factors in retinal photoreceptor cells. Keio J Med 45:140-154

Hirsch N, Harris WA (1997) *Xenopus Pax-6* and retinal development. J Neurobiol 32:45-61

Hitchcock PF, Macdonald RE, VanDeRyt JT, Wilson SW (1996) Antibodies against Pax6 immunostain amacrine and ganglion cells and neuronal progenitors, but not rod precursors, in the normal and regenerating retina of the goldfish. J Neurobiol 29:399-413

Hofmann HD (1988a) Ciliary neuronotrophic factor stimulates choline acetyltransferase activity in cultured chicken retina neurons. J Neurochem 51:109-113

Hofmann HD (1988b) Development of cholinergic retinal neurons from embryonic chicken in monolayer cultures: stimulation by glial cell-derived factors. J Neurosci 8:1361-1369

Hogan BL, Horsburgh G, Cohen J, Hetherington CM, Fisher G, Lyon MF (1986) Small eyes (*Sey*): a homozygous lethal mutation on chromosome 2 which affects the differentiation of both lens and nasal placodes in the mouse. J Embryol Exp Morphol 97:95-110

Holt CE, Bertsch TW, Ellis HM, Harris WA (1988) Cellular determination in the *Xenopus* retina is independent of lineage and birth date. Neuron 1:15-26

Hoover F, Seleiro EA, Kielland A, Brickell PM, Glover JC (1998) Retinoid X receptor γ gene transcripts are expressed by a subset of early generated retinal cells and eventually restricted to photoreceptors. J Comp Neurol 391:204-213

Huang S, Moody SA (1997) Three types of serotonin-containing amacrine cells in tadpole retina have distinct clonal origins. J Comp Neurol 387:42-52

Hunter DD, Murphy MD, Olsson CV, Brunken WJ (1992) S-laminin expression in adult and developing retinae: a potential cue for photoreceptor morphogenesis. Neuron 8:399-413

Hyatt GA, Schmitt EA, Fadool JM, Dowling JE (1996) Retinoic acid alters photoreceptor development in vivo. Proc Natl Acad Sci USA 93:13298-13303

Ishibashi M, Moriyoshi K, Sasai Y, Shiota K, Nakanishi S, Kageyama R (1994) Persistent expression of helix-loop-helix factor HES-1 prevents mammalian neural differentiation in the central nervous system. Embo J 13:1799-1805

Ishibashi M, Ang SL, Shiota K, Nakanishi S, Kageyama R, Guillemot F (1995) Targeted disruption of mammalian hairy and Enhancer of split homolog-1 (MASH-1) leads to up-regulation of neural helix-loop-helix factors, premature neurogenesis, and severe neural tube defects. Genes Dev 9:3136-3148

Jacquemin E, Jonet L, Oliver L, Bugra K, Laurent M, Courtois Y, Jeanny JC (1993) Developmental regulation of acidic fibroblast growth factor (aFGF) expression in bovine retina. Int J Dev Biol 37:417-423

Jasoni CL, Reh TA (1996) Temporal and spatial pattern of MASH-1 expression in the developing rat retina demonstrates progenitor cell heterogeneity. J Comp Neurol 369:319-327

Jensen AM, Wallace VA (1997) Expression of Sonic hedgehog and its putative role as a precursor cell mitogen in the developing mouse retina. Development 124:363-371

Kanekar S, Perron M, Dorsky R, Harris WA, Jan LY, Jan YN, Vetter ML (1997) *Xath5* participates in a network of bHLH genes in the developing *Xenopus* retina. Neuron 19:981-994

Karlsson M, Clary DO, Lefcort FB, Reichardt LF, Karten HJ, Hallbook F (1998) Nerve growth factor receptor TrkA is expressed by horizontal and amacrine cells during chicken retinal development. J Comp Neurol 400:408-416

Kastner P, Mark M, Chambon P (1995) Nonsteroid nuclear receptors: what are genetic studies telling us about their role in real life? Cell 83:859-869

Kelley MW, Turner JK, Reh TA (1994) Retinoic acid promotes differentiation of photoreceptors in vitro. Development 120:2091-2102

Kelley MW, Turner JK, Reh TA (1995a) Regulation of proliferation and photoreceptor differentiation in fetal human retinal cell cultures. Invest Ophthalmol Visual Sci 36:1280–1289

Kelley MW, Turner JK, Reh TA (1995b) Ligands of steroid/thyroid receptors induce cone photoreceptors in vertebrate retina. Development 121:3777–3785

Kirsch M, Fuhrmann S, Wiese A, Hofmann HD (1996) CNTF exerts opposite effects on in vitro development of rat and chick photoreceptors. Neuroreport 7:697–700

Kirsch M, Lee MY, Meyer V, Wiese A, Hofmann HD (1997) Evidence for multiple, local functions of ciliary neurotrophic factor (CNTF) in retinal development: expression of CNTF and its receptors and in vitro effects on target cells. J Neurochem 68:979–990

Kirsch M, Schulz-Key S, Wiese A, Fuhrmann S, Hofmann H (1998) Ciliary neurotrophic factor blocks rod photoreceptor differentiation from postmitotic precursor cells in vitro. Cell Tissue Res 291:207–216

Kljavin IJ, Lagenaur C, Bixby JL, Reh TA (1994) Cell adhesion molecules regulating neurite growth from amacrine and rod photoreceptor cells. J Neurosci 14:5035–5049

Kochhar DM, Jiang H, Penner JD, Johnson AT, Chandraratna RA (1998) The use of a retinoid receptor antagonist in a new model to study vitamin A-dependent developmental events. Int J Dev Biol 42:601–608

Krauss S, Johansen T, Korzh V, Moens U, Ericson JU, Fjose A (1991) Zebrafish pax[zf-a]: a paired box-containing gene expressed in the neural tube. Embo J 10:3609–3619

Lake N (1994) Taurine and GABA in the rat retina during postnatal development. Visual Neurosci 11:253–260

Lee JE (1997) NeuroD and neurogenesis. Dev Neurosci 19:27–32

Levine EM, Roelink H, Turner J, Reh TA (1997) Sonic hedgehog promotes rod photoreceptor differentiation in mammalian retinal cells in vitro. J Neurosci 17:6277–6288

Li HS, Yang JM, Jacobson RD, Pasko D, Sundin O (1994) Pax-6 is first expressed in a region of ectoderm anterior to the early neural plate: implications for stepwise determination of the lens. Dev Biol 162:181–194

Libby RT, Hunter DD, Brunken WJ (1996) Developmental expression of laminin β2 in rat retina. Further support for a role in rod morphogenesis. Invest Ophthalmol Visual Sci 37:1651–1661

Libby RT, Xu Y, Selfors LM, Brunken WJ, Hunter DD (1997) Identification of the cellular source of laminin β2 in adult and developing vertebrate retinae. J Comp Neurol 389:655–667

Lillien L, Cepko C (1992) Control of proliferation in the retina: temporal changes in responsiveness to FGF and TGF α. Development 115:253–266

Lillien L (1995) Changes in retinal cell fate induced by overexpression of EGF receptor. Nature 377:158–162

Lillien L, Wancio D (1998) Changes in epidermal growth factor receptor expression and competence to generate glia regulate timing and choice of differentiation in the retina. Mol Cell Neurosci 10:296–308

Liu IS, Chen JD, Ploder L, Vidgen D, van der Kooy D, Kalnins VI, McInnes RR (1994) Developmental expression of a novel murine homeobox gene (Chx10): evidence for roles in determination of the neuroretina and inner nuclear layer. Neuron 13:377–393

Mangelsdorf DJ, Borgmeyer U, Heyman RA, Zhou JY, Ong ES, Oro AE, Kakizuka A, Evans RM (1992) Characterization of three RXR genes that mediate the action of 9-cis retinoic acid. Genes Dev 6:329–344

Mascarelli F, Raulais D, Counis MF, Courtois Y (1987) Characterization of acidic and basic fibroblast growth factors in brain, retina and vitreous chick embryo. Biochem Biophys Res Commun 146:478–486

McCaffery P, Lee MO, Wagner MA, Sladek NE, Drager UC (1992) Asymmetrical retinoic acid synthesis in the dorsoventral axis of the retina. Development 115:371–382

McCaffery P, Drager UC (1993) Retinoic acid synthesis in the developing retina. Adv Exp Med Biol 328:181–190

McEvilly RJ, Erkman L, Luo L, Sawchenko PE, Ryan AF, Rosenfeld MG (1996) Requirement for Brn-3.0 in differentiation and survival of sensory and motor neurons. Nature 384:574–577

McFarlane S, McNeill L, Holt CE (1995) FGF signaling and target recognition in the developing *Xenopus* visual system. Neuron 15:1017–1028

McFarlane S, Zuber ME, Holt CE (1998) A role for the fibroblast growth factor receptor in cell fate decisions in the developing vertebrate retina. Development 125:3967–3975

McWhirter JR, Goulding M, Weiner JA, Chun J, Murre C (1997) A novel fibroblast growth factor gene expressed in the developing nervous system is a downstream target of the chimeric homeodomain oncoprotein E2A-Pbx1. Development 124:3221–3232

Mey J, McCaffery P, Drager UC (1997) Retinoic acid synthesis in the developing chick retina. J Neurosci 17:7441–7449

Meyer D, Birchmeier C (1995) Multiple essential functions of neuregulin in development. Nature 378:386–390

Morrow EM, Belliveau MJ, Cepko CL (1998) Two phases of rod photoreceptor differentiation during rat retinal development. J Neurosci 18:3738–3748

Morrow EM, Furukawa T, Lee JE, Cepko CL (1999) NeuroD regulates multiple functions in the developing neural retina in rodent. Development 126:23–36

Nag TC, Jotwani G, Wadhwa S (1998) Immunohistochemical localization of taurine in the retina of developing and adult human and adult monkey. Neurochem Int 33:195–200

Negishi K, Teranishi T, Kato S (1982) New dopaminergic and indoleamine-accumulating cells in the growth zone of goldfish retinas after neurotoxic destruction. Science 216:747–749

Negishi K, Teranishi T, Kato S (1985) Growth rate of a peripheral annulus defined by neurotoxic destruction in the goldfish retina. Brain Res 352:291–295

Neophytou C, Vernallis AB, Smith A, Raff MC (1997) Muller-cell-derived leukaemia inhibitory factor arrests rod photoreceptor differentiation at a postmitotic pre-rod stage of development. Development 124:2345–2354

Nicotra CM, Gueli MC, de Luca G, Bono A, Pintaudi AM, Paganini A (1994) Retinoid dynamics in chicken eye during pre- and postnatal development. Mol Cell Biochem 132:45–55

Ohuchi H, Yoshioka H, Tanaka A, Kawakami Y, Nohno T, Noji S (1994) Involvement of androgen-induced growth factor (FGF-8) gene in mouse embryogenesis and morphogenesis. Biochem Biophys Res Commun 204:882–888

Passini MA, Levine EM, Canger AK, Raymond PA, Schechter N (1997) *Vsx-1* and *Vsx-2*: differential expression of two paired-like homeobox genes during zebrafish and goldfish retinogenesis. J Comp Neurol 388:495–505

Patstone G, Pasquale EB, Maher PA (1993) Different members of the fibroblast growth factor receptor family are specific to distinct cell types in the developing chicken embryo. Dev Biol 155:107–123

Pittack C, Grunwald GB, Reh TA (1997) Fibroblast growth factors are necessary for neural retina but not pigmented epithelium differentiation in chick embryos. Development 124:805–816

Prati M, Calvo R, Morreale G, Morreale de Escobar G (1992) L-thyroxine and 3,5,3'-triiodothyronine concentrations in the chicken egg and in the embryo before and after the onset of thyroid function. Endocrinology 130:2651–2659

Puschel AW, Gruss P, Westerfield M (1992) Sequence and expression pattern of pax-6 are highly conserved between zebrafish and mice. Development 114:643–651

Quiring R, Walldorf U, Kloter U, Gehring WJ (1994) Homology of the *eyeless* gene of *Drosophila* to the *Small eye* gene in mice and Aniridia in humans. Science 265:785–789

Reh TA, Tully T (1986) Regulation of tyrosine hydroxylase-containing amacrine cell number in larval frog retina. Dev Biol 114:463–469

Reh TA (1987) Cell-specific regulation of neuronal production in the larval frog retina. J Neurosci 7:3317–3324

Reh TA, Kljavin IJ (1989) Age of differentiation determines rat retinal germinal cell phenotype: induction of differentiation by dissociation. J Neurosci 9:4179–4189

Reh TA (1992) Cellular interactions determine neuronal phenotypes in rodent retinal cultures. J Neurobiol 23:1067–1083

Rickman DW, Brecha NC (1995) Expression of the proto-oncogene, trk, receptors in the developing rat retina. Visual Neurosci 12:215–222

Rickman DW, Rickman CB (1996) Suppression of trkB expression by antisense oligonucleotides alters a neuronal phenotype in the rod pathway of the developing rat retina. Proc Natl Acad Sci USA 93:12564–12569

Riou JF, Delarue M, Mendez AP, Boucaut JC (1998) Role of fibroblast growth factor during early midbrain development in Xenopus. Mech Dev 78:3–15

Rossant J, Zirngibl R, Cado D, Shago M, Giguere V (1991) Expression of a retinoic acid response element-hsplacZ transgene defines specific domains of transcriptional activity during mouse embryogenesis. Genes Dev 5:1333–1344

Roztocil T, Matter-Sadzinski L, Alliod C, Ballivet M, Matter JM (1997) NeuroM, a neural helix-loop-helix transcription factor, defines a new transition stage in neurogenesis. Development 124:3263–3272

Ruberte E, Dolle P, Chambon P, Morriss-Kay G (1991) Retinoic acid receptors and cellular retinoid binding proteins. II. Their differential pattern of transcription during early morphogenesis in mouse embryos. Development 111:45–60

Sasai Y, Kageyama R, Tagawa Y, Shigemoto R, Nakanishi S (1992) Two mammalian helix-loop-helix factors structurally related to Drosophila hairy and Enhancer of split. Genes Dev 6:2620–2634

Seleiro EA, Darling D, Brickell PM (1994) The chicken retinoid-X-receptor-γ gene gives rise to two distinct species of mRNA with different patterns of expression. Biochem J 301:283–288

Sjoberg M, Vennstrom B, Forrest D (1992) Thyroid hormone receptors in chick retinal development: differential expression of mRNAs for α and N-terminal variant β receptors. Development 114:39–47

Smith SM, Dickman ED, Power SC, Lancman J (1998) Retinoids and their receptors in vertebrate embryogenesis. J Nutr 128:467S–470S

Snow RL, Robson JA (1994) Ganglion cell neurogenesis, migration and early differentiation in the chick retina. Neuroscience 58:399–409

Sommer L, Ma Q, Anderson DJ (1996) Neurogenins, a novel family of atonal-related bHLH transcription factors, are putative mammalian neuronal determination genes that reveal progenitor cell heterogeneity in the developing CNS and PNS. Mol Cell Neurosci 8:221–241

Stenkamp DL, Gregory JK, Adler R (1993) Retinoid effects in purified cultures of chick embryo retina neurons and photoreceptors. Invest Ophthalmol Visual Sci 34:2425–2436

Takabatake T, Ogawa M, Takahashi TC, Mizuno M, Okamoto M, Takeshima K (1997) Hedgehog and patched gene expression in adult ocular tissues. FEBS Lett 410:485–489

Takebayashi K, Sasai Y, Sakai Y, Watanabe T, Nakanishi S, Kageyama R (1994) Structure, chromosomal locus, and promoter analysis of the gene encoding the mouse helix-loop-helix factor HES-1. Negative autoregulation through the multiple N box elements. J Biol Chem 269:5150–5156

Taylor M, Reh TA (1990) Induction of differentiation of rat retinal germinal neuroepithelial cells by cAMP. J Neurobiol 21:470–481

Tcheng M, Fuhrmann G, Hartmann MP, Courtois Y, Jeanny JC (1994a) Spatial and temporal expression patterns of FGF receptor genes type 1 and type 2 in the developing chick retina. Exp Eye Res 58:351–358

Tcheng M, Oliver L, Courtois Y, Jeanny JC (1994b) Effects of exogenous FGFs on growth, differentiation, and survival of chick neural retina cells. Exp Cell Res 212:30–35

Tomita K, Nakanishi S, Guillemot F, Kageyama R (1996a) MASH-1 promotes neuronal differentiation in the retina. Genes Cells 1:765–774

Tomita K, Ishibashi M, Nakahara K, Ang SL, Nakanishi S, Guillemot F, Kageyama R (1996b) Mammalian hairy and Enhancer of split homolog 1 regulates differentiation of retinal neurons and is essential for eye morphogenesis. Neuron 16:723–734

Truslove GM (1962) A gene causing ocular retardation in the mouse. J Embryol Exp Morphol 10:652–660

Turner DL, Cepko CL (1987) A common progenitor for neurons and glia persists in rat retina late in development. Nature 328:131–136

Turner DL, Snyder EY, Cepko CL (1990) Lineage-independent determination of cell type in the embryonic mouse retina. Neuron 4:833–845

Turner EE, Jenne KJ, Rosenfeld MG (1994) Brn-3.2: a Brn-3-related transcription factor with distinctive central nervous system expression and regulation by retinoic acid. Neuron 12:205–218

Waid DK, McLoon SC (1995) Immediate differentiation of ganglion cells following mitosis in the developing retina. Neuron 14:117–124

Waid DK, McLoon SC (1998) Ganglion cells influence the fate of dividing retinal cells in culture. Development 125:1059–1066

Walther C, Gruss P (1991) Pax-6, a murine paired box gene, is expressed in the developing CNS. Development 113:1435–1449

Wanaka A, Milbrandt J, Johnson EM, Jr. (1991) Expression of FGF receptor gene in rat development. Development 111:455–468

Watanabe T, Raff MC (1990) Rod photoreceptor development in vitro: intrinsic properties of proliferating neuroepithelial cells change as development proceeds in the rat retina. Neuron 4:461–467

Watanabe T, Raff MC (1992) Diffusible rod-promoting signals in the developing rat retina. Development 114:899–906

Wetts R, Fraser SE (1988) Multipotent precursors can give rise to all major cell types of the frog retina. Science 239:1142–1145

Wexler EM, Berkovich O, Nawy S (1998) Role of the low-affinity NGF receptor (p75) in survival of retinal bipolar cells. Visual Neurosci 15:211–218

Wilke TA, Gubbels S, Schwartz J, Richman JM (1997) Expression of fibroblast growth factor receptors (FGFR1, FGFR2, FGFR3) in the developing head and face. Dev Dyn 210:41–52

Xiang M, Zhou L, Peng YW, Eddy RL, Shows TB, Nathans J (1993) Brn-3b: a POU domain gene expressed in a subset of retinal ganglion cells. Neuron 11:689–701

Xiang M, Zhou L, Macke JP, Yoshioka T, Hendry SH, Eddy RL, Shows TB, Nathans J (1995) The Brn-3 family of POU-domain factors: primary structure, binding specificity, and expression in subsets of retinal ganglion cells and somatosensory neurons. J Neurosci 15:4762–4785

Xiang M, Gan L, Zhou L, Klein WH, Nathans J (1996) Targeted deletion of the mouse POU domain gene Brn-3a causes selective loss of neurons in the brainstem and trigeminal ganglion, uncoordinated limb movement, and impaired suckling. Proc Natl Acad Sci USA 93:11950–11955

Yan RT, Wang SZ (1998) NeuroD induces photoreceptor cell overproduction in vivo and de novo generation in vitro. J Neurobiol 36:485–496

Cell Fate Specification in the *Drosophila* Retina

Justin P. Kumar and Kevin Moses[1]

Introduction

The compound eye of the fruit fly, *Drosophila melanogaster*, is a useful model system for investigating the issues that underlie cell fate specification. More than a decade ago, Tomlinson and Ready proposed that the cells within the fly eye make use of a combinatorial code of signals in adopting specific cellular identities. This concept has driven our effort to understand how the individual facets of the fly eye are assembled (Ready et al. 1976; Tomlinson 1985). The attraction of the fly eye as an experimental system also rests within its simple adult structure along with its precise and stereotyped development. This is, of course, coupled to the extraordinary power of *Drosophila* genetic techniques to discover genes (the genes mentioned in this chapter are summarized in Table 1). We will attempt to summarize what is known about the molecular development of the first (R8) and last (R7) photoreceptors to join the developing unit eye or ommatidia. We are focusing on these two cells, in part, because the molecules and mechanisms that underlie their development are the best understood. The mechanisms used by these two cells, a proneural one for R8 and an inductive one for R7, are repeatedly used during development. Lessons learned from the development of these two cells have considerable implications for other experimental systems.

Arranged in a nearly crystalline array are nearly 800 identical unit eyes or ommatidia that comprise the compound eye of *Drosophila*. Assembled within each ommatidium are just 20 differing cells, each occupying a precise position within the facet (Fig. 1). At the heart of the ommatidium lie eight photoreceptor neurons (labeled R1-8) that are arranged in an asymmetric trapezoidal pattern. Above the photoreceptor core lie four cone cells that secrete the overlying lens. Surrounding the ommatidial core is a sheath of pigment cells that optically insulate one ommatidium from another.

The adult compound eye is derived from a monolayer epithelium, the eye-antennal imaginal disk. During the third larval instar, pattern formation

[1] Department of Cell Biology, Emory University School of Medicine, 1648 Pierce Drive, Atlanta, Georgia 30322-3030, USA

Results and Problems in Cell Differentiation, Vol. 31
M. E. Fini (Ed.): Vertebrate Eye Development
© Springer-Verlag Berlin Heidelberg 2000

Table 1. Summary of genes treated in this chapter

Gene name	Abbreviation	Synonyms	Type of protein
anterior open	*aop*	*yan, pokkuri*	Ets domain nuclear
atonal	*ato*		Basic helix-loop-helix
bride of sevenless	*boss*	*F*	G-protein coupled receptor
corkscrew	*csw*	*E(sev)1A, l(1)2Db*	Tyrosine phosphatase
daughter of sevenless	*dos*	*E(csw)3A, Su(sev)3A*	PH domain cytoplasmic
Draf1	*raf*	*l(1)pole hole, polyhomeotic*	MAP kinase kinase kinase
Delta	*Dl*	*Overflow*	EGF repeat containing transmembrane
Disabled	*Dab*		Phosphotyrosine interaction domain
Downstream of raf1	*Dsor1*	*MEK, MAPKK*	MAP kinase kinase
downstream of receptor kinase	*drk*	*E(sev)2B, crkl, Su(sev)R1*	SH2-SH3-SH2
Enhancer of Split Complex	*ES-C*		Nine nuclear factors
Epidermal growth factor receptor	*Egfr*	*DER, c-erbB, flb, top, Elp*	Receptor tyrosine kinase
Gap1	*Gap*	*mip, sextra*	GTPase
Notch	*N*	*fa, spl, Ax, Chp, Co, nd, shd, swb*	EGF repeat transmembrane receptor
phyllopod	*phyl*	*Su(Raf)2C*	Novel nuclear
pointed	*pnt*	*D-ets-2*	Ets containing nuclear
ras	*ras*	*IMPDH, raspberry-lethal*	Small GTPase signal
rolled	*rl*	*DmERK-A, MAPK, Dsor2*	MAP kinase
rough	*ro*		Homeobox nuclear
scabrous	*sca*		Secreted
sevenless	*sev*		Receptor tyrosine kinase
seven in absentia	*sina*		Ring finger transcription factor
seven-up	*svp*		Nuclear hormone receptor
Son of sevenless	*Sos*	*l(2)br24, br25, 34Ea, E(sev)2A, Su(tor)2-2*	Guanine nucleotide exchange factor
spitz	*spi*		Transforming growth factor-alpha
Suppressor of Hairless	*Su(H)*	*l(2)br7, dRBP-JK*	Transcription factor
tramtrack	*ttk*	*oss, FTZ-F2, oversensitive, E(yan)100D, schmal*	Zinc finger transcription factor

begins with the initiation of a wave of differentiation at the posterior margin of the disk that sweeps anteriorly. The leading edge of the wave is visualized by an indentation in the epithelium, the morphogenetic furrow [1]. The earliest events in eye development, including the initiation and progression of the morphogenetic furrow along with the molecules that govern its movement, are considered in the preceding chapter by U. Heberlein and

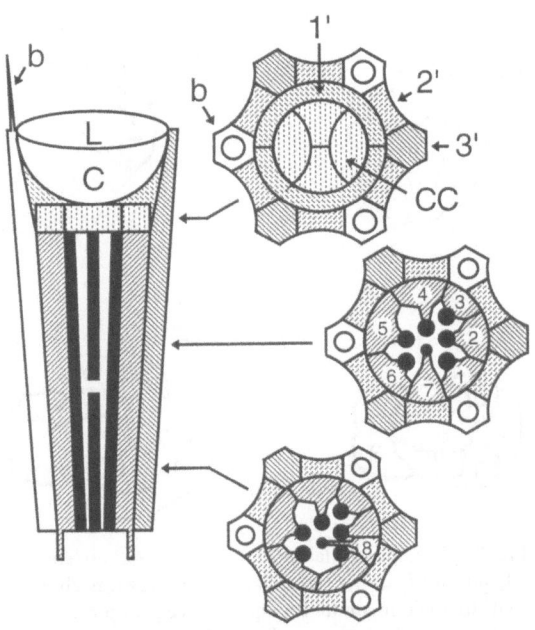

Fig. 1. The structure of the ommatidium. To the *left* is a diagram of the long section of one facet or ommatidium taken from the dorsal half of the right eye. Anterior is shown to the *right*. To the *right* are three cross-sections cut at the levels indicated. Cells and structures are indicated as follows: *b* bristle (two cells, *unshaded*); *L* lens (*unshaded*); *C* crystalline cone (*unshaded*); *1* primary pigment cell (*left diagonal broken shading*); *2* secondary pigment cell (*right diagonal broken shading*); *3* tertiary pigment cell (*left diagonal shading*); *CC* cone cell (*vertical broken shading*). The eight photoreceptor cells are *numbered* (*right diagonal shading*). Only one example of each cell type is labeled, others are shaded similarly. Note: the convention is to count 20 cells: 8 photoreceptors, 4 cone cells, 2 primary pigment cells, 3 secondary pigment cells (there are 6, but each is shared between 2 ommatidia, so the total counted is 3), 1 tertiary pigment cell (there are 3, but each is shared between 3 ommatidia, so the total counted is 1) and 2 cells forming the mechanosensory bristle (there are 3 bristles, but each is shared between 3 ommatidia, so the total counted is 1 bristle group of 2 cells). Thus 8 + 4 + 2 + 3 + 1 + 2 = 20

J. Treisman. Our discussion will focus on the events of ommatidial assembly, a process that begins within the morphogenetic furrow.

The first steps in ommatidial assembly begin with the formation of large periodic cell groupings termed rosettes within the morphogenetic furrow (see Fig. 2). Within the furrow, the rosette undergoes a process of maturation that reduces it to just five cells which are the precursors to the first five photoreceptors, R8, the R2/5 pair, and the R3/4 pair. Collectively, these five precursor cells are called the precluster (Ready et al. 1976; Wolff et al. 1993). The first column of evenly spaced preclusters can be seen lying just posterior to the morphogenetic furrow. One cell within the precluster will differentiate first and become the R8 neuron. The remaining four cells will differentiate in pairs with the R2/5 pair first and the R3/4 pair following. A tight wave of mitosis follows just behind the morphogenetic furrow and gives birth to a

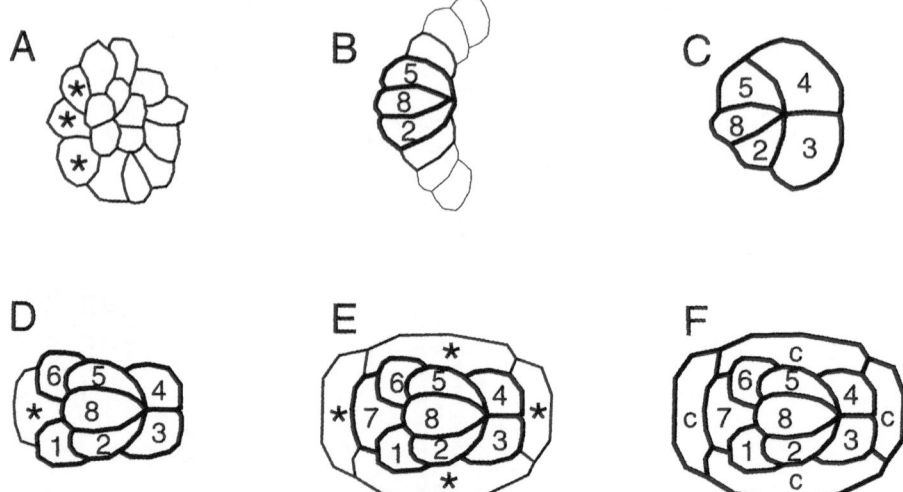

Fig. 2A–F. Ommatidial assembly. Six stages are shown from a third instar eye imaginal disk. In all, anterior is to the *right*. Each stage shown is about 4 to 6 h after the last. **A** The rosette stage. Cells that will form the R8 equivalence group are indicated with *asterisks*. **B** The arc stage. Cells that are specified are *numbered* (by future photoreceptor cell fate) and are in *heavy lines*, the next two cells are in *mid-level lines*, arc-terminal cells that will be excluded at the next stage are in the *light line*. **C** The five cell precluster. **D** Cells 1 and 6 have been specified, the cell in the next position (*asterisk*) will normally be specified as the R7, but will become a cone cell if Sevenless pathway signals are blocked. **E** The R7 has been specified, and the next four outer cells have not (*asterisks*). While these unspecified cells do express the Sevenless receptor at this stage, they are occluded from the R8 cell, and thus receive no Boss ligand. If Sevenless signaling is ectopically activated in these cells at this stage, all will become R7s. **F** The four cells that did not "see" Boss become cone cells (*c*). The specification of the pigment and bristle cells follows later

pool of cells from which the final three photoreceptors are recruited, the R1/6 first and ultimately the R7 neuron. The cone and pigment cells are also borne in the second mitotic wave and are added to the cluster later in development.

The developing fly ommatidium is constructed using two distinct molecular and developmental mechanisms. Its birth begins with the formation of the R8 photoreceptor neuron, the founder cell. Its appearance is brought about by interactions between proneural and neurogenic genes within the morphogenetic furrow. R8 photoreceptor and embryonic neuroblast specification are similar in that both depend on proneural and neurogenic gene signaling pathways and are not lineage-dependent. However, the R8 cell, once specified, is postmitotic, while the embryonic neuroblasts give rise to specific and stereotyped lineages (Ready et al. 1976; Wolff and Ready 1991; Lawrence and Green 1979). Following R8 specification, the remaining ommatidial cells are recruited into the cluster by inductive mechanisms. Each precursor cell receives cues from neighboring cells that have already differentiated and the combination of cues received by the precursor cell determines its ultimate cellular fate. The recruitment of the R7 neuron, the last photoreceptor to be added to the ommatidium, is one of the best understood examples of an inductive mechanism in metazoan development (Tomlinson 1988; Ready 1989; Rubin 1989; Zipursky 1989; Banerjee and Zipursky 1990; Lawrence and Tomlinson 1991; Cagan and Zipursky 1992; Krämer and Cagan 1994; Freeman 1997; Dickson 1998).

2
The R8 Photoreceptor Cell

Within the morphogenetic furrow individual cells from a larger epithelial field are singled out to become R8 cell precursors. This requires the function of a single proneural gene, *atonal*, whose actions define a developmental state in which clusters of cells are competent to become neuronal precursors (Ghysen and Dambly-Chaudiere 1989; Romani et al. 1989). Proneural genes have been best studied in the development of sensory organs of the peripheral nervous system (PNS) and neuroblast lineages of the central nervous system (CNS Jan and Jan 1993; Goodman and Doe 1993). The *atonal* gene encodes a basic Helix-Loop-Helix (bHLH) transcription factor, which has as a proneural function and determines the R8 photoreceptor cell fate (Jarman et al. 1993). In *atonal* mutants there is a complete failure of R8 cells to appear behind the furrow, resulting in the complete deletion of the compound eyes (Jarman et al. 1994, 1995). In contrast ectopic expression of Atonal results in the formation of extra R8 cells in each ommatidium (Dokucu et al. 1996). All cells just anterior to the furrow express Atonal (the prefurrow phase) but within the furrow Atonal expression is refined first to groups of 15–20 cells (the intermediate phase) and then down to an equivalence group of 2–3 cells (Dokucu et al. 1996; Baker et al. 1996). From this equivalence group, one cell continues to express Atonal and becomes the R8 photoreceptor precursor. A developing wild-type eye disk reveals three columns of single, evenly spaced, putative R8 cells staining positively for this proneural protein (Jarman et al. 1994; Dokucu et al. 1996; Baker et al. 1996).

It is becoming clear that very similar processes are involved in the specification of retinal precursor cells in vertebrates. In the vertebrate retina the first neural cell type to differentiate is the retinal ganglion cell (RGC, reviewed in [25]). The specification of the RGCs depends on the expression of homologues of *Drosophila* Atonal, known as *Xath* in *Xenopus* and *Math* in mouse (Kanekar et al. 1997; Brown et al. 1998; Tsuda et al. 1998). The *Math3* gene is expressed early in much of the developing nervous system, but later becomes restricted to the retina, and this is mediated at the transcriptional level (Tsuda et al. 1998). *Math5* expression is largely restricted to the developing eye and begins before neural differentiation and persists later in progenitor cells. When Math5 is ectopically expressed in *Xenopus* the retinal is expanded (Brown et al. 1998). Targeted ectopic expression of Xath5 in *Xenopus* retinal progenitor cells increases the percentage of RGCs in the retina (Kanekar et al. 1997). Thus it seems likely that Atonal homologues function in the developing vertebrate retina, as Atonal does in the fly, to specify the founding retinal neuronal cell type.

Notch is a transmembrane receptor protein containing a number of EGF-like and ankyrin repeats and cysteine-rich domains (Wharton et al. 1985; Kidd et al. 1986; Artavanis-Tsakonas et al. 1995). Loss of *Notch* function within the morphogenetic furrow results in the neuralization of all cells, and many of these extra neurons adopt the R8 cell fate (Cagan and Ready 1989; Baker et al. 1990). Neural hypertrophy in *Notch* loss-of-function mutants is a familiar phenotype, first seen in the development of the embryonic nervous system (Poulson 1937). This is seen also in mutants for any of a set of neurogenic loci that includes *Delta, mastermind, bigbrain, neuralized*, and the *Enhancer of split Complex* (*ES-C*) (Lehmann et al. 1983). Together, the neurogenic genes' products function to single out individual cells from proneural domains and facilitate their progression into neuronal precursor cells. In neurogenic gene loss-of-function mutants, extra neural precursors are made, a phenotype that contrasts with that of the proneural genes. In addition to *Notch*, mutations in *Delta, mastermind, neuralized*, and *ES-C* disrupt ommatidial development (Dietrich and Campos-Ortega 1984; Campos-Ortega and Knust 1990; Parody and Muskavitch 1993; Parks et al. 1995; Ligoxygakis et al. 1998; Fischer-Vize et al. 1992).

Cagan and Ready clearly showed that Notch is required for the recruitment of all cells within the developing eye and also that loss of Notch often has opposing effects on cells that are in differing developmental states. This led to the proposition that Notch played a permissive rather than an instructive role in cell fate choices within the retina (Cagan and Ready 1989). A series of recent papers has demonstrated that Notch can both promote and repress R8 formation by affecting the *atonal* expression pattern differently in two phases. First, *Notch* is required to induce the broad Atonal expression domain ahead of the furrow. Later, signaling via Notch is required to restrict *atonal* expression to the intermediate group stage within the furrow and finally to the single Atonal expressing R8 cells that lie posterior to the furrow (Baker et al. 1996; Ligoxygakis et al. 1998; Baker and Zitron 1995; Baker and Yu 1997). These results have largely explained the apparent contradiction that *Notch* is both a neurogenic and proneural gene in the developing eye.

Removing *Notch* function (with a conditional allele) for a short period of time leads to a neurogenic phenotype in which large fractions of cells within the morphogenetic furrow are neuralized (Cagan and Ready 1989). The newly neuralized cells are derived from the intermediate group of Atonal-expressing cells. In wild type, Notch functions in all but one of the intermediate group cells to prevent neurogenesis by repressing Atonal expression. This cell is then singled out to become the future R8 photoreceptor and will continue to express Atonal. Without *Notch*, Atonal expression is not further restricted and a single cell is not selected to become the founder R8. Instead, multiple cells retain Atonal expression and adopt the R8 neuronal fate (Baker and Yu 1997). In ectopic expression experiments in which Notch is rendered ligand-independent and constitutively active (*Notch^{Act}*), Atonal expression is lost from the intermediate groups and photoreceptor development is abolished (Baker et al. 1996).

A source of some confusion has been the apparent proneural function that *Notch* played in eye development. Photoreceptors can be lost as well as gained in Notch depletion experiments. In normal eye development, Atonal expression rises dramatically as the broad zone of Atonal-expressing cells ahead of the furrow is focused down to the intermediate group stage within the furrow (Baker and Yu 1997). As Atonal expression rises, transcription of its downstream targets is activated. It is believed that at least one of these targets (*scabrous*) is responsible for lateral inhibition between or within intermediate groups. If *Notch* function is removed at this stage, Atonal levels never rise to the levels required to activate *scabrous* transcription and the initial refinement of Atonal expression never occurs (Baker and Yu 1997). This results in the loss of ommatidial preclusters and thus the apparent proneural phenotype.

The multiple effects of *Notch* loss-of-function on eye development can be mimicked by loss of *Delta* (Knust and Campos-Ortega 1990; Parody and Muskavitch 1993; Parks et al. 1995; Baker and Zitron 1995). Delta, like Notch, is a transmembrane protein that contains a number of EGF-like repeats and also contains a cysteine-rich motif common to all known Notch ligands (Vässin and Campos-Ortega 1987; Kopczynski et al. 1988; Muskavitch 1994). Keeping in line with its role as a ligand for Notch, Delta exerts its effects nonautonomously, and Delta and Notch have been shown to mediate adhesion when they are expressed on the surface of cells in culture (Heitzler and Simpson 1991; Fehon et al. 1990; Rebay et al. 1991). Like *Notch*, loss of *Delta* within the morphogenetic furrow leads to the formation of extra R8 cells within each ommatidium (Parks et al. 1995; Baker and Zitron 1995). Also like *Notch*, *Delta* also has proneural functions. Large mosaic clones containing populations of *Delta* null cells show no neural differentiation at the center of the clone (Baker and Yu 1997). It is interesting to note that another Notch ligand (Serrate) is expressed within the morphogenetic furrow but no proneural or neurogenic phenotypes result from its absence (Bakerf and Yu 1997; Sun and Artavanis-Tsakonas 1996). This may not be surprising, as *Serrate* mutants do not have neurogenic phenotypes in other tissues and *Serrate* has never been placed within the neurogenic group of genes.

The *Notch* mutant effects on neural precursor formation are thought to occur through its downstream targets, the *ES-C* (Jennings et al. 1994; Bailey and Posakony 1995). The activation of at least one *ES-C* gene has been shown conclusively to occur through *Suppressor of Hairless* (*Su(H)*) (Bailey and Posakony 1995; Lecourtois and Schweisguth 1998). It is likely that the other *ES-C* genes are regulated in a similar manner. Within *ES-C* lie at least nine genes, of which seven encode bHLH transcription factors (*mδ, mγ, mβ, m3, m5, m7*, and *m8*). The remaining two (*m4* and *groucho*) encode non-bHLH nuclear proteins (Delidakis et al. 1991; Delidakis and Artavanis-Tsakonas 1992; Knust et al. 1992; Schrons et al. 1992). Mutations that remove most or all seven bHLH transcription factors result in the neuralization of cells within the furrow (Campos-Ortega and Knust 1990; Ligoxygakis et al. 1998; Treisman et al. 1997). Loss of *Su(H)* gives a near-identical phenotype further placing it as a regulator of *ES-C* genes downstream to *Notch*. It is interesting to note that in mosaic clones that lack *ES-C* and *Su(H)* function there is only a neurogenic component, unlike clones lacking *Notch* or *Delta*, in which both neurogenic and proneural phenotypes are seen (Ligoxygakis et al. 1998). This suggests that the proneural functions of Notch and Delta are controlled in a mechanism independent of the ES-C and Su(H) proteins. These functions may be mediated by the ankyrin repeats that lie within the intracellular domain of Notch. The Su(H)-binding site in Notch lies outside the ankyrin repeat domain (Tamura et al. 1995). Transgenic flies in which the ankyrin repeats alone are expressed have phenotypes different from activated Notch or Su(H) and this is also consistent with the action of two independent pathways (Matsuno et al. 1997).

Within the morphogenetic furrow, several *ES-C* bHLH factors (Mγ, Mβ, Mδ, M7, and M8) are expressed in distinct patterns (Dokucu et al. 1996; Baker et al. 1996, de Celis et al. 1996). The precise localization of the Mγ and Mδ proteins is known in the greatest detail: cells just anterior to the furrow contain Atonal along with the Mγ and Mδ. However, as the intermediate groups of Atonal-expressing cells is established within the furrow, Mγ and Mδ are restricted to clusters of cells that interdigitate between the Atonal-expressing groups (Dokucu et al. 1996; Baker et al. 1996). This clearly suggests that the *ES-C* genes act to repress *atonal* in some cells within the morphogenetic furrow. Overexpression of the Mδ protein leads to almost complete deletion of Atonal expression and the resulting adults consequently have very few ommatidia (Ligoxygakis et al. 1998).

The selection of R8 cannot be the sole responsibility of these proneural and neurogenic genes for two reasons. First, the absence of neurogenic genes or the ectopic expression of proneural genes should transform all newly neuralized cells into R8 neurons, which is clearly not the case (Dokucu et al. 1996; Baker et al. 1990; Parks et al. 1995). Second, in two additional mutations (*scabrous* and *rough*) the selection of R8 or the expression of Atonal is also affected (Baker et al. 1990; van Vactor et al. 1991; Heberlein et al. 1991).

A new twist to R8 specification came with the observation that mutations within the homeobox containing transcription factor Rough gave rise to multiple R8 cells per developing ommatidial cluster (van Vactor et al. 1991; Heberlein et al. 1991); but it was not until the expression pattern of Rough and Atonal were compared that a role for *rough* in R8 specification was suggested (Dokucu et al. 1996). Rough was first reported to be broadly expressed in the furrow and then within the R2/5 and R3/4 cell pairs

(Tomlinson et al. 1988). Closer examination of Rough expression within the furrow revealed that it is expressed in a pattern complementary to that of Atonal. This pattern is identical to some of the ES-C proteins (Dokucu et al. 1996). It has been suggested that the extra R8 cells in *rough* null mutants are derived from the R8 equivalence group, two to three cells from which a single R8 precursor is selected (Cagan 1993). Thus, Rough may mediate the final step of R8 precursor selection, the resolution of a single R8 cell per cluster. Ectopic expression of Rough blocks the initiation of Atonal expression within the furrow and may explain the furrow arrest seen in these transgenic flies (Dokucu et al. 1996; Basler et al. 1990; Kimmel et al. 1991).

scabrous mutants were first recognized because of their recessive rough eye phenotype (Morgan et al. 1938). Among the many abnormalities in developing *scabrous* eyes are defects in the spacing between preclusters and the specification of multiple R8 neurons within a single precluster (Baker et al. 1990; Baker and Zitron 1995; Ellis et al. 1994). Despite this phenotype, *scabrous* is not classified as a neurogenic gene, since its function is dispensable in the embryonic CNS (Mlodzik et al. 1990). Scabrous is a secreted and diffusible signaling protein that contains a fibrinogen-related domain (like Fibrinogen and Tenascin (Mlodzik et al. 1990; Lee et al. 1996)). Scabrous protein is found within a subset of Atonal-expressing cells: anterior to the furrow, where all cells express Atonal, Scabrous is absent. Scabrous expression then becomes coincident with that of Atonal within the furrow and persists in the R8 precursor cells that lie in three columns posterior to the furrow (Baker et al. 1996; Lee et al. 1996).

A key insight came from the phenotype of a double mutant for *scabrous* and *Notch*: the refinement of Atonal expression never occurs and all the cells within the morphogenetic furrow adopt an R8 cell fate (Baker and Zitron 1995; Lee et al. 1996). This *scabrous/Notch* double mutant phenotype is different from that of any other known genotype. In contrast, loss of *Notch* alone still allows for the initial phase of Atonal refinement. It is the singling out of an individual cell to continue Atonal expression and adopt a R8 precursor fate which is affected (Baker and Yu 1997). These results suggest an interesting two-pronged mode of inhibiting R8 formation. Short-range inhibition within the proneural group of cells occurs through the Notch signaling pathway to restrict R8 precursor formation to a single individual cell. This is consistent with Notch and Delta as surface-bound molecules that require sending and receiving cells to be in contact with each other (Artavanis-Tsakonas et al. 1995). Longer-range inhibition, from proneural groups to the non-Atonal-expressing cells, may be mediated by Scabrous, which is a secreted molecule (Baker and Yu 1998; Lee et al. 1998).

Consistent with the selection of the founder cell being independent of inductive cues, development of the R8 neuron occurs independently of the EGF receptor signaling cascade which has been implicated in cell fate choices of nearly all the cells in the fly ommatidium (Freeman 1997, 1996). The EGF receptor lies at the top of the ras pathway and is thought to modulate cell fate choices through the phosphorylation of transcription factors by MAP Kinase (the product of the rolled gene in *Drosophila*, (Schweitzer and Shilo 1997)). $Egfr^{Elp}$ is a gain of function mutation for the EGF receptor (Baker and Rubin 1989; Lesokhin et al. 1999). Although fewer ommatidial facets are present, these ommatidia are normally constructed (Baker and Rubin 1989). Developing $Egfr^{Elp}$ homozygotes do not have clusters with extra R8 cells as one would expect if the EGF receptor functions in R8 selection (Baker and Rubin 1992). Recently, a temperature-sensitive *Egfr* mutation was isolated that mimics the null phenotype at restrictive temperatures. Removing *Egfr* function during eye development does not affect R8 development (Kumar et al. 1998). Early attempts to recover retinal mosaic clones for *Egfr* nulls were difficult to interpret due to an early cell proliferation function of the receptor (Baker and Rubin 1989, 1992; Xu and Rubin 1993). Clones are significantly smaller than their wild-type twin spot (Xu and Rubin 1993). However, recently, the Minute technique has been used to recover large *Egfr* null clones and, consistent with the temperature-sensitive experiments, R8 development is largely normal (Dominguez et al. 1998; Lawrence et al. 1979).

3
The R7 Photoreceptor Cell

The *sevenless* mutant was isolated in a screen for mutants that failed to phototax in counter-current distribution experiments (Benzer 1973). The defects in phototaxis were subsequently shown to be due to the absence of UV-sensitive rhodopsins that are expressed specifically in the R7 photoreceptors (Harris et al. 1976; Fryxell and Meyerowitz 1987; Zuker et al. 1987). The rhodopsins are absent since the R7 itself is deleted in *sevenless* homozygotes (Harris et al. 1976) in which the R7 precursor is transformed into an accessory cell (Tomlinson and Ready 1986, 1987). *sevenless* encodes a receptor tyrosine kinase (Banerjee et al. 1987; Basler and Hafen 1988; Bowtell et al. 1988; Hafen et al. 1987) and this strongly suggested that interactions between neighboring cells play a role in the development of the R7.

Interestingly, the Sevenless protein is not restricted to just the presumptive R7 cell: its expression is dynamic during development and can be detected at the apical surface in a majority of ommatidial cells at differing times and levels (the R3/4 pair, the R1/6 pair, R7 and the four cone cells), suggesting that many cells have the potential to adopt the R7 cell fate (Banerjee et al. 1987; Tomlinson et al. 1987). A constitutively active Sevenless receptor expressed in its normal domain transforms all these cells into R7 neurons (Basler et al. 1989, 1991; Basler and Hafen 1989; Bowtell et al. 1989; Mullins and Rubin 1991). More importantly, however the expression pattern suggested that a mechanism of communication must exist within the ommatidia to restrict the R7 cell fate to a single cell. Transient expression of

Sevenless in all ommatidial cells (in a *sevenless* mutant background) only restores the R7 in a subset of ommatidial columns, suggesting that within the eye, ommatidia were only competent to use the *sevenless* pathway for a restricted period of time (Basler and Hafen 1989; Bowtell et al. 1989; Mullins and Rubin 1991).

A second gene that affects the specification of R7 is *bride of sevenless (boss)* (Reinke and Zipursky 1988). In *boss* mutants, the R7 cells are deleted from the ommatidia, but unlike *sevenless*, *boss* is required in the R8 photoreceptor and not R7 (Reinke and Zipursky 1988; Hart et al. 1990). Boss is a seven-membrane-spanning protein and is expressed solely on the surface of the R8 photoreceptor neuron (Krämer et al. 1991). Immunostains of cultured transfected cells and wild-type eye imaginal disks showed that Boss is internalized by the R7 precursor (Krämer et al. 1991; Krämer 1993; Hart et al. 1993; Tomlinson 1991; Cagan et al. 1992). The presence of the Hook protein is required for the internalization of boss into multivesicular bodies within the R7 precursor (Krämer and Phistry 1996). In normal development, the R8 cell comes into direct contact with the R7 precursor providing a cellular basis for R8 to R7 communication (Tomlinson and Ready 1986).

However, like the R7 precursor, four of the developing outer photoreceptors (the R1/6 and R3/4 pairs) also come into direct contact with R8 and express the Boss receptor Sevenless (Tomlinson and Ready 1986; Tomlinson et al. 1987). Activation of the receptor in these cells can force them to adopt an R7-like cellular fate (Basler et al. 1991). However, these same cells, when exposed to ectopic Boss protein, are refractory to this signal and do not adopt an R7 cell fate (van Vactor et al. 1991). Preventing ectopic R7 cell fate in these cells is, in part, due to the *seven-up* gene, which encodes a steroid receptor and is expressed in the R1/6 and R3/4 pairs. Mutations in *seven-up* result in the transformation of these cells into R7-like photoreceptors (Mlodzik et al. 1990). Conversely, ectopic expression of Seven-up in the R7 precursor causes it to fore go the R7 cell fate and it becomes an outer photoreceptor instead (Hiromi et al. 1993). Interestingly, *seven-up* requires a functioning ras signal transduction pathway to suppress transformations in outer photoreceptors to R7 cells (Kramer et al. 1995; Begemann et al. 1995).

Like both *boss* and *sevenless*, mutations in any Ras pathway component results in the loss of the R7 photoreceptor. Large-scale genetic screens have identified seven downstream components of the Sevenless cascade, of which four also lie downstream of the EGF receptor tyrosine kinase (Simon et al. 1991; Karim et al. 1996; Dickson et al. 1996; Rogge et al. 1991). Two of these loci encoded a Ras protein and a putative guanine nucleotide exchange factor, Sos. Placing Ras downstream of the Sevenless receptor tyrosine kinase was a significant advance in the study of the R7 photoreceptor. Mutations within the Ras GTPase-activating protein, Gap1, added to our understanding of how Ras shuttles between GTP and GDP bound forms (Gaul et al. 1992). Isolation of the SH3-SH2-SH3 adapter protein, Drk, physically linked the receptor to the Sos-ras-Gap1 complex (Oliver et al. 1993; Simon et al. 1993). From other systems it is known that downstream of Ras lie three protein kinases, Raf, Mek and Erk (Marshall 1994). Screens in both the eye and embryo quickly isolated all three fly homologues (Dickson et al. 1996; Biggs et al. 1994; Brunner et al. 1994; Tsuda et al. 1993; Dickson et al. 1992). All of these proteins have vertebrate homologues in the Ras pathway and constitute a conserved "cassette" that also functions downstream of other major *Drosophila* receptor tyrosine kinases such as the EGF receptor, the FGF receptor (Breathless), and Torso (Wassarman et al. 1995; Diaz-Benjumea and Hafen 1994). These functions of Ras pathway proteins are well understood at the genetic and biochemical levels. They represent a linear path directly from the reception of an extracellular signal to the execution of that signal via the modulation of transcription factor activities.

At the bottom of the biochemical cascade lies Erk (Marshall 1994; Cano and Mahadevan 1995), encoded by the *rolled* locus, and a key player in R7 development. While loss of *rolled* can affect the development of both the R1/6 pair and the R7 photoreceptor cell types, a gain of function mutation in *rolled* converts all Sevenless-expressing cells into R7s (Brunner et al. 1994). Activation of Erk has been shown to occur through its phosphorylation by the Raf and Dsor1 (Mek) kinases (Tsuda et al. 1993; Howe et al. 1992). Upon activation, Erk is translocated to the nucleus, where it modulates transcription by phosphorylating nuclear factors (Brunner et al. 1994; O'Neill et al. 1994). The activity of Erk is thought to be tightly regulated by phosphatases that reside within the nucleus (Nebreda 1994; Muda et al. 1997).

Except for the R8 cell, all developing photoreceptors require the Ras pathway for their specification (Freeman 1996, 1997; Kumar et al. 1998; Dominguez et al. 1998). Furthermore, only two receptor tyrosine kinases are known to function in the eye, Sevenless and the EGF receptor. So how do cells with a common developmental history interpret an identical signal yet execute different developmental programs? This, in part, rests within the combination of transcription factors that are expressed in the developing eye (Kumar, Moses 1997). It may also rest on the recent discovery that downstream of receptor tyrosine kinases lie multiple signaling pathways and not just the single linear Ras to Erk cassette (Allard et al. 1996; Le and Simon 1998; Herbst et al. 1996; Raabe et al. 1996; Mac Dougall and Waterfield 1996).

A picture of which nuclear factors lie downstream of the Ras pathway and how they regulate R7 development has emerged in recent years. It began with the isolation of a number of transcription factors that, when absent, delete the R7 photoreceptor solely or in part. Genetic screens identified *seven in absentia* (*sina*), *tramtrack, pointed, anterior open, Djun*, and *phyllopod* (Dickson 1998; O'Neill et al. 1994; Carthew and Rubin 1990; Rogge et al. 1992; Xiong and Montell 1993; Lai and Rubin 1992; Dickson et al. 1995; Dickson 1995; Chang et al. 1995; Bohmann et al. 1994).

One of the first mutations isolated in transcription factors that affected the R7 cell was that of *sina*. In *sina* mutants, like *sevenless*, the putative R7 precursor becomes a non-neuronal cone cell. Also like Sevenless, while Sina protein is expressed in many cells within the developing ommatidium, it is required autonomously in only one cell, the presumptive R7 cell precursor (Carthew and Rubin 1990). Its role in the determination of R7 had remained enigmatic until the discovery of a role in these cells for both *phyllopod* and *tramtrack*. Phyllopod is a novel nuclear protein that is required for the development of the photoreceptors borne in the second mitotic wave (the R1/6 pair and the R7 cells). Like all components in the Ras pathway, ectopic expression of Phyllopod in Sevenless-expressing cells can transform these cells into R7-like photoreceptors, suggesting that Phyllopod is regulated by the Ras signaling cascade (Xiong and Montell 1993; Dickson et al. 1995; Chang et al. 1995; Lai et al. 1997). Tramtrack is a negative regulator of neuronal differentiation and lies downstream of Erk (Xiong and Montell 1993; Lai et al. 1997). Ommatidia in *tramtrack* mutants contain supernumerary R7 cells (Xiong and Montell 1993). Tramtrack is normally expressed in the undifferentiated and non-neuronal cone cells (Lai et al. 1996) and ectopic expression of Tramtrack blocks photoreceptor development in the R3/4 pair and the R7 (which express Sevenless). First evidence for a connection between these three proteins came from observations that ectopically expressed Phyllopod downregulates Tramtrack in the cone cells. Also, Tramtrack is ectopically expressed in photoreceptors if both *sina* and *phyllopod* are absent (Li et al. 1997; Tang et al. 1997). In vitro binding assays and yeast two-hybrid screens suggest that the three proteins physically interact to form a complex. A reasonable explanation is that in a subset of neuronal precursors (the R1/6 pair and the R7 cells) Sina and Phyllopod posttranscriptionally downregulate *tramtrack* function, thereby releasing the block on photoreceptor development (Dickson 1998). Since Phyllopod is not expressed within the five precluster cells (R8, and the R2/5 and R3/4 pairs), another nuclear factor may be present that interacts with Sina in those cells to downregulate the activity of *tramtrack*.

Anterior open (also known as Yan) is an ETS domain transcription factor and is the second known general barrier to photoreceptor differentiation (O'Neill et al. 1994; Lai and Rubin 1992). Like *tramtrack*, mutations in *anterior open* yield ommatidia containing extra photoreceptors, the majority of which are R7-like cells (Lai and Rubin 1992). Anterior open appears to lie directly downstream of Erk: within the C-terminal of Anterior open reside eight putative Erk phosphorylation sites which are essential for Anterior open function (O'Neill et al. 1994; Lai and Rubin 1992; Rebay and Rubin 1995). Removal of these phosphorylation sites impedes the ability for Erk to act and results in an Anterior open isoform that blocks photoreceptor determination. This suggests that the Ras-Erk pathway normally negatively regulates Anterior open and thus promotes neuronal differentiation in the eye (Rebay and Rubin 1995). Anterior open is thought to repress neuronal cell fate determination through a mechanism that includes Tramtrack (Lai et al. 1997). By genetic epistasy these three genes lie in the regulatory order: *anterior open* > *sina* > *tramtrack*, so it is not expected that any interaction between Anterior open and Tramtrack will be direct (Lai and Rubin 1992; Lai et al. 1996; Yamamoto et al. 1996).

The other ETS-containing protein known to function in the eye is the product of the *pointed* gene, which is required for normal photoreceptor development. Like Anterior open, Pointed contains an Erk phosphorylation site and is thought to lie downstream of the Ras pathway (Brunner et al. 1994; Lai and Rubin 1992); but unlike Anterior open, Pointed is a positive regulator of neuronal differentiation: in *pointed* null mosaic clones, cells never express neuronal markers (O'Neill et al. 1994). The activity of Pointed seems to be dependent upon Djun. Loss of Djun (mimicked by dominant negative isoforms) can suppress the development of ectopic R7 cells that are induced by activated Ras (Bohmann et al. 1994). In contrast, the R7 cell fate is promoted in the R7 equivalence group by constitutively active forms of Djun. These constitutively active isoforms of Djun are unable to promote the R7 cell fate in the absence of pointed, suggesting that Djun and Pointed cooperate to establish the R7 cell fate for precursor cells. There is some additional biochemical evidence that suggests that both proteins are required to activate transcription of certain promoters (Treier et al. 1995). As expected of negative regulators of photoreceptor determination, Anterior open acts antagonistically to repress Djun-Pointed-mediated transcriptional activation in vitro and to suppress the ectopic R7 development seen in constitutively active Djun mutants in vivo (Treier et al. 1995).

The straightforward linear path from the receptor down the Ras pathway to these transcription factors is surprisingly not the end of the story for the R7. Abolition of Drk binding to Sevenless does not completely block the induction of the R7 cell fate (Raabe et al. 1996). This suggests that pathways downstream of Sevenless might exist that are independent of the canonical Ras pathway. Indeed, the identification of three new components that act downstream to *sevenless* but not in the Ras cascade (*daughter of sevenless, corkscrew,* and *Disabled*) suggests that other signal transduction pathways do function in parallel to the Ras signaling module (Allard et al. 1996; Le 1998; Herbst et al. 1996; Raabe et al. 1996; MacDougall and Waterfield 1996). What additional components lie downstream of Sevenless and/or other receptor tyrosine kinase receptors is an open question.

4
Concluding Remarks

To assume that the resources of the *Drosophila* eye have already been exhausted would be a considerable miscalculation. Significant questions about how cells adopt their primary cell fate remain. This Chapter has dealt with the development of just two cells within the eye, both of which have considerably different histories from one another. How other cells with common developmental histories still adopt differing cellular identities is still an open question worthy of continued exploration. We are considerably closer today than in the early days of the sevenless story to understanding how cells adopt their cellular fates; but we are still a long way from completely understanding the mechanisms that underlie the combinatorial code for ommatidial assembly insightfully and brilliantly proposed by Tomlinson and Ready more than a decade ago. The processes which underlie ommatidial founder cell formation in *Drosophila* (chiefly the focusing of Atonal expression) appear to be very similar in vertebrates. This similarity extends that seen earlier in the role of Pax6 in eye specification (see the chapter by Treisman and Heberlein in this Volume) and all these data taken together suggest that in evolutionary terms there may be only one eye after all.

Acknowledgements. We thank the National Eye Institute and National Science Foundation for supporting our work through the funding of the following grants: RO1 EY 0299, F32 EY 06763, and IBN-9507857.

References

Allard JD, Chang HC, Herbst R, McNeill H, Simon MA (1996) The SH2-containing tyrosine phosphatase Corkscrew is required during signaling by Sevenless, Ras1 and Raf. Development 122:1137–1146

Artavanis-Tsakonas S, Matsuno K, Fortini ME (1995) Notch signalling. Science 268:225–232

Bailey AM, Posakony JW (1995) Suppressor of Hairless directly activates transcription of *Enhancer of split* complex genes in response to Notch receptor activity. Genes Dev 9:2609–2622

Baker NE, Mlodzik M, Rubin GM (1990) Spacing differentiation in the developing *Drosophila* eye: a fibrinogen-related lateral inhibitor encoded by *scabrous*. Science 250:1370–1377

Baker NE, Rubin GM (1989) Effect on eye development of dominant mutations in *Drosophila* homologue of the EGF receptor. Nature 340:150–153

Baker NE, Rubin GM (1992) *Ellipse* mutations in the *Drosophila* homologue of the EGF receptor affect pattern formation, cell division, and cell death in eye imaginal disc. Dev Biol 150:381–396

Baker NE, Yu SY (1997) Proneural function of neurogenic genes in the developing *Drosophila* eye. Curr Biol 7:122–132

Baker NE, Yu SY (1998) The R8-photoreceptor equivalence group in *Drosophila*: fate choice precedes regulated *Delta* transcription and is independent of *Notch* gene dose. Mech Dev 74:3–14

Baker NE, Yu S, Han D (1996) Evolution of proneural Atonal expression during distinct regulatory phases in the developing *Drosophila* eye. Curr Biol 6:1290–1301

Baker NE, Zitron AE (1995) *Drosophila* eye development: Notch and Delta amplify a neurogenic pattern conferred on the morphogenetic furrow by Scabrous. Mech Dev 49:173–189

Banerjee U, Renfranz PJ, Hinton DR, Rabin BA, Benzer S (1987) The *sevenless*+ protein is expressed apically in cell membranes of developing *Drosophila* retina; it is not restricted to R7. Cell 51:151–158

Banerjee U, Renfranz PJ, Pollock JA, Benzer S (1987) Molecular characterization and expression of *sevenless*, a gene involved in neuronal pattern formation in the *Drosophila* eye. Cell 49:281–291

Banerjee U, Zipursky SL (1990) The role of cell–cell interaction in the development of the *Drosophila* visual system. Neuron 4:177–187

Basler K, Christen B, Hafen E (1991) Ligand-independent activation of the Sevenless receptor tyrosine kinase changes the fate of cells in the developing *Drosophila* eye. Cell 64:1069–1081

Basler K, Hafen E (1988) Control of photoreceptor cell fate by the *sevenless* protein requires a functional tyrosine kinase domain. Cell 54:299–311

Basler K, Hafen E (1989) Dynamics of *Drosophila* eye development and temporal requirements of Sevenless expression. Development 107:723–731

Basler K, Hafen E (1989) Ubiquitous expression of *sevenless*: position-dependent specification of cell fate. Science 243:931–937

Basler K, Siegrist P, Hafen E (1989) The spatial and temporal expression pattern of *sevenless* is exclusively controlled by gene-internal elements. EMBO J 8:2381–2386

Basler K, Yen D, Tomlinson A, Hafen E (1990) Reprogramming cell fate in the developing *Drosophila* retina: transformation of R7 cells by ectopic expression of *rough*. Genes Dev 4:728–739

Begemann G, Michon AM, v.d.Voorn L, Wepf R, Mlodzik M (1995) The *Drosophila* orphan nuclear receptor Seven-up requires the Ras pathway for its function in photoreceptor determination. Development 121:225–235

Benzer S (1973) Genetic dissection of behavior. Sci Am 229:24–37

Biggs WHr, Zavitz KH, Dickson B, van der Straten A, Brunner D, Hafen E, Zipursky SL (1994) The *Drosophila rolled* locus encodes a MAP kinase required in the *sevenless* signal trandsuction pathway. EMBO J 13:1628–1635

Bohmann D, Ellis MC, Staszewski LM, Mlodzik M (1994) *Drosophila* Jun mediates Ras-dependent photoreceptor determination. Cell 78:973–986

Bowtell DDL, Simon MA, Rubin GM (1988) Nucleotide sequence and structure of the *sevenless* gene of *Drosophila melanogaster*. Genes Dev 2:620–634

Bowtell DDL, Simon MA, Rubin GM (1989) Ommatidia in the developing *Drosophila* eye require and can respond to *sevenless* for only a restricted period. Cell 56:931–936

Brown NL, Kanekar S, Vetter ML, Tucker PK, Gemza DL, Glaser T (1998) *Math5* encodes a murine basic helix-loop-helix transcription factor expressed during early stages of retinal neurogenesis. Development 125:4821–4833

Brunner D, Oellers N, Szabad J, Biggs WHr, Zipursky SL, Hafen E (1994) A gain-of-function mutation in *Drosophila* MAP kinase activates multiple receptor tyrosine kinase signaling pathways. Cell 76:875–888

Cagan RL (1993) Cell fate specification in the developing *Drosophila* retina. Development Suppl:19–28

Cagan RL, Krämer H, Hart AC, Zipursky SL (1992) The Bride of Sevenless and Sevenless interaction: internalization of a transmembrane ligand. Cell 69:393–399

Cagan RL, Ready DF (1989) Notch is required for successive cell decisions in the developing *Drosophila* retina. Genes Dev 3:1099–1112

Cagan RL, Zipursky SL (1992) Cell choice and patterning in the *Drosophila* retina. In: Shankland M, Macaguo ER (eds) Academic Press, San Diego, pp 189–224

Campos-Ortega JA, Knust E (1990) Defective ommatidial cell assembly leads to defective morphogenesis: a phenotypic analysis of the *E(spl)D* mutation of *Drosophila melanogaster*. Wilhelm Roux's Arch Dev Biol 198:286–294

Cano E, Mahadevan LC (1995) Parallel signal processing among mammalian MAPKs. Trends Biochem Sci 20:117–122

Carthew RW, Rubin GM (1990) *Seven in absentia*, a gene required for specification of R7 cell fate in the *Drosophila* eye. Cell 63:561–577

Cepko C, Austin C, Yang X, Alexiades M, Ezzeddine D (1996) Cell fate determination in the vertebrate retina. Proc Natl Acad Sci USA 93:589–895

Chang HC, Solomon NM, Wassarman DA, Karim FD, Therrien M, Rubin GM, Wolff T (1995) *Phyllopod* functions in the fate determination of a subset of photoreceptors in *Drosopihila*. Cell 80:463–472 de Celis JF, de Celis J, Ligoxygakis P, Preiss A, Delidakis C, Bray S (1996) Functional relationships between *Notch*, *Su(H)* and the bHLH genes of the *E(spl)* complex: the *E(spl)* genes mediate only a subset of *Notch* activities during imaginal development. Development 122:2719–2728

Delidakis C, Artavanis-Tsakonas S (1992) The *Enhancer of split [E(spl)]* locus of *Drosophila* encodes seven independent helix-loop-helix proteins. Proc Natl Acad Sci USA 89:8731–8735

Delidakis C, Preiss A, Hartley DA, Artavanis-Tsakonas S (1991) Two genetically and molecularly distinct functions involved in early neurogenesis reside within the *Enhancer of split* locus of *Drosophila melanogaster*. Genetics 129:803–823

Diaz-Benjumea FJ, Hafen E (1994) The Sevenless signalling cassette mediates *Drosophila* EGF receptor function during epidermal development. Development 120:569–578

Dickson B (1995) Nuclear factors in Sevenless signalling. Trends Genet 11:106–111

Dickson B, Sprenger F, Morrison D, Hafen E (1992) Raf functions downstream of Ras1 in the Sevenless signal transduction pathway. Nature 360:600–603

Dickson BJ (1998) Photoreceptor development: breaking down the barriers. Curr Biol 8:90–92

Dickson BJ, Dominguez M, van der Straten A, Hafen E (1995) Control of *Drosophila* photoreceptor cell fates by Phyllopod, a novel nuclear protein acting downstream of the raf kinase. Cell 80:453–462

Dickson BJ, van der Straten A, Dominguez M, Hafen E (1996) Mutations modulating Raf signaling in *Drosophila* eye development. Genetics 142:163–171

Dietrich U, Campos-Ortega JA (1984) The expression of neurogenic loci in imaginal epidermal cells of *Drosophila* melanogaster. J Neurogenet 1:315–332

Dokucu ME, Zipursky SL, Cagan RL (1996) Atonal, Rough and the resolution of proneural clusters in the developing *Drosophila* retina. Development 122:4139–4147

Dominguez M, Wasserman JD, Freeman M (1998) Multiple functions of the EGF receptor in *Drosophila* eye development. Curr Biol 8:1039–1048

Ellis MC, Weber U, Wiersdorff V, Mlodzik M (1994) Confrontation of *scabrous* expressing and non-expressing cells is essential for normal ommatidial spacing in the *Drosophila* eye. Development 120:1959–1969

Fehon RG, Kooh PJ, Rebay I, Regan CL, Xu T, Muskavitch MAT, Artavanis-Tsakonas S (1990) Molecular interactions between the protein products of the neurogenic loci *Notch* and *Delta*, two EGF-homologous genes in *Drosophila*. Cell 61:523–534

Fischer-Vize JA, Vize PD, Rubin GM (1992) A unique mutation in the *Enhancer of split* gene complex affects the fates of the mystery cells in the developing *Drosophila* eye. Development 115:89–101

Freeman M (1996) Reiterative use of the EGF Receptor triggers differentiation of all cell types in the *Drosophila* eye. Cell 87:651–660

Freeman M (1997) Cell determination strategies in the *Drosophila* eye. Development 124:261–270

Fryxell KJ, Meyerowitz EM (1987) An opsin gene that is expressed only in the R7 photoreceptor cell in *Drosophila*. EMBO J 6:443–451

Gaul U, Mardon G, Rubin GM (1992) A putative Ras GTPase activating protein acts as a negative regulator of signaling by the Sevenless receptor tyrosine kinase. Cell 68:1007–1019

Ghysen A, Dambly-Chaudiere C (1989) Genesis of the *Drosophila* peripheral nervous system. Trends Genet 5:251–255

Goodman CS, Doe CQ (1993) Embryonic development of the *Drosophila* central nervous system. In: Bate M, Martinez Arias A (eds) Cold Spring Harbor Laboratory Press, Cold Spring Harbor, New York, pp 1131–1206

Hafen E, Basler K, Edstroem JE, Rubin GM (1987) *Sevenless*, a cell specific homeotic gene of *Drosophila*, encodes a putative transmembrane receptor with a tyrosine kinase domain. Science 236:55–63

Harris WA, Stark WS, Walker JA (1976) Genetic dissection of the photoreceptor system in the compound eye of *Drosophila* melanogaster. J Physiol 256:415–439

Hart AC, Krämer H, van Vactor DLJ, Paidhungat M, Zipursky SL (1990) Induction of cell fate in the *Drosophila* retina: the Bride of Sevenless protein is predicted to contain a large extracellular domain and seven transmembrane segments. Genes Dev 4:1835-1847

Hart AC, Krämer H, Zipursky SL (1993) Extracellular domain of the Boss transmembrane ligand acts as an antagonist of the Sev receptor. Nature 361:732-736

Heberlein U, Mlodzik M, Rubin GM (1991) Cell-fate determination in the developing *Drosophila* eye: role of the *rough* gene. Development 112:703-712

Heitzler P, Simpson P (1991) The choice of cell fate in the epidermis of *Drosophila*. Cell 64:1083-1092

Herbst R, Carroll PM, Allard JD, Schilling J, Raabe T, Simon MA (1996) Daughter of Sevenless is a substrate of the phosphotyrosine phosphatase Corkscrew and functions during Sevenless signaling. Cell 85:899-909

Hiromi Y, Mlodzik M, West SR, Rubin GM, Goodman CS (1993) Ectopic expression of Seven-up causes cell fate changes during ommatidial assembly. Development 118:1123-1135

Howe LR, Leevers SJ, Gomez N, Nakielny S, Cohen P, Marshall CJ (1992) Activation of the MAP Kinase pathway by the protein kinase Raf. Cell 71:335-342

Jan YN, Jan LY (1993) The peripheral nervous system. In: Bate M, Martinez Arias A (eds) Cold Spring Harbor Laboratory Press, Cold Spring Harbor, New york, pp 1207-1244

Jarman AP, Grau Y, Jan LY, Jan YN (1993) *Atonal* is a proneural gene that directs chordotonal organ formation in the *Drosophila* peripheral nervous system. Cell 73:1307-1321

Jarman AP, Grell EH, Ackerman L, Jan LY, Jan YN (1994) *Atonal* is the proneural gene for *Drosophila* photoreceptors. Nature 369:398-400

Jarman AP, Sun Y, Jan LY, Jan YN (1995) Role of the proneural gene, *atonal*, in formation of *Drosophila* chordotonal organs and photoreceptors. Development 121:2019-2030

Jennings B, Preiss A, Delidakis C, Bray S (1994) The Notch signaling pathway is required for *Enhancer of split* bHLH protein expression during neurogenesis in the *Drosophila* embryo. Development 120:3537-3548

Kanekar S, Perron M, Dorsky R, Harris WA, Jan LY, Jan YN, Vetter ML (1997) Xath5 participates in a network of bHLH genes in the developing *Xenopus* retina. Neuron 19:981-994

Karim FD, Chang HC, Therrien M, Wassarman DA, Laverty T, Rubin GM (1996) A screen for genes that function downstream of Ras1 during *Drosophila* eye development. Genetics 143:315-329

Kidd S, Kelley MR, Young MW (1986) Sequence of the *Notch* locus of *Drosophila melanogaster*: relationship of the encoded protein to mammalian clotting and growth factors. Mol Cell Biol 6:3094-3108

Kimmel BE, Heberlein U, Rubin GM (1991) The homeo domain protein Rough is expressed in a subset of cells in the developing *Drosophila* eye where it can specify photoreceptor cell subtype. Genes Dev 4:712-727

Knust E, Schrons H, Grawe F, Campos-Ortega JA (1992) Seven genes of the *Enhancer of split* complex of *Drosophila melanogaster* encode helix-loop-helix proteins. Genetics 132:505-518

Kopczynski CC, Alton AK, Fechtel K, Kooh PJ, Muskavitch MAT (1988) *Delta*, a *Drosophila* neurogenic gene, is transcriptionally complex and encodes a protein related to blood coagulation factors and epidermal growth factor of vertebrates. Genes Dev 2:1723-1735

Krämer H (1993) Patrilocal cell–cell interactions: Sevenless captures its bride. Trends Cell Biol 3:103-105

Krämer H, Cagan RL (1994) Determination of photoreceptor cell fate in the *Drosophila* retina. Curr Opin Neurobiol 4:14-20

Krämer H, Cagan RL, Zipursky SL (1991) Interaction of *bride of sevenless* membrane-bound ligand and the *sevenless* tyrosine kinase receptor. Nature 352:207-213

Krämer H, Phistry M (1996) Mutations in the *Drosophila hook* gene inhibits endocytosis of the Boss transmembrane ligand into multivesicular bodies. J Cell Biol 133:1205-1215

Kramer S, West SR, Hiromi Y (1995) Cell fate control in the *Drosophila* retina by the orphan receptor Seven-up: its role in the decisions mediated by the Ras signalling pathway. Development 121:1361-1372

Kumar JP, Moses K (1997) Transcription factors in eye development: a gorgeous mosaic? Genes Dev 11:2023–2028

Kumar JP, Tio M, Hsiung F, Akopyan S, Gabay L, Seger R, Shilo BZ, Moses K (1998) Dissecting the roles of the *Drosophila* EGF receptor in eye development and MAP kinase activation. Development 125:3875–3885

Lai ZC, Fetchko M, Li Y (1997) Repression of *Drosophila* photoreceptor cell fate through cooperative action of two transcriptional repressors Yan and Tramtrack. Genetics 147:1131–1137

Lai ZC, Harrison SD, Karim F, Li Y, Rubin GM (1996) Loss of *tramtrack* gene activity results in ectopic R7 cell formation, even in a *sina* mutant background. Proc Natl Acad Sci USA 93:5025–5030

Lai ZC, Rubin GM (1992) Negative control of photoreceptor development in *Drosophila* by the product of the *yan* gene, an ETS domain protein. Cell 70:609–620

Lawrence PA, Green SM (1979) Cell lineage in the developing retina of *Drosophila*. Dev Biol 71:142–152

Lawrence PA, Struhl G, Morata G (1979) Bristle patterns and compartment boundaries in the tarsi of *Drosophila*. J Embryol Exp Morphol 51:195–208

Lawrence PA, Tomlinson A (1991) A marriage is consummated. Nature 352:193

Le N, Simon MA (1998) Disabled is a putative adaptor protein that functions during signaling by the Sevenless receptor tyrosine kinase. Mol Cell Biol 18:4844–4854

Lecourtois M, Schweisguth F (1998) Indirect evidence for *Delta*-dependent intracellular processing of Notch in *Drosophila* embryos. Curr Biol 8:771–774

Lee EC, Hu X, Yu SY, Baker NE (1996) The *scabrous* gene encodes a secreted glycoprotein dimer and regulates proneural development in *Drosophila* eyes. Mol Cell Biol 16:1179–1188

Lee E-C, Yu S-Y, Hu X, Mlodzik M, Baker NE (1998) Functional analysis of the fibrinogen-related *scabrous* gene from *Drosophila melanogaster* identifies potential effector and stimulatory protein domains. Genetics 150:663–673

Lehmann R, Jimenez F, Dietrich U, Campos-Ortega JA (1983) On the phenotype and development of mutants of early neurogenesis in *Drosophila* melanogaster. Wilhelm Roux's Arch Dev Biol 192:62–74

Lesokhin AM, Yu SY, Katz J, Baker NE (1999) Several levels of EGF receptor signaling during photoreceptor specification in wild-type, *Ellipse*, and null mutant *Drosophila*. Dev Biol 205:129–144

Li S, Li Y, Carthew RW, Lai Z-C (1997) Photoreceptor cell differentiation requires regulated proteolysis of the transcriptional repressor Tramtrack. Cell 90:469–478

Ligoxygakis P, Yu SY, Delidakis C, Baker NE (1998) A subset of Notch functions during *Drosophila* eye deveeloopment require Su(H) and the E(spl) gene complex. Development 125:2893–2900

MacDougall LK, Waterfield MD (1996) Receptor signaling: *To* Sevenless, a daughter. Curr Biol 6:1250–1253

Marshall CJ (1994) MAP kinase kinase kinase, MAP kinase kinase and MAP kinase. Curr Opin Genet Dev 4:82–89

Matsuno K, Go MJ, Sun X, Eastman DS, Artavanis-Tsakonas S (1997) Suppressor of Hairless-independent events in Notch signaling imply novel pathway elements. Development 124:4265–4273

Mlodzik M, Baker NE, Rubin GM (1990) Isolation and expression of *scabrous*, a gene regulating neurogenesis in *Drosophila*. Genes Dev 4:1848–1861

Mlodzik M, Hiromi Y, Weber U, Goodman CS, Rubin GM (1990) The *Drosophila seven-up* gene, a member of the steroid receptor gene superfamily controls photoreceptor cell fates. Cell 60:211–224

Morgan TH, Bridges CB, Schultz J, Schultz J (1938) Constitution of the germinal material in relation to heredity. Yearb Carnegie Inst Wash 37:304–310

Muda M, Boschert U, Smith A, Antonsson B, Gillieron C, Chabert C, Camps M, Martinou I, Ashworth A, Arkinstall S (1997) Molecular cloning and functional characterization of a novel mitogen-activated protein kinase phosphatase, MKP-4. J Biol Chem 272:5141–5151

Mullins MC, Rubin GM (1991) Isolation of temperature-sensitive mutations of the tyrosine kinase receptor Sevenless (Sev) in *Drosophila* and their use in determining its time of action. Proc Natl Acad Sci USA 88:9387–9391

Muskavitch MAT (1994) Delta-Notch signaling and *Drosophila* cell fate choice. Dev Biol 166:415–430

Nebreda AR (1994) Inactivation of MAP kinases. Trends Biochem Sci 19:1–2

Oliver JP, Raabe T, Henkemeyer M, Dickson B, Mbamalu G, Margolis B, Schlessinger J, Hafen E, Pawson T (1993) A *Drosophila* SH2-SH3 adaptor protein implicated in coupling the Sevenless tyrosine kinase to an activator of Ras guanine nucleotide exchange, Sos. Cell 73:179–191

O'Neill EM, Rebay I, Tijan R, Rubin GM (1994) The activities of two ETS-related transcription factors required for *Drosophila* eye development are modulated by the Ras/MAPK pathway. Cell 78:137–147

Parks AL, Turner FR, Muskavitch MAT (1995) Relationships between complex Delta expression and the specification of retinal cell fates during *Drosophila* eye development. Mech Dev 50:201–216

Parody TR, Muskavitch MAT (1993) The pleiotropic function of *Delta* during postembyonic development of *Drosophila melanogaster*. Genetics 135:527–539

Poulson D (1937) Chromosomal deficiencies and embryonic development of *Drosophila* melanogaster. Proc Natl Acad Sci USA 23:133–137

Raabe T, Riesgo-Escovar J, Liu X, Bausenwein BS, Deak P, Maröy P, Hafen E (1996) DOS, a novel pleckstrin homology domain-containing protein required for signal transduction between Sevenless and Ras1 in *Drosophila*. Cell 85:911–920

Ready DF (1989) A multifaceted approach to neural development. Trends Neurosci 12:102–110

Ready DF, Hanson TE, Benzer S (1976) Development of the *Drosophila* retina, a neurocrystalline lattice. Dev Biol 53:217–240

Rebay I, Fleming RJ, Fehon RG, Cherbas L, Cherbas P, Artavanis-Tsakonas S (1991) Specific EGF repeats of Notch mediate interactions with Delta and Serrate: implications for Notch as a multifunctional receptor. Cell 67:687–699

Rebay I, Rubin GM (1995) Yan functions as a general inhibitor of differentiation and is negatively regulated by activation of Ras1/MAPK pathway. Cell 81:857–866

Reinke R, Zipursky SL (1988) Cell–cell interaction in the *Drosophila* retina: the *bride of sevenless* gene is required in photoreceptor R8 for R7 cell development. Cell 55:321–330

Rogge R, Cagan R, Majumdar A, Dulaney T, Banerjee U (1992) Neuronal development in the *Drosophila* retina: the *sextra* gene defines an inhibitory component in the developmental pathway of R7 photoreceptor cells. Proc Natl Acad Sci USA 89:5271–5275

Rogge RD, Karlovich CA, Banerjee U (1991) Genetic dissection of a neurodevelopmental pathway: Son of Sevenless functions downstream of the Sevenless and EGF receptor tyrosine kinases. Cell 64:39–48

Romani S, Camuzano S, Macagno ER, Modolell J (1989) Expression of *achaete* and *scute* genes in *Drosophila* imaginal discs and their function in sensory organ development. Genes Dev 3:997–1007

Rubin GM (1989) Development of the *Drosophila* retina: inductive events studied at single cell resolution. Cell 57:519–520

Schrons H, Knust E, Campos-Ortega JA (1992) The *Enhancer of split* complex and adjacent genes in the 96F region of *Drosophila melanogaster* are required for segregation of neural and epidermal progenitor cells. Genetics 132:481–503

Schweitzer R, Shilo B-Z (1997) A thousand and one roles for the *Drosophila* EGF receptor. Trends Genet 13:191–196

Simon MA, Bowtell DDL, Dodson GS, Laverty TR, Rubin GM (1991) Ras1 and a putative guanine nucleotide exchange factor perform crucial steps in signaling by the Sevenless protein tyrosine kinase. Cell 67:701–716

Simon MA, Dodson GS, Rubin GM (1993) An SH3-SH2-SH3 protein is required for p21 Ras1 activation and binds to Sevenless and Sos proteins in vitro. Cell 73:169–177

Sun X, Artavanis-Tsakonas S (1996) The intracellular deletions of *Delta* and *Serrate* define dominant negative forms of the *Drosophila* Notch ligands. Development 122:2465-2474

Tamura K, Taniguchi Y, Minoguchi S, Sakai T, Tun T, Furukawa T, Honjo T (1995) Physical interaction between a novel domain of the receptor Notch and the transcription factor RBP-J k/Su(H). Curr Biol 5:1416-1423

Tang AH, Neufeld TP, Kwan E, Rubin GM (1997) PHYL acts to down-regulate TTK88, a transcriptional repressor of neuronal cell fates, by a SINA dependent mechanism. Cell 90:459-467

Tomlinson A (1985) The cellular dynamics of pattern formation in the eye of *Drosophila*. J Embryol Exp Morphol 89:313-331

Tomlinson A (1988) Cellular interactions in the developing *Drosophila* eye. Development 104:183-193

Tomlinson A (1991) Developmental mechanisms. Sevenless and its bride: a marriage of molecules? Curr Biol 1:132-134

Tomlinson A, Bowtell DDL, Hafen E, Rubin GM (1987) Localization of the *sevenless* protein, a putative receptor for positional information in the eye imaginal disc of *Drosophila*. Cell 51:143-150

Tomlinson A, Kimmel BE, Rubin GM (1988) *Rough*, a *Drosophila* homeobox gene required in photoreceptors R2 and R5 for inductive interaction in the developing eye. Cell 55:771-784

Tomlinson A, Ready DF (1986) *Sevenless*: a cell specific homeotic mutation of the *Drosophila* eye. Science 231:400-402

Tomlinson A, Ready DF (1987) Neuronal differentiation in the *Drosophila* ommatidium. Dev Biol 120:366-376

Treier M, Bohmann D, Mlodzik M (1995) JUN cooperates with the ETS domain protein Pointed to induce photoreceptor R7 fate in the *Drosophila* eye. Cell 83:753-760

Treisman JE, Luk A, Rubin GM, Heberlein U (1997) *Eyelid* antagonizes *wingless* signaling during *Drosophila* development and has homology to the Bright family of DNA-binding proteins. Genes Dev 11:1949-1962

Tsuda H, Takebayashi K, Nakanishi S, Kageyama R (1998) Structure and promoter analysis of *Math3* gene, a mouse homolog of *Drosophila* proneural gene *atonal*. Neural-specific expression by dual promoter elements. J Biol Chem 273:6327-6333

Tsuda L, Inoue YH, Yoo M-A, Mizuno M, Hata M, Lim Y-M, Adachi-Yamada T, Ryo H, Masamune Y, Nishida Y (1993) A protein kinase similar to MAP kinase activator acts downstream of the Raf kinase in *Drosophila*. Cell 72:407-414

van Vactor DL, Cagan RL, Krämer H, Zipursky SL (1991) Induction in the developing compound eye of *Drosophila*: multiple mechanisms restrict R7 induction to a single retinal precursor. Cell 67:1145-1155

Vässin H, Campos-Ortega JA (1987) Genetic analysis of *Delta*: a neurogenic gene of *Drosophila melanogaster*. Genetics 116:433-445

Wassarman DA, Therrien M, Rubin GM (1995) The Ras signaling pathway in *Drosophila*. Curr Opin Genet Dev 5:44-50

Wharton KA, Johansen KM, Xu T, Artavanis-Tsakonas S (1985) Nucleotide sequence from the neurogenic locus Notch implies a gene product that shares homology with proteins containing EGF-like repeats. Cell 43:567-581

Wolff T, Ready DF (1991) The beginning of pattern formation in the *Drosophila* compound eye: the morphogenetic furrow and the second mitotic wave. Development 113:841-850

Wolff T, Ready DF (1993) Chapter 22: Pattern formation in the *Drosophila* retina. In: Bate M, Martinez-Arias A (eds) The development of *Drosophila melanogaster*. Cold Spring Harbor Laboratory Press, Cold Spring Harbor, New York, pp 1277-1325

Xiong WC, Montell C (1993) *tramtrack* is a transcriptional repressor required for cell fate determination in the *Drosophila* eye. Genes Dev 7:1085-1096

Xu T, Rubin GM (1993) Analysis of genetic mosaics in developing and adult *Drosophila* tissues. Development 117:1223-1237

Yamamoto D, Nihonmatsu I, Matsuo T, Miyamoto H, Kondo S, Hirata K, Ikegami Y (1996) Genetic interactions of *pokkuri* with *seven in absentia*, *tramtrack* and downstream components of the *sevenless* pathway in R7 photoreceptor induction in *Drosophila melanogaster*. Wilhelm Roux's Arch Dev Biol 205:215–224

Zipursky SL (1989) Molecular and genetic analysis of *Drosophila* eye development: *sevenless*, *bride of sevenless* and *rough*. Trends Neurosci 12:183–189

Zuker CS, Montell C, Jones K, Laverty T, Rubin GM (1987) A rhodopsin gene expressed in photoreceptor cell R7 of the *Drosophila* eye: homologies with other signal-transducing molecules. J Neurosci 7:1550–1557

Roles of the Extracellular Matrix in Retinal Development and Maintenance

Richard T. Libby[1], William J. Brunken[2,3], and Dale D. Hunter[4]

1
Introduction

The extracellular matrix (ECM) is a highly organized meshwork of secreted macromolecules that consists of proteins, proteoglycans, and polysaccharides. Components of the ECM provide cells with different signals that are involved in many aspects of the development of multicellular organisms, such as: (1) when to start, continue, or stop dividing; (2) when to differentiate; (3) where to migrate; (4) how to polarize; (5) where to form a synapse; and (6) whether to die or survive. Thus, ECM molecules may be playing many different roles in the development of a complex tissue such as the retina.

When discussing the ECM, particularly its function, it is important to go beyond the traditional components of the ECM (i.e., the large macromolecules such as collagen), as over the past several years many new components of the ECM have been identified; moreover, many of these appear to have functions in development. Also, a complete description of ECM function must include diffusible factors that may be sequestered by ECM components and the numerous receptors that are now known to be activated by ECM molecules.

In the vertebrate retina, there are many known and hypothesized functions of the ECM, including those in such major developmental processes as axon outgrowth, cell differentiation, cell division, and cell survival/death. To facilitate these diverse roles, distinct ECMs are present within the retina and, in several instances, their compositions change during development. Retinal ECMs can be divided broadly into two groups: basement membrane and nonbasement membrane ECMs. There are three basement membranes associated with the retina: the inner limiting membrane (ILM), Bruch's membrane, and the basement membrane associated with the retinal vasculature.

[1] MRC Institute of Hearing Research, Nottingham, UK
[2] Department of Biology, Boston College, Chestnut Hill, Massachusetts, USA
[3] Cutaneous Biology Research Center, Massachussetts General Hospital, Harvard Medical School, Charlestown, Massachusetts, USA
[4] Departments of Neuroscience, Anatomy and Cell Biology, and Ophthalmology, Tufts University School of Medicine, Boston, Massachussetts 02111, USA

Results and Problems in Cell Differentiation, Vol. 31
M. E. Fini (Ed.): Vertebrate Eye Development
© Springer-Verlag Berlin Heidelberg 2000

These basement membranes serve to delimit the retina from the surrounding non-neuronal tissues. Nonbasement membrane ECMs extend between the ILM and the retinal pigmented epithelium (RPE); based on differences in the distribution of ECM components, both the adult and developing retina could be subdivided into many different areas. Here, we will discuss the proven and potential roles ECM molecules play in retinal development, concentrating primarily on the effects one family of ECM molecules – the laminins – exert on photoreceptor development. Laminins, and the molecules with which they interact, have been implicated as contributing factors in most stages of development, and thus provide a good model to examine the ways in which the ECM can affect the development of a tissue.

2
Multiple Extrinsic Cues are Involved in Retinal Development

Many aspects of retinal development are controlled by extrinsic signals; these include not only ECM molecules, but also soluble and membrane-bound molecules. ECM molecules have two ways of affecting retinal development: (1) direct signaling; (2) indirect signaling by interacting with other types of extracellular molecules. The latter, indirect, effect of ECM molecules is often overlooked as one of their functions. ECM molecules can sequester other molecules in biologically appropriate locations and at physiologically relevant concentrations, and particular ECM molecules may be important facilitators of functions of other extracellular molecules.

Several groups have shown that soluble factors may affect many aspects of retinal development. The fibroblast growth factor (FGF) family of proteins is involved in the decision in the optic cup to become neural retina or RPE (Guillemot and Cepko 1992; Pittack et al. 1997; Hyer et al. 1998). Another major patterning event, the establishment of the dorsal/ventral axis, may be controlled by retinoic acid (Hyatt et al. 1992; Hyatt et al. 1996). Proliferation of retinal precursors can be affected by several soluble growth factors including members of the FGF family (Gensburger et al. 1987; Lillien and Cepko 1992) transforming growth factor (TGF)α (Lillien and Cepko 1992), and TGFβ3 (Anchan and Reh 1995). Many different soluble molecules have been implicated in the cell fate decision or differentiation of retinal cells, including FGF-1 (Guillemot and Cepko 1992; Désiré et al. 1998) FGF-2 (formerly bFGF) (Désiré et al. 1998; Hicks and Courtois 1992; Zhao and Barnstable 1996), taurine (Altshuler et al. 1993), retinoic acid (Kelley et al. 1994; Hyatt et al. 1996), leukemia inhibitory factor (LIF; Neophytou et al. 1997), and ciliary neurotrophic factor (CNTF; Neophytou et al. 1997; Fuhrmann et al. 1995; Ezzeddine et al. 1997). In fact, extrinsic, soluble factors are thought to be involved in almost every aspect of retinal development, and ECM molecules may be necessary for them to function properly. For instance, members of the

FGF family are known to bind to glycosaminoglycans (one of the major components of the ECM); in some instances, this binding is necessary for their function (reviewed in Faham et al. 1998; McKeechan et al. 1998).

Several membrane-bound molecules have also been shown to be involved during retinal development. Cell–cell interactions are thought to be important in establishing neuronal phenotypes in the retina (Reh 1992) and in vivo studies have shown that the notch-delta signaling pathway may be instrumental in early cell fate decisions in the retina (Austin et al. 1995; Henrique et al. 1997). Overexpression of the EGF receptor promotes the expression of a retinal cell type (Lillien 1995) and disruption of FGF receptor signaling affects the cell fate decision of retinal progenitors (McFarlane et al. 1998).

Several molecules involved in cell adhesion are also involved in retinal development. Antibodies to N-cadherin prevent proper histogenesis of retinal explant cultures (Matsunaga et al. 1988) and, similarly, gicerin was shown to be important in the histogenesis of the retina (Tsukamoto et al. 1997). At least one member of the integrin family of ECM receptors (α6) is required for proper lamination of the retina (Georger-Labouesse et al. 1998). Neuregulins (which can be soluble or membrane-bound) have been shown to promote retinal cell survival and neurite outgrowth (Bermingham-McDonogh et al. 1996). Neurite outgrowth from retinal cells is also promoted by N-cadherin (Matsunaga et al. 1988) and members of the immunoglobin family of cell adhesion molecules (Hankin and Lagenaur 1994). Together, these data demonstrate that many extrinsic factors are involved in retinal development. Although some of these interact indirectly with the ECM, others have direct interactions with retinal ECM components.

3
Laminins

Laminins are large heterotrimeric glycoproteins of the ECM consisting of an α, a β, and a γ chain (reviewed in Burgeson et al. 1994); disruption of many of the genes encoding laminin chains are known to cause pathologies in mammals (Table 1). Presently, in humans and rodents, five α chains, three β chains, and three γ chains have been identified and these chains are known to combine to form at least 12 different laminins (Table 2). It is critical to distinguish between the different laminins, and not simply use the term laminin, as the biological activity of the laminins are the result of the emergent properties of the component chains forming a given heterotrimer. Given the large number of chains (11) and even larger number of possible trimers (over 45), the number of unique biological roles in the developing and adult animal are indeed vast.

Laminins are a major component of all basement membranes and they have primarily been studied only in basement membranes. Recently, laminins

Table 1. Laminin chains and disease

Chain	Other names[a]	Species	Disease	Phenotype	Reference for phenotypes
α1	A, Ae				
α2	Merosin, Am, m chain	Human	Congenital muscular dystrophy	Congenital muscular dystrophy	[137]
		Mouse	dy/dy mouse	Dystrophic mouse	[138]
			Murine muscular dystrophy	Muscular dystrophy	[139]
		Mouse	Knockout/null mutant	Severe muscular dystrophy	[140]
α3	165kd, Kalinin A	Human	Junctional epidermolysis bullosa	Junctional epidermolysis bullosa (JEB)	[113, 141]
		Mouse	Transgenic/null mutant	JEB, skin blistering, postnatal lethality	[142]
α4					
α5		Mouse	Transgenic/null mutant	Embryonic lethality	[143]
β1	B1, B1e				
β2	s chain, B1s	Mouse	Transgenic/null mutant	Abnormal neuromuscular junction	[51]
			Transgenic/null mutant	Nephrodic syndrome	[144]
β3	140 kd, B1k, Kalinin B1		Junctional epidermolysis bullosa	Junctional epidermolysis bullosa (JEB)	[145]
γ1	B2, B2e	Mouse	Transgenic/null mutant	Embryonic lethality	[146]
γ2	B2t		Junctional Epidermolysis Bullosa	Junctional epidermolysis bullosa (JEB)	[110, 147, 148]
γ3					

The laminin family currently consists of 5α, 3β, and 3γ chains. In humans, mutations in several laminin chains (α2, α3, β3, and γ2) are known to cause disease, and, in mice, mutations (naturally occurring and artificially created) in some of the same laminin genes cause similar abnormalities. Targeted ablations of the laminin α5 and γ1 genes in mouse result in embryonic lethality, demonstrating that these chains are necessary for normal development.
[a] The laminin nomenclature has been standardized by Burgeson et al. (1994).

have been found in ECMs that are not associated with basement membranes (Hunter et al. 1992; Koch et al. 1999) and these observations have greatly extended the numbers and types of cells that laminins may be interacting with during development.

In the developing mammalian retina, laminins are expressed in many different locations. These include sites that are associated with important developmental events: the ventricular surface of the retina (Libby et al. 1996; Libby et al., submitted); the vitreal border (McLoon et al. 1988; Sarthy and Fu 1990; Morissette and Carbonetto 1995; Lentz et al. 1997); the ECM associated

Table 2. Known laminin trimers[a]

	Composition	Previous names	Reference
Laminin-1	$\alpha 1 \beta 1 \gamma 1$	EHS laminin	[149]
Laminin-2	$\alpha 2 \beta 1 \gamma 1$	Merosin	[150]
Laminin-3	$\alpha 1 \beta 2 \gamma 1$	s-Laminin	[150]
Laminin-4	$\alpha 2 \beta 2 \gamma 1$	Merosin, s-merosin	[150]
Laminin-5	$\alpha 3 \beta 3 \gamma 2$	Kalinin, nicein, epiligrin	[151]
Laminin-6	$\alpha 3 \beta 1 \gamma 1$	k-Laminin	[152]
Laminin-7	$\alpha 3 \beta 2 \gamma 1$	ks-Laminin	[153]
Laminin-8	$\alpha 4 \beta 1 \gamma 1$		[154]
Laminin-9	$\alpha 4 \beta 2 \gamma 1$		[154]
Laminin-10	$\alpha 5 \beta 1 \gamma 1$		[154]
Laminin-11	$\alpha 5 \beta 2 \gamma 1$		[154]
Laminin-12	$\alpha 2 \beta 1 \gamma 3$		[32]
Laminin-13	$\alpha 3 \beta 2 \gamma 3$		Libby et al., submitted
Laminin-14	$\alpha 4 \beta 2 \gamma 3$		Libby et al., submitted
Laminin-15	$\alpha 5 \beta 2 \gamma 3$		Libby et al., submitted

Fifteen laminin heterotrimers have been hypothesized, each its own combination of an α, a β, and a γ chain. Although there is the potential for the 11 known laminin chains to form 45 distinct laminins, it is unlikely that all are produced, as some may be thermodynamically unstable; however, it is likely that more laminins will be found.
[a] The laminin nomenclature has been standardized by Burgeson et al. (1994).

with the developing retinal synaptic layers (Libby et al. submitted); and Bruch's membrane (Libby et al. 1996; Sarthy and Fu 1990; Dong and Chung 1991) (Fig. 1a). It is important to note that these structures do not all express precisely the same complement of laminins. Furthermore, in several locations, there is a developmental change in the types of laminin expressed. Laminins are also present in the adult mammalian retina. They are localized in the interphotoreceptor matrix (IPM; Hunter et al. 1992; Libby et al. submitted), the retinal synaptic layers (Libby et al. 1999), and in the inner limiting membrane (Fig. 1b). Thus, the presence of so many distinct laminins, expressed in many different locations, suggests a diverse role for laminins in retinal development and function.

3.1
Photoreceptor Development

Photoreceptor cells are responsible for the sensory transduction which converts light into neural code; thus, for normal vision to occur it is necessary to have proper development and maintenance of photoreceptors. Photoreceptor development can be broken down into four developmental processes: (1) the progenitor cell's withdrawal from the cell cycle and adoption of the photoreceptor fate; (2) morphogenesis of the inner and outer segments; (3) synaptogenesis with retinal interneurons; (4) continued maintenance of the photoreceptor phenotype (survival). Although some of their development is

A   **B**

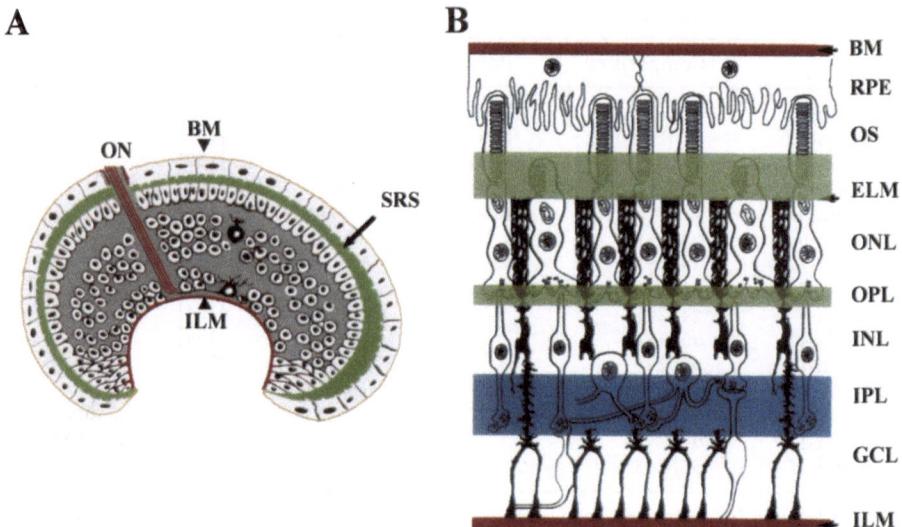

Fig. 1A, B. Different laminin compartments in the adult and developing retina. In retinal development (A) there are at least three ECMs that contain laminins: Bruch's membrane (*BM yellow*); the subretinal space (*SRS green*) and the internal limiting membrane (*ILM* including the optic nerve, *ON red*). Also, as the inner and outer plexiform layers develop (not shown), they begin to contain laminins. In the adult retina (B), laminins have been identified within five distinct ECMs: *BM* (*red*); the interphotoreceptor matrix surrounding the inner segments (*green*); the outer plexiform layer (*OPL green*); the inner plexiform layer (*IPL blue*); and the *ILM* (*red*). Interestingly, the ECMs associated with photoreceptors (the interphotoreceptor matrix and the ECM at their synapses in the OPL) appear to contain the same laminins; similarly, the two retinal basement membranes (BM and the ILM) appear to contain the same laminins. *OS* Outer segments; *ELM* external limiting membrane; *ONL* outer nuclear layer; *INL* inner plexiform layer; *GCL* ganglion cell layer

controlled intrinsically (Cepko 1999; Belliveau and Cepko 1999; Morrow et al. 1998), it appears that for proper photoreceptor development and function, many extrinsic factors are necessary.

Extrinsic factors have been hypothesized to be involved in photoreceptor development with most of these factors being small diffusible molecules (mainly growth factors). In fact, studies carried out in vitro suggest that a wide array of these molecules can have both positive and negative effects on rod photoreceptor development. FGF-2 (Hicks and Courtois 1992), taurine (Altshuler et al. 1993), and retinoic acid (Kelley et al. 1994) all promote rod photoreceptor development. TGF-α inhibits rod production while promoting the production of another cell that is simultaneously being generated by the retinal progenitor, the Müller cell (Lillien 1995). LIF (Neophytou et al. 1997) and CNTF (Neophytou et al. 1997; Ezzeddine et al. 1997) have also been shown to inhibit rod photoreceptor differentiation. LIF appears to arrest differentiating rod cells somewhere between cell fate commitment and the expression of its visual pigment (Neophytou et al. 1997). CNTF also appears to arrest photoreceptor differentiation prior to the expression of visual pigment, and may respecify the fate of the presumptive rod photoreceptor cell to another fate (Ezzeddine et al. 1997). Many different factors appear to modulate the commitment of a progenitor cell to the rod photoreceptor fate and subsequently to permit and/or promote its differentiation in vitro, suggesting that many different factors may modulate rod photoreceptor development in vivo. Furthermore, these studies suggest that different factors may affect different stages of rod photoreceptor development.

4
Laminins and Photoreceptor Development

In rodents, the expression of laminins is spatially and temporally associated with the major developmental and morphogenetic events in photoreceptor development. Laminins are present at the ventricular surface of the retina – the surface where photoreceptors develop and through which they extend their inner and outer segments – prior to rod photoreceptor development and continue to be expressed in the subretinal space (SRS) throughout rod photoreceptor morphogenesis (Fig. 2). In fact, laminins appear to fill the SRS until outer segment morphogenesis begins, at which point laminin expression is restricted to the IPM surrounding the inner segments; this pattern of distribution is continued in the adult retina in all mammals examined (Hunter et al. 1992; Libby et al. 1996), including humans (Libby et al. submitted).

4.1
Photoreceptor Fate Determination

Retinal progenitors are capable of giving rise to more than one cell type (Turner and Cepko 1987; Holt et al. 1988; Wetts and Fraser 1988; Wetts et al. 1989; Turner et al. 1990), suggesting that extrinsic cues may be involved in retinal cell fate determination. The molecular mechanisms which specify cell fate might be generally programmed and extrinsically controlled by environmental factors. In fact, it has been shown that both intrinsic and extrinsic mechanisms are involved in retinal cell fate determination (reviewed in Cepko 1999). Many extrinsic factors have been hypothesized or shown to be involved in retinal cell fate determination and differentiation (see above), including laminins.

Several in vitro studies have suggested that laminins are involved in cell fate determinations and differentiation in the retina, particularly with respect to photoreceptors. Laminin-1 was shown to powerfully alter the development of Y79 cells, a retinoblastoma-derived cell line that is thought to be a pluripotent precursor cell. Grown on laminin-1 (composed of chains $\alpha 1 \beta 1 \gamma 1$), these cells take on neuronal-like characteristics, whereas, in the

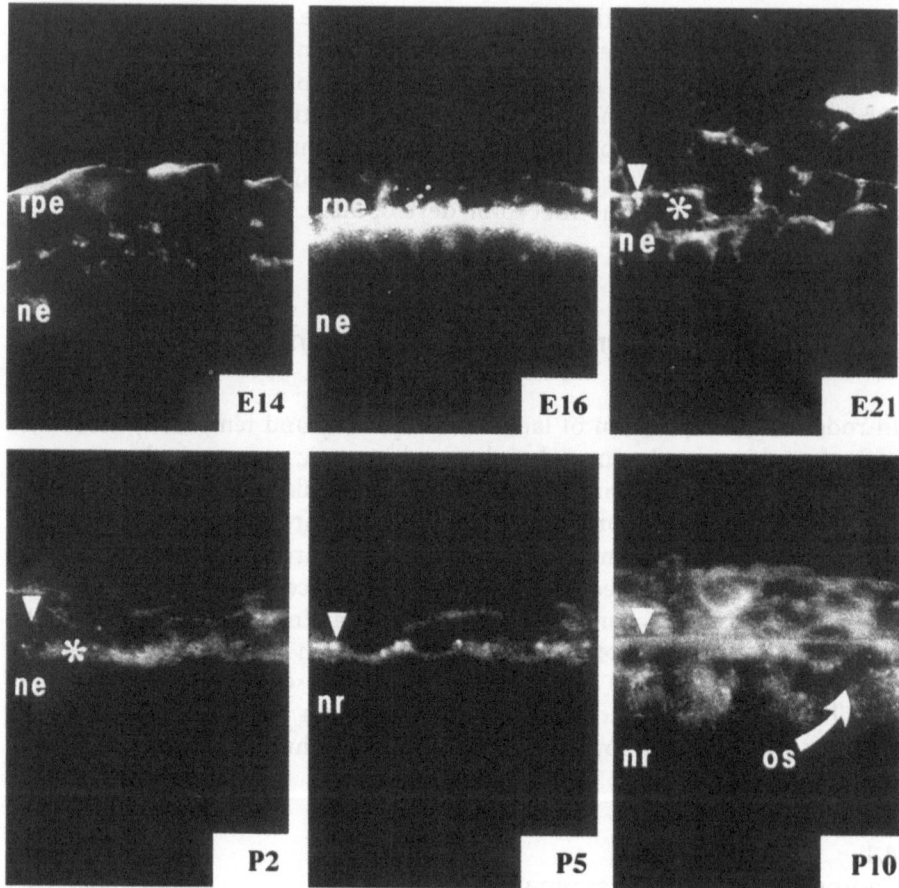

Fig. 2. Some laminins are temporally and spatially expressed in a position to interact with photoreceptors throughout their development. Here, immunohistochemistry for laminin β2 is shown on transverse rat retinal sections during photoreceptor development. At embryonic day 14 (*E14*), laminin β2 is present at the two surfaces which border the subretinal space: the apical surface of the neuroepithelium (*ne*) and the apical surface of the retinal pigmented epithelium (*rpe*). By E16, laminin β2 appears to fill the subretinal space; this pattern of expression is continued through postnatal day 10 (*P10*). *Arrowheads* point to Bruch's membrane and *asterisks* mark RPE cells. *os* Developing outer segments; *nr* neural retina. (After Libby et al. (1996))

absence of laminin-1, they express more photoreceptor-like characteristics (Campbell and Chader 1988). Frade et al. (1996) have also shown that laminin-1 can induce neuronal characteristics in an in vitro system of retinal development: laminin-1 greatly enhanced the number of retinal neuroepithelial cells differentiating into neurons. They also found matrices that contained a different laminin complement (including the α2 chain) did not have a similar, neuron-promoting effect. We have shown that laminins containing the laminin β2 chain (formerly s-laminin or the s chain) are required for the expression of the rod photoreceptor phenotype in an in vitro model system for rod photoreceptor development (Hunter et al. 1992). Furthermore, in vitro, laminin β2 can modulate the expression between two retinal cell fates, the rod photoreceptor and the rod bipolar cell (Hunter and Brunken 1997; Fig. 3). These data show that laminins can stimulate undifferentiated retinal cells to differentiate, and suggest that the composition of the laminin heterotrimer may be important in mediating the differentiation of retinal cell types: a laminin-1-rich and laminin β2 chain-poor environment would be permissive for retinal neuronal differentiation, whereas a laminin β2 chain-rich and a laminin-1-poor environment would be permissive for photoreceptor differentiation.

During development, different laminins are expressed at the vitreal and ventricular borders of the retinal neuroepithelium. Retinal neuroepithelial cells contact these two surfaces at different points in their cell cycle; the ventricular surface is where retinal progenitors undergo mitosis. Hinds and Hinds (1979) observed that terminally mitotic retinal progenitor cells which adopt a photoreceptor fate: maintain a junctional complex with the ventricular surface (eventually this surface becomes the external limiting membrane); remain in the ventricular layer; grow only a short vitread-directed process. In contrast, those cells that are destined to become retinal neurons

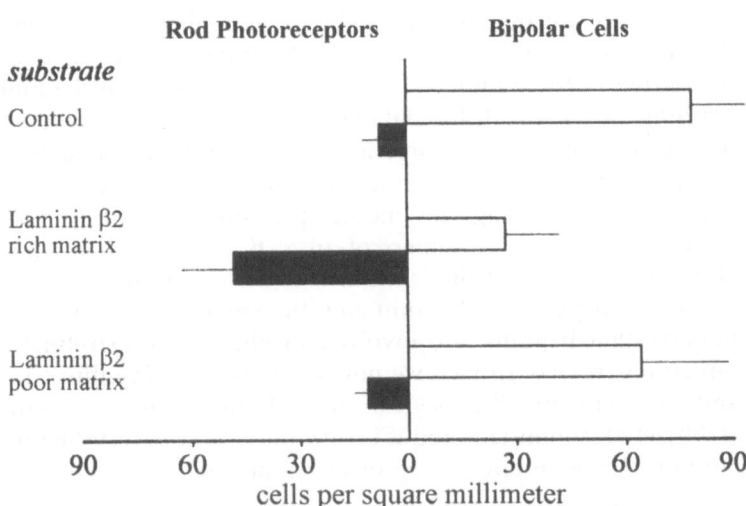

Fig. 3. In vitro, laminin β2 affects the differentiation of retinal neuroepithelial cells. When grown on glass coverslips at low density (control) only a few retinal neuroepithelial cells differentiate as rod photoreceptors (expressing rhodopsin [RetP1]), whereas many cells differentiate as bipolar cells (expressing protein kinase C [PKC]). However, when these cells are grown on a matrix rich in the laminin β2 chain, many rods and few bipolar cells differentiate. Retinal neuroepithelial cells grown on a matrix from which the laminin β2 chain has been removed differentiate in a manner indistinguishable from controls. For details of these experiments see Hunter and Brunken (1997). (After Hunter and Brunken (1997))

(nonphotoreceptors): lose their junctional contact with the ventricular surface; migrate out of the ventricular layer; and grow a long vitread-directed
process. Thus, it is possible that molecules at these surfaces help to guide
retinal progenitor development; in fact, Hinds and Hinds (1979) suggested
that differences in vitreal and ventricular surfaces are involved in retinal cell
differentiation. As mentioned above, the development of retinal cells is
strikingly different on different laminin matrices, and, in fact, the laminin
complements at the vitreal border (McLoon et al. 1988; Sarthy and Fu 1990;
Morissette and Carbonetto 1995; Lentz et al. 1997) and at the ventricular
surface (Hunter et al. 1992; Libby et al. 1996; Libby et al. submitted) are
different. Taken together, anatomical studies analyzing the movements of
retinal progenitor cells, studies of laminins' effects in vitro, and the differences between laminin matrices at the two surfaces of the retinal neuroepithelium suggest that laminins are involved in the differentiation, and
potentially the cell fate decisions, of retinal progenitors, with one laminin
environment supporting neuronal differentiation and another supporting
photoreceptor differentiation.

As mentioned above, laminins containing the β2 chain are involved in
photoreceptor differentiation in vitro (Hunter et al. 1992; Hunter and
Brunken 1997) and may even be involved with the cell fate decision to become a photoreceptor instead of a retinal neuron (Hunter and Brunken
1997). A laminin β2 chain-null mutant mouse (Noakes et al. 1995) has allowed us to test the role of β2 chain-containing laminins in photoreceptor
development in vivo (Libby et al. 1999). Interestingly, there appears to be no
disruption in the number of photoreceptor cells or in the number of retinal
neurons, at odds with the data generated in vitro. However, this is perhaps
not surprising, as the expression of the laminin α and γ chains that are
thought to form trimers with the laminin β2 chain are still expressed (Libby
et al. 1999), indicating that another β chain is substituting for β2 (laminins
are always secreted as trimers of an α, β, and γ). Furthermore, Frade et al.
(1996) suggested that the laminin α chain was responsible for the neuronal
differentiating effect of laminin on retinal neuroepithelial cells. Thus, it is still
possible that laminins are involved in photoreceptor/retinal cell fate determinations in vivo. In fact, we now know that the laminin α3, α4, α5, β2, γ2,
and γ3 chains are all present in the SRS during photoreceptor development
(Libby et al. submitted) and it is now important to test the function of all the
laminin chains in photoreceptor development.

4.2
Photoreceptor Morphogenesis: Inner and Outer Segment Development

The ultrastructural anatomy of the developing photoreceptor has been well
characterized. Numerous electron microscopic studies have described the
time course of rod photoreceptor morphogenesis in rodents (De Roberts

1956; De Roberts 1960; Weidman and Kuwabara 1968; Weidman and Kuwabara 1969; Galbavy and Olson 1979; Nir et al. 1984; Obata and Usukara 1992). These studies showed that specialized apical processes of photoreceptors could be detected around the time of birth, and these processes began to protrude through the ventricular surface into the SRS, forming rudimentary inner segments. The rudimentary inner segments continue to expand longitudinally and at about the 5th postnatal day the first signs of outer segments can be detected. Lengthening of outer segments continues until approximately postnatal day 20, with most growth occurring between the 10th and 18th postnatal days. Extracellular molecules positioned at the ventricular surface and within the SRS – such as laminins – are excellent candidates for providing positional information to differentiating rods and positive signals for segment morphogenesis.

Laminins are expressed in a position to interact with the developing photoreceptor segments (see above; Fig. 2). Laminins do appear to be involved in the morphogenesis of inner and outer segments, as inner and outer segments do not reach their normal length in laminin β2-null mice. Inner and outer segment morphogenesis appears to begin normally; however, by the end of the second postnatal week, the mutants' inner and outer segment length lags behind that of normal littermates and, by the third postnatal week, they are only 50 % of the length of those in normal littermates (Libby et al. 1999). Thus, laminins appear to be involved in photoreceptor morphogenesis, although laminins containing the laminin β2 chain do not provide positional information to photoreceptors, nor are they necessary for the initial stages of segment formation. It will be of interest to determine the roles of the other laminins, and their component chains, that we know are in a position to interact with developing inner and outer segment. Interestingly, there are laminins that are not expressed in the adult IPM that are present surrounding developing inner and outer segments, suggesting unique roles for laminins in the developing IPM (Libby et al., submitted).

4.3
Photoreceptor Synaptogenesis

The anatomy of photoreceptor synaptogenesis in mammals has been well defined by several groups (e.g., Olney 1968; Olney 1968; Blanks et al. 1974). First, two horizontal processes oppose the rod terminal. Second, synapse-specific specializations (including the synaptic ribbon) appear in the rod terminal, forming an immature release site; finally, the immature synaptic region is invaded by a bipolar cell dendrite, which enters between two flanking horizontal cell processes, forming the mature triad typical of adult rod photoreceptor synapses.

The molecular organization of the developing photoreceptor synapse, in contrast, is not well studied. However, several of the molecular components involved in the development of another synpase, the neuromuscular junction, are defined, and at least two of them are ECM molecules: a β2 chain-containing laminin (Noakes et al. 1995) and agrin (Gautam et al. 1996). At the neuromuscular junction, these molecules are thought to interact with α-dystroglycan, which itself interacts with β-dystroglycan; this complex is thought to anchor dystrophin, which interacts with the cytoskeleton, to the membrane. This overall complex is thought to maintain the structural integrity of the nerve/muscle interaction (reviewed in Sunada and Campbell 1996; Fig. 4). Recently, laminin (Libby et al. 1999) and agrin (Kröger et al. 1996; Kröger 1997) have been detected in the ECM of retinal plexiform layers, in the adult and during synaptogenesis. Dystrophin (e.g., Pillers et al. 1993; Schmitz et al. 1993; Schmitz and Drenckhahn 1997; Ueda et al. 1995; Ueda et al. 1997) and dystroglycan (e.g., Montanaro et al. 1995; Drenckhahn et al. 1996) are present in the outer plexiform layer, where they are found in photoreceptor terminals (Schmitz and Drenckhahan 1997; Montanaro et al. 1995; Drenckhahn et al. 1996; Koulen et al. 1998; Ueda et al. 1998; Blank et al. 1999; Claudepierre et al. 1999). Thus, it appears that the complex that is important for the development and maintenance of the neuromuscular

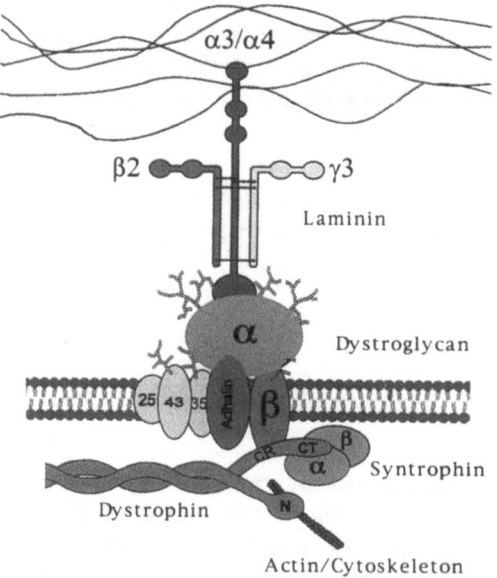

Fig. 4. Laminin, dystroglycan and dystrophin are known to form a complex that links the ECM to the cytoskeleton in muscle. In the retina, dystrophin and dystroglycan are present in photoreceptor terminals and laminins are present in the ECM of the outer plexiform layer (OPL), thus, this same complex may help to anchor the photoreceptor terminal in the OPL. The structural details of the molecules depicted are for diagrammatic purposes; the α4 chain is truncated in its shortarm and the dystrophin molecule is a monomer. (After Sunada and Campbell (1996))

junction is also present at photoreceptor synapses. However, the developmental expression of dystroglycan lags behind the expression of dystrophin and the laminins in the outer plexiform layer, suggesting that it is not the prime mover in target recognition and synapse stabilization (Libby et al. 1999).

Mutations in one component of the complex described above are known to affect retinal physiology in both humans and mice. Duchenne and Becker muscular dystrophies and the pathologies associated with the mdx mouse are caused by mutations in the dystrophin gene. Duchenne and Becker patients and mice carrying certain alleles of *mdx* also have defects in their electroretinograms (ERGs): a disruption of the rod-mediated b-wave amplitude and an increase in its implicit time, while the a-wave and cone-mediated b-waves are spared in both human and mice (Pillers et al. 1993; Pillers et al. 1995; Girlanda et al. 1997; Pillers et al. 1999; Fitzgerald et al. 1994; Lenk et al. 1996; Kameya et al. 1997). The physiological data indicate a failure of synaptic transmission from the photoreceptor cells to the second-order neurons in the INL, which correlates with the localization of dystrophin in photoreceptor terminals; however, no anatomical disruption of the photoreceptor synapse has been found (Blank et al. 1999). It should be noted that different isoforms of dystrophin are expressed in the retina, Dp427, Dp260, Dp140, Dp71 (Pillers et al. 1993; Kameya et al. 1997; D'Souza et al. 1995), and they have different expression patterns; analysis of different *mdx* alleles with mutations that disrupt different dystrophin isoforms has led to the idea that two of the isoforms are important in retinal physiology. Recently, Howard et al. (1998) and Pillers et al. (1999) have suggested that dystrophin in the photoreceptor terminal as well as dystrophin isoforms around the ILM and retinal vasculature are both involved in maintaining proper retinal physiology. Thus, dystrophin and the ECM complex with which it indirectly interacts are directly involved maintaining proper photoreceptor synaptic function.

The laminin α3, α4, α5, β2, and γ3 chains are all expressed throughout photoreceptor synaptogenesis, and their expression is maintained in the adult. At least one of these, the laminin chain, is necessary for proper photoreceptor synaptic development. Electron micrographic analysis of the photoreceptor synapses in the laminin β2-null mouse shows marked disorganization; ERGs from these mice demonstrate that, like the mdx mouse, there is synaptic disruption between photoreceptors and bipolar cells (Libby et al. 1999). As mutations in dystrophin and laminin β2 cause similar physiological phenotypes, and dystrophin and laminin (as well as the complex that is capable of linking them) are present in the OPL, it is tempting to speculate that these two molecules are acting in concert during the differentiation and stabilization of photoreceptor synapses. This would be analogous to the neuromuscular junction, where disruptions in both laminin β2 and dystrophin result in neuromuscular pathologies (Noakes et al. 1995; Hoffman et al. 1987). Furthermore, laminins are present in the matrix of the adult and developing ILM and retinal vasculature, where they could be interacting with the dystrophin complex and affecting retinal physiology in as-yet-undefined ways.

5
Role of the Interphotoreceptor Matrix in Maintaining Photoreceptor Viability

The IPM is a specialized ECM that surrounds the inner and outer segments of photoreceptors. Molecules within the IPM are essential for photoreceptor function and in maintaining photoreceptor viability. Many studies have demonstrated the presence of proteoglycans and polysaccharides in the IPM (e.g. Bach and Berman 1971; Berman and Bach 1968; Adler and Klucznik 1982; and see the series of papers from J.G. Hollyfield and colleagues), and an overview of all the matrix molecules in the IPM is a review unto itself. Importantly, the IPM is not a homogenous matrix: for example, there are differences in the matrix around rods and cones (e.g. Varner et al. 1987; Hageman and Johnson 1987; Mieziewska et al. 1994). Furthermore, at least with respect to one matrix component, there is a difference between the matrix surrounding inner segments and that surrounding outer segments (Hunter et al. 1992). Here, we will concentrate on a few of the IPM components that may be important in several different aspects of photoreceptor survival.

Acharya et al. (1998) have suggested that the "basic molecular scaffold" of the IPM is created by interactions between two of its components, the glycosaminoglycan hyaluronan (Hollyfield et al. 1998) and SPACR (sialoprotein associated with cones and rods; Tien et al. 1992; Acharya et al. 1998). While the biological roles of SPACR are not well understood, hyaluronan is known to have many important roles within the ECM, including an involvement in ECM organization (reviewed in Knudson and Knudson 1993).

Laminins are present in both the adult and developing IPM (Hunter et al. 1992; Libby et al. 1996; Libby et al. 1997; Libby et al., submitted). The presence of laminins within the IPM and the ability of some laminins to form polymers that are important in providing structural integrity to basement membranes (Yurchenco et al. 1992) suggests that they are playing a major structural role within the IPM. Recently, we have established that 5 of the 11 known laminin chains are produced by the Müller cell in both rodent and human retinae ($\alpha 3$, $\alpha 4$, $\alpha 5$, $\beta 2$, and $\gamma 3$; Libby et al. 1997; Libby et al. submitted) and that these chains can potentially form three different laminins (heterotrimers containing $\alpha 3\beta 2\gamma 3$, $\alpha 4\beta 2\gamma 3$, and $\alpha 5\beta 2\gamma 3$). Interestingly, these laminins are only found surrounding inner segments in mammals, suggesting (perhaps not surprisingly) that the basic structure of the IPM surrounding the inner segments is very different from that surrounding the outer segments.

Presently, laminins and the SPACR-hyaluronan complex appear to be the major structural components of the IPM; however, it would not be surprising, as additional isoforms of traditional basement membrane components are identified, that more structural components will be identified. Laminins and the SPACR-hyaluronan complex probably do more than just supply the macromolecular structure of the IPM as, along with numerous other molecules that have been identified within the IPM, they are involved in many of the functions that are necessary for vision, including retinal adhesion, cycling of visual components in the retina, and possibly providing trophic support for photoreceptors. Thus, any discussion of maintaining photoreceptor survival must include the components of the IPM. The clinician should be mindful of these gene products in considering etiologies of the human disease.

5.1
Retinal Adhesion

Normal photoreceptor function and survival relies on the neural retina remaining closely adherent to the RPE. Loss of retinal adhesion, known as retinal detachment, is a serious clinical condition that has multiple etiologies. Interestingly, retinal adhesion does not rely on any obvious anatomical junctions, but rather on a variety of other mechanisms (reviewed in Marmor 1993), including interactions between molecular elements of the IPM and the retinal cells. For example, enzymatic disruption of glycoconjugates (Yao et al. 1992; Yao et al. 1994) and proteoglycans (Lazarus and Hageman 1992) within the IPM causes a temporary loss of retinal adhesion, suggesting that ECM components are important determinants for normal retinal adhesion. Little is known about retinal adhesion in the adult and even less is known about adhesion during development.

Cellular surfaces that appose the IPM contain several molecules that are known to be involved in cellular adhesion through their interactions with ECM components. For example, members of the integrin family – many of which function in cell adhesion (reviewed in Hass and Plow 1994) – are present on the apical surface of the RPE and photoreceptor inner and outer segments (Anderson et al. 1995; Chen et al. 1997; Lin and Clegg 1998). In addition, the neural cell adhesion molecule (NCAM) is present on the apical surface of the RPE and photoreceptor outer segments (Gunderson et al. 1993). Another cell adhesion molecule, CD44, has been localized on the microvilli of the Müller cell, which project from the outer limiting membrane into the IPM (Chaitin et al. 1994). CD44 is a known receptor for hyaluronan (reviewed by Lesley and Hyman 1998) and thus may be an important link between the SPACR-hyaluronan complex within the IPM and the proximal border of the IPM.

Laminins are powerful mediators of cell adhesion (e.g. Terranova et al. 1983; Timpl et al. 1983) and mutations in some laminin chains are responsible for some of the cases of the adhesive disorder of the skin known as junctional epidermolysis bullosa (e.g. Aberdam et al. 1994; Pulkkinen et al. 1994; Pulkkinen et al. 1994; Kivirikko et al. 1995) in which there is a dramatic blistering and eventual dissolution of the skin. The major transmembrane receptor for laminins are the integrins (reviewed by Aumailley et al. 1996) which are dimers consisting of an α and β subunit. Studies of integrin expression in retina have generated a large literature outside the scope of this chapter. However, some integrin receptors are present on the surface of the cells bordering the IPM (Anderson et al. 1995; Lin et al. 1998). Interestingly, the vitronectin receptor (αVβ5) has a pattern of expression around cone photoreceptors similar to that of IPM laminins: from the ELM to the base of the outer segments (Anderson et al. 1995). However, αVβ5 integrin's ability to bind to IPM laminins is unknown. It seems unlikely that photoreceptor vitronectin receptors would recognize these laminins, as vitronectin receptors (αVβ3 and αVβ5) rely on RGD sequence recognition for binding to its substrates.

Further support for the laminins in the IPM contributing to retinal adhesion is that photoreceptors adhere to at least one laminin chain: photoreceptors have been shown to adhere to a recombinant fragment of laminin β2 (Hunter et al. 1992). Thus, it appears that the structural components of the IPM are directly involved in retinal adhesion; however, the exact mechanisms of retinal adhesion remain elusive.

5.2
Photoreceptor Survival

There are numerous diseases (with etiologies numbering in the dozens) that affect photoreceptor survival and can be collectively classified as photoreceptor degenerations. Photoreceptors undergo degeneration in response to overstimulation (LaVail et al. 1992) and as the result of many genetic abnormalities affecting proteins that are important in phototransduction and photoreceptor/RPE cell physiology (Farber and Danciger 1997). Furthermore, it appears that normal healthy photoreceptors require constant signals from their environment for continued viability (Campochiaro et al. 1996). Together, these data suggest that photoreceptors have an intrinsic cell death program that is poised to be activated as the result of any internal or external abnormalities. As many retinal diseases in humans are characterized by photoreceptor degenerations, it is of obvious clinical importance to characterize what factors contribute to maintaining photoreceptor viability; several factors thought to be important mediators of photoreceptor survival are present in the IPM.

Hewitt et al. (1990) identified a component of the IPM that selectively enhances the survival of photoreceptors and not other neurons in culture. Similarly, another group found that insoluble components of the IPM contain neurotrophic activities (Tombran Tink et al. 1992). Recently, this same group reported that neuron-specific enolase (NSE) was present in the IPM and had some neurotrophic activity. NSE was localized to the basal area of the IPM – a pattern similar to that shown for laminins – and it was also found to be tightly associated with the IPM.

Several well-characterized diffusible growth factors are also known to be in the IPM. A member of the transforming growth factor β superfamily, inhibin, has been shown to be in the IPM (Ying et al. 1995). FGF-2 is perhaps the best-studied diffusible factor that is present in the IPM (Gao and Hollyfield 1992); FGF-2 is known to affect the survival of photoreceptors in vitro (Carwile et al. 1998; Fontaine et al. 1998), to delay the degeneration of photoreceptors in rodent models of photoreceptor degeneration in vivo (e.g. Uteza et al. 1999; Faktorovich et al. 1990), and to protect photoreceptors from light-induced degeneration (e.g., LaVail et al. 1992; Faktorovich et al. 1992; Masuda et al. 1995). Furthermore, photoreceptors have receptors for FGF-2 in their outer segments (Morimoto et al. 1993), and when FGF-2 receptors are functionally removed, photoreceptors degenerate (Campochiaro et al. 1996). These data suggest that diffusible growth factors are necessary to maintain photoreceptor viability even in a normal healthy animal; thus, an important role of the IPM is to bind and sequester these soluble components in a position where they can interact with photoreceptors.

Both FGF-2 and NSE have been shown to be tightly associated with the IPM (Hageman et al. 1991; Li et al. 1995). However, it is unknown how these soluble molecules are sequestered in the IPM. Interestingly, an ECM molecule, heparin, is known to bind FGF-2; in some systems, heparin is required for FGF-2 function (reviewed in Faham et al. 1998; McKeehan et al. 1998). In the retina, FGF-2 has a more potent effect in protecting photoreceptors from light damage in the presence of heparin both in vitro (Carwile et al. 1998) and in vivo (LaVail et al. 1992). Thus, the IPM's involvement with diffusible factors may go beyond the simple role of sequestering them around photoreceptors at a physiologically relevant concentration.

Laminins in the IPM may also be contributing to photoreceptor survival. It has been shown that laminins can act in cooperation with growth factors and thereby contribute to neuronal survival (Millaruelo et al. 1988). In addition, laminin (presumably laminin-1) enables FGF-2 to promote neuronal survival (Schmidt and Kater 1995); this synergistic activity is blocked by an antibody against an integrin receptor, suggesting that laminins are acting directly through a receptor-mediated event. The authors hypothesized that laminin, through an integrin receptor, was enabling FGF-2 to promote neuronal survival by affecting how a cell responds to FGF-2. Thus, it is possible that the laminins in the IPM are directly stimulating photoreceptors and enabling them to respond to environmental signals that are necessary in preventing photoreceptor degeneration.

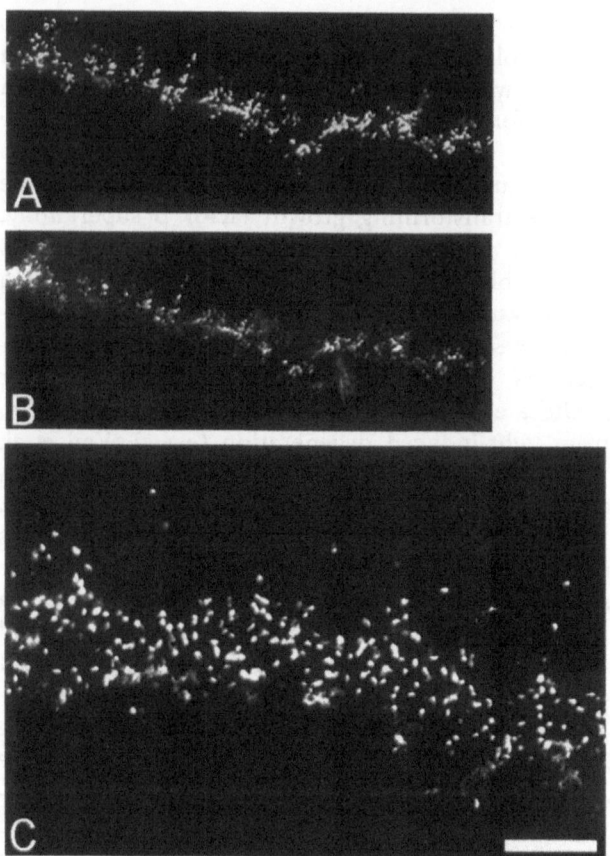

Fig. 5A–C. Laminin is expressed in the outer plexiform layer along with elements of the dystrophin-dytroglycan complex. A marker of the ribbons in the synaptic terminal, B16, is present in arc-shaped structures in the outer plexiform layer of the rat retina (A); on the same section, laminin β2 chain immunoreactivity colocalizes with the ribbon antigen, suggesting synaptic localization (B). Dystrophin is also confined to discrete sites in the OPL (C). *Bar* 10 μm (A, B); 5 μm (C). (Unpubl. work of the authors)

6
Conclusion

The field of matrix biology is experiencing a profound explosion in the identification of novel ECM molecules and their receptors. It is important to continue the search for new matrix components, their receptors, and molecules which modify both (matrix metalloproteases, MMPs, and tissue inhibitors of MMPs, TIMPs), as many of these molecules will most likely be important in retinal diseases (e.g., Fariss et al. 1998); for instance, recently a novel ECM molecule has been found to cause retinitis pigmentosa (Eudy et al. 1998). Perhaps more important will be to incorporate these biologically active molecules into the surgical paradigms which are currently being developed to implant retinal prostheses (Humayun et al. 1999; Zrenner et al. 1999), not only to increase the biocompatability of artificial prostheses but also to improve the integration between the biological tissue and the electronic interface. It will also be important, in the case of transplantation of fetal photoreceptors or neural stem cells, to incorporate those molecules into the stem cell populations which will achieve the desired outcome: photoreceptor-promoting molecules to repair retinal degenerations; neurotrophic molecules to repair optic nerve injuries, etc. While these goals seem remote, the recent success in the use of matrices and matrix-derived molecules in cancer therapies and wound healing suggests that neuronal repair may be an achievable goal.

References

Aberdam D, Galliano MF, Vailly J, Pulkkinen L, Bonifas J, Christiano AM, Tryggvason K, Uitto J, Epstein EH Jr, Ortonne JP et al. (1994) Herlitz's junctional epidermolysis bullosa is linked to mutations in the gene (LAMC2) for the gamma 2 subunit of nicein/kalinin (LAMININ-5). Nat Genet 6:299-304

Acharya S, Rayborn ME, Hollyfield JG (1998) Characterization of SPACR, a sialoprotein associated with cones and rods present in the interphotoreceptor matrix of the human retina: immunological and lectin binding analysis. Glycobiol 8:997-1006

Acharya S, Rodriguez IR, Moreira EF, Midura RJ, Misono K, Todres E, Hollyfield JG (1998) SPACR, a novel interphotoreceptor matrix glycoprotein in human retina that interacts with hyaluronan. J Biol Chem 273:31599-31606

Adler AJ, Klucznik KM (1982) Proteins and glycoproteins of the bovine interphotoreceptor matrix: composition and fractionation. Exp Eye Res 34:423-434

Altshuler D, Lo Turco JJ, Rush J, Cepko C (1993) Taurine promotes the differentiation of a vertebrate retinal cell type in vitro. Development 119:1317-1328

Anchan RM, Reh TA (1995) Transforming growth factor-β-3 is mitogenic for rat retinal progenitor cells in vitro. J Neurobiol 28:133-145

Anderson DH, Johnson LV, Hageman GS (1995) Vitronectin receptor expression and distribution at the photoreceptor-retinal pigment epithelial interface. J Comp Neurol 360:1-16

Aumailley M, Gimond C, Patricia R (1996) Integrin-mediated cellular interactions with laminins. In: Ekblom P, Timpl R (eds) The laminins. Harwood Academic Publishers, Amsterdam, pp 127-158

Austin CP, Feldman DE, Ida JA, Jr, Cepko CL (1995) Vertebrate retinal ganglion cells are selected from competent progenitors by the action of Notch. Development 121:3637-3650

Bach G, Berman ER (1971) Amino sugar-containing compounds of the retina. I. Isolation and identification. Biochim Biophys Acta 252:453-461

Baudoin C, Miquel C, Gagnoux Palacios L, Pulkkinen L, Christiano AM, Uitto J, Tadini G, Ortonne JP, Meneguzzi G (1994) A novel homozygous nonsense mutation in the LAMC2 gene in patients with the Herlitz junctional epidermolysis bullosa. Hum Mol Genet 3:1909-1910

Belliveau MJ, Cepko CL (1999) Extrinsic and intrinsic factors control the genesis of amacrine and cone cells in the rat retina. Development 126:555-566

Berman ER, Bach G (1968) The acid mucopolysaccharides of cattle retina. Biochem J 108:75-88

Bermingham-McDonogh O, McCabe KL, Reh TA (1996) Effects of GGF/neuregulins on neuronal survival and neurite outgrowth correlate with erbB2/neu expression in developing rat retina. Development 122:1427-1438

134   R. T. Libby et al.

Blank M, Koulen P, Blake DJ, Kröger S (1999) Dystrophin and beta-dystroglycan in photoreceptor terminals from normal and mdx3Cv mouse retinae. Eur J Neurosci 11:2121-2133

Blanks JC, Adinolfi AM, Lolley RN (1974) Synaptogenesis in the photoreceptor terminal of the mouse retina. J Comp Neurol 156:81-93

Burgeson RE, Chiquet M, Deutzmann R, Ekblom P, Engel J, Kleinman H, Martin GR, Meneguzzi G, Paulsson M, Sanes J, Timpl R, Tryggvason K, Yamada Y, Yurcheno PD (1994) A new nomenclature for the laminins. Matrix Biol 14:209-211

Campbell M, Chader GJ (1988) Retinoblastoma cells in tissue culture. Ophthalmic Paediatr Genet 9:171-199

Campochiaro PA, Chang M, Ohsato M, Vinores SA, Nie Z, Hjelmeland L, Mansukhani A, Basilico C, Zack DJ (1996) Retinal degeneration in transgenic mice with photoreceptor-specific expression of a dominant-negative fibroblast growth factor receptor. J Neurosci 16:1679-1688

Carwile ME, Culbert RB, Sturdivant RL, Kraft TW (1998) Rod outer segment maintenance is enhanced in the presence of bFGF, CNTF and GDNF. Exp Eye Res 66:791-805

Cepko CL (1999) The roles of intrinsic and extrinsic cues and bHLH genes in the determination of retinal cell fates. Curr Opin Neurobiol 9:37-46

Chaitin MH, Wortham HS, Brun-Zinkernagel AM (1994) Immunocytochemical localization of CD44 in the mouse retina. Exp Eye Res 58:359-365

Champliaud MF, Lunstrum GP, Rousselle P, Nishiyama T, Keene DR, Burgeson RE (1996) Human amnion contains a novel laminin variant, laminin 7, which like laminin 6, covalently associates with laminin 5 to promote stable epithelial-stromal attachment. J Cell Biol 132:1189-1198

Chen W, Joos TO, Defoe DM (1997) Evidence for beta 1-integrins on both apical and basal surfaces of Xenopus retinal pigment epithelium. Exp Eye Res 64:73-84

Claudepierre T, Rodius F, Frasson M, Fontaine V, Picaud S, Dreyfus H, Mornet D, Rendon A (1999) Differential distribution of dystrophins in rat retina. Invest Ophthalmol Visual Sci 40:1520-1529

De Roberts E (1956) Electron microscope observations on the submicroscopic organization of retinal rods. J Biophys Biochem Cytol 2:319-334

De Roberts E (1960) Some observations on the ultrastructure and morphogenesis of photoreceptors. J Gen Physiol 43:1-13

Désiré L, Courtois Y, Jeanny JC (1998) Suppression of fibroblast growth factors 1 and 2 by antisense oligonucleotides in embryonic chick retinal cells in vitro inhibits neuronal differentiation and survival. Exp Cell Res 241:210-221

D'Souza VN, Nguyen TM, Morris GE, Karges W, Pillers DA, Ray PN (1995) A novel dystrophin isoform is required for normal retinal electrophysiology. Hum Mol Genet 4:837-842

Dong LJ, Chung AE (1991) The expression of the genes for entactin, laminin A, laminin B1 and laminin B2 in murine lens morphogenesis and eye development. Differentiation 48:157-172

Drenckhahn D, Holbach M, Ness W, Schmitz F, Anderson LV (1996) Dystrophin and the dystrophin-associated glycoprotein, beta-dystroglycan, co-localize in photoreceptor synaptic complexes of the human retina. Neuroscience 73:605-612

Engvall E, Earwicker D, Haaparanta T, Ruoslahti E, Sanes JR (1990) Distribution and isolation of four laminin variants; tissue-restricted distribution of heterotrimers assembled from five different subunits. Cell Regul 1:731-740

Eudy JD, Weston MD, Yao S, et al (1998) Mutation of a gene encoding a protein with extracellular matrix motifs in Usher syndrome type IIa. Science 280:1753-1757

Ezzeddine ZD, Yang X, DeChiara T, Yancopoulos G, Cepko CL (1997) Postmitotic cells fated to become rod photoreceptors can be respecified by CNTF treatment of the retina. Development 124:1055-1067

Faham S, Linhardt RJ, Rees DC (1998) Diversity does make a difference: fibroblast growth factor-heparin interactions. Curr Opin Struct Biol 8:578-586

Faktorovich EG, Steinberg RH, Yasumura D, Matthes MT, LaVail MM (1990) Photoreceptor degeneration in inherited retinal dystrophy delayed by basic fibroblast growth factor. Nature 347:83-86

Faktorovich EG, Steinberg RH, Yasumura D, Matthes MT, LaVail MM (1992) Basic fibroblast growth factor and local injury protect photoreceptors from light damage in the rat. J Neurosci 12:3554-3567

Farber DB, Danciger M (1997) Identification of genes causing photoreceptor degenerations leading to blindness. Curr Opin Neurobiol 7:666-673

Fariss RN, Apte SS, Luthert PJ, Bird AC, Milam AH (1998) Accumulation of tissue inhibitor of metalloproteinases-3 in human eyes with Sorsby's fundus dystrophy or retinitis pigmentosa. Br j Ophthalmol 82:1329-1334

Fitzgerald KM, Cibis GW, Giambrone SA, Harris DJ (1994) Retinal signal transmission in Duchenne muscular dystrophy: evidence for dysfunction in the photoreceptor/deplorizing bipolar cell pathway. J Clin Invest 93:2425-2430

Fontaine V, Kinkl N, Sahel J, Dreyfus H, Hicks D (1998) Survival of purified rat photoreceptors in vitro is stimulated directly by fibroblast growth factor-2. J Neurosci 18:9662-9672

Frade JM, Martinez Morales JR, Rodriguez Tebar A (1996) Laminin-1 selectively stimulates neuron generation from cultured retinal neuroepithelial cells. Exp Cell Res 222:140-149

Fuhrmann S, Kirsch M, Hofmann HD (1995) Ciliary neurotrophic factor promotes chick photoreceptor development in vitro. Development 121:2695-2706

Galbavy ES, Olson MD (1979) Morphogenesis of rod cells in the retina of the albino rat: a scanning electron microscopic study. Anat Rec 195:707-717

Gao H, Hollyfield JG (1992) Basic fibroblast growth factor (bFGF) immunolocalization in the rodent outer retina demonstrated with an anti-rodent bFGF antibody. Brain Res 585:355-360

Gautam M, Noakes PG, Moscoso L, Rupp F, Scheller RH, Merlie JP, Sanes JR (1996) Defective neuromuscular synaptogenesis in agrin-deficient mutant mice. Cell 85:525-535

Gensburger C, Labourdette G, Sensenbrenner M (1987) Brain basic fibroblast growth factor stimulates the proliferation of rat neuronal precursor cells in vitro. FEBS Lett 217:1-5

Georges-Labouesse E, Mark M, Messaddeq N, Gansmüller A (1998) Essential role of alpha 6 integrins in cortical and retinal lamination. Curr Biol 8:983-986

Girlanda P, Quartarone A, Buceti R, Sinicropi S, Macaione V, Saad FA, Messina L, Danieli GA, Ferreri G, Vita G (1997) Extra-muscle involvement in dystrophinopathies: an electroretinography and evoked potential study. J Neurol Sci 146:127-132

Guillemot F, Cepko CL (1992) Retinal fate and ganglion cell differentiation are potentiated by acidic FGF in an in vitro assay of early retinal development. Development 114:743-754

Gundersen D, Powell SK, Rodriguez Boulan E (1993) Apical polarization of N-CAM in retinal pigment epithelium is dependent on contact with the neural retina. J Cell Biol 121:335-343

Haas TA, Plow EF (1994) Integrin-ligand interactions: a year in review. Curr Opin Cell Biol 6:656-662

Hageman GS, Johnson LV (1987) Chondroitin 6-sulfate glycosaminoglycan is a major constituent of primate cone photoreceptor matrix sheaths. Curr Eye Res 6:639-646

Hageman GS, Kirchoff Rempe MA, Lewis GP, Fisher SK, Anderson DH (1991) Sequestration of basic fibroblast growth factor in the primate retinal interphotoreceptor matrix. Proc Natl Acad Sci USA 88:6706-6710

Hankin MH, Lagenaur CF (1994) Cell adhesion molecules in the early developing mouse retina: retinal neurons show preferential outgrowth in vitro on L1 but not N-CAM. J Neurobiol 25:472-487

Helbling-Leclerc A, Zhang X, Topaloglu H, Cruaud C, Tesson F, Weissenbach J, Tome FM, Schwartz K, Fardeau M, Tryggvason K, et al. (1995) Mutations in the laminin alpha 2-chain gene (LAMA2) cause merosin-deficient congenital muscular dystrophy. Nat Genet 11:216-218

Henrique D, Hirsinger E, Adam J, Le Roux I, Pourquie O, Ish-Horowicz D, Lewis J (1997) Maintenance of neuroepithelial progenitor cells by delta-notch singalling in the embryonic chick retina. Curr Biol 7:661-670

Hewitt AT, Lindsey JD, Carbott D, Adler R (1990) Photoreceptor survival-promoting activity in interphotoreceptor matrix preparations: characterization and partial purification. Exp Eye Res 50:79-88

Hicks D, Courtois Y (1992) Fibroblast growth factor stimulates photoreceptor differentiation in vitro. J Neurosci 12:2022-2033

Hinds JW, Hinds PL (1979) Differentiation of photoreceptors and horizontal cells in the embryonic mouse retina: an electron microscopic, serial section analysis. J Comp Neurol 187:495–511

Hoffman EP, Brown RH, Jr, Kunkel LM (1987) Dystrophin: the protein product of the Duchenne muscular dystrophy locus. Cell 51:919–928

Hollyfield JG, Rayborn ME, Tammi M, Tammi R (1998) Hyaluronan in the interphotoreceptor matrix of the eye: species differences in content, distribution, ligand binding and degradation. Exp Eye Res 66:241–248

Holt CE, Bertsch TW, Ellis HM, Harris WA (1988) Cellular determination in the *Xenopus* retina is independent of lineage and birth date. Neuron 1:15–26

Howard PL, Dally GY, Wong MH, Ho A, Weleber RG, Pillers DA, Ray PN (1998) Localization of dystrophin isoform Dp71 to the inner limiting membrane of the retina suggests a unique functional contribution of Dp71 in the retina. Hum Mol Genet 7:1385–1391

Humayun MS, deJaun E, Weiland JD, Dagnelie G, Katona S, Greenberg R, Suzuke S (1999) Pattern electrical stimulation of the human retina. Vision Res 39:2569–2576

Hunter DD, Brunken WJ (1997) Beta 2 laminins modulate neuronal phenotype in the rat retina. Mol Cell Neurosci 10:7–15

Hunter DD, Murphy MD, Olsson CV, Brunken WJ (1992) S-laminin expression in adult and developing retinae: a potential cue for photoreceptor morphogenesis. Neuron 8:399–413

Hyatt GA, Schmitt EA, Fadool JM, Dowling JE (1996) Retinoic acid alters photoreceptor development in vivo. Proc Natl Acad Sci USA 93:13298–13303

Hyatt GA, Schmit, EA, Marsh Armstrong NR, Dowling JE (1992) Retinoic acid-induced duplication of the zebrafish retina. Proc Natl Acad Sci USA 89:8293–8297

Hyatt GA, Schmitt EA, Marsh Armstrong N, McCaffery P, Drager UC, Dowling JE (1996) Retinoic acid establishes ventral retinal characteristics. Development 122:195–204

Hyer J, Mima T, Mikawa T (1998) FGF1 patterns the optic vesicle by directing the placement of the neural retina domain. Development 125:869–877

Kameya S, Araki E, Katsuki M, Mizota A, Adachi E, Nakahara K, Nonaka I, Sakuragi S, Takeda S, Nabeshima Y (1997) Dp260 disrupted mice revealed prolonged implicit time of the b-wave in ERG and loss of accumulation of beta-dystroglycan in the outer plexiform layer of the retina. Hum Mol Genet 6:2195–2203

Kelley MW, Turner JK, Reh TA (1994) Retinoic acid promotes differentiation of photoreceptors in vitro. Development 120:2091–2102

Kivirikko S, McGrath JA, Baudoin C, Aberdam D, Ciatti S, Dunnill MG, McMillan JR, Eady RA, Ortonne JP, Meneguzzi G, et al. (1995) A homozygous nonsense mutation in the alpha 3 chain gene of laminin 5 (LAMA3) in lethal (Herlitz) junctional epidermolysis bullosa. Hum Mol Genet 4:959–962

Koch M, Olson PF, Albus A, Jin W, Hunter DD, Brunken WJ, Burgeson RE, Champliaud MF (1999) Characterization and expression of the laminin gamma3 chain: a novel, non-basement membrane-associated, laminin chain. J Cell Biol 145:605–618

Koulen P, Blank M, Kröger S (1998) Differential distribution of beta-dystroglycan in rabbit and rat retina. J Neurosci Res 51:735–747

Kröger S (1997) Differential distribution of agrin isoforms in the developing and adult avian retina. Mol Cell Neurosci 10:149–161

Kröger S, Horton SE, Honig LS (1996) The developing avian retina expresses agrin isoforms during synaptogenesis. J Neurobiol 29:165–182

LaVail MM, Unoki K, Yasumura D, Matthes MT, Yancopoulos GD, Steinberg RH (1992) Multiple growth factors, cytokines, and neurotrophins rescue photoreceptors from the damaging effects of constant light. Proc Nat Acad Sci USA 89:11249–11253

Lazarus HS, Hageman GS (1992) Xyloside-induced disruption of interphotoreceptor matrix proteoglycans results in retinal detachment. Invest Ophthalmol Visual Sci 33:364–376

Lenk U, Oexle K, Voit T, Ancker U, Hellner KA, Speer A, Hubner C (1996) A cysteine 3340 substitution in the dystroglycan-binding domain of dystrophin associated with Duchenne muscular dystrophy, mental retardation and absence of the ERG b-wave. Hum Mol Genet 5:973–975

Lentz SI, Miner JH, Sanes JR, Snider WD (1997) Distribution of the ten known laminin chains in the pathways and targets of developing sensory axons. J Comp Neurol 378:547–561

Lesley J, Hyman R (1998) CD44 structure and function. Front Biosci 3:D616–630

Li A, Lane WS, Johnson LV, Chader GJ, Tombran-Tink J (1995) Neuron-specific enolase: a neuronal survival factor in the retinal extracellular matrix? J Neurosci 15:385–393

Libby RT, Hunter DD, Brunken WJ (1996) Developmental expression of laminin β2 in rat retina. Further support for a role in rod morphogenesis. Invest Ophthalmol Visual Sci 37:1651–1661

Libby RT, Yin X, Selfors LM, Brunken WJ, Hunter DD (1997) Identification of the cellular source of laminin β2 in adult and developing vertebrate retinae. J Comp Neurol 389:355–367

Libby RT, Lavallee CR, Balkema GW, Brunken WJ, Hunter DD (1999) Disruption laminin beta 2 chain production causes alterations in morphology and function in the cns. J Neurosci 19:9399–9411

Lillien L (1995) Changes in retinal cell fate induced by overexpression of EGF receptor. Nature 377:158–162

Lillien L, Cepko C (1992) Control of proliferation in the retina: temporal changes in responsiveness to FGF and TGF alpha. Development 115:253–266

Lin H, Clegg DO (1998) Integrin alphavbeta5 participates in the binding of photoreceptor rod outer segments during phagocytosis by cultured human retinal pigment epithelium. Invest Ophthalmol Visual Sci 39:1703–1712

Marinkovich MP, Lunstrum GP, Keene DR, Burgeson RE (1992) The dermal-epidermal junction of human skin contains a novel laminin variant. J Cell Biol 119:695–703

Masuda K, Watanabe I, Unoki K, Ohba N, Muramatsu T (1995) Functional Rescue of photoreceptors from the damaging effects of constant light by survival-promoting factors in the rat. Invest Ophthalmol Visual Sci 36:2142–2146

Matsunaga M, Hatta K, Takeichi M (1988) Role of N-cadherin cell adhesion molecules in the histogenesis of neural retina. Neuron 1:289–295

McFarlane S, Zuber ME, Holt CE (1998) A role for the fibroblast growth factor receptor in cell fate decisions in the developing vertebrate retina. Development 125:3967–3975

McKeehan WL, Wang F, Kan M (1998) The heparan sulfate-fibroblast growth factor family: diversity of structure and function. Prog Nucleic Acid Res Mol Biol 59:135–176

McLoon SC, McLoon LK, Palm SL, Furcht LT (1988) Transient expression of laminin in the optic nerve of the developing rat. J Neurosci 8:1981–1990

Mieziewska K, Szel A, Van Veen T, Aguirre GD, Philp N (1994) Redistribution of insoluble interphotoreceptor matrix components during photoreceptor differentiation in the mouse retina. J Comp Neurol 345:115–124

Millaruelo AI, Nieto Sampedro M, Cotman CW (1988) Cooperation between nerve growth factor and laminin or fibronectin in promoting sensory neuron survival and neurite outgrowth. Brain Res 466:219–228

Miner JH, Cunningham J, Sanes JR (1998) Roles for laminin in embryogenesis: exencephaly, syndactyly, and placentopathy in mice lacking the laminin alpha5 chain. J Cell Biol 143:1713–1723

Miner JH, Patton BL, Lentz SI, Gilbert DJ, Snider WD, Jenkins NA, Copeland NG, Sanes JR (1997) The laminin alpha chains: expression, developmental transitions, and chromosomal locations of alpha1–5, identification of heterotrimeric laminins 8–11, and cloning of a novel alpha3 isoform. J Cell Biol 137:685–701

Miyagoe Y, Hanaoka K, Nonaka I, Hayasaka M, Nabeshima Y, Arahata K, Takeda S (1997) Laminin alpha2 chain-null mutant mice by targeted disruption of the Lama2 gene: a new model of merosin (laminin 2)-deficient congenital muscular dystrophy. FEBS Lett 415:33–39

Montanaro F, Carbonetto S, Campbell KP, Lindenbaum M (1995) Dystroglycan expression in the wild type and mdx mouse neural retina: synaptic colocalization with dystrophin, dystrophin-related protein but not laminin. J Neurosci Res 42:528–538

Morimoto A, Matsuda S, Uryu K, Fujita H, Okumura N, Sakanaka M (1993) Light- and electron-microscopic localization of basic fibroblast growth factor in adult rat retina. Okajimas Folia Anat Jpn 70:7–12

Morissette N, Carbonetto S (1995) Laminin α2 chain (M chain) is found within the pathway of avian and murine retinal projections. J Neurosci 15:8067-8082

Morrow EM, Belliveau MJ, Cepko CL (1998) Two phases of rod photoreceptor differentiation during rat retinal development. J Neurosci 18:3738-3748

Neophytou C, Vernallis AB, Smith A, Raff MC (1997) Muller-cell-derived leukaemia inhibitory factor arrests rod photoreceptor differentiation at a postmitotic pre-rod stage of development. Development 124:2345-2354

Nir I, Cohen D, Papermaster DS (1984) Immunocytochemical localization of opsin in the cell membrane of developing rat retinal photoreceptors. J Cell Biol 98:1788-1795

Noakes PG, Gautam M, Mudd J, Sanes JR, Merlie JP (1995) Aberrant differentiation of neuromuscular junctions in mice lacking s-laminin/laminin β2. Nature 374:258-262

Noakes PG, Miner JH, Gautam M, Cunningham JM, Sanes JR, Merlie JP (1995) The renal glomerulus of mice lacking s-laminin/laminin β2: nephrosis despite molecular compensation by laminin β1. Nat Genet 10:400-406

Obata S, Usukura J (1992) Morphogenesis of the photoreceptor outer segment during postnatal development in the mouse (BALB/c) retina. Cell Tissue Res 269:39-48

Olney JW (1968) An electron microscopic study of synapse formation, receptor outer segment development, and other aspects of developing mouse retina. Invest Ophthalmol 7:250-268

Olney JW (1968) Centripetal sequence of appearance of receptor-bipolar synaptic structures in developing mouse retina. Nature 218:281-282

Pillers DA, Weleber RG, Green DG, Rash SM, Dally GY, Howard PL, Powers MR, Hood DC, Chapman VM, Ray PN, Woodward WR (1999) Effects of dystrophin isoforms on signal transduction through neural retina: genotype-phenotype analysis of duchenne muscular dystrophy mouse mutants. Mol Genet Metab 66:100-110

Pillers DM, Bulman DE, Weleber RG, Sigesmund DA, Musarella MA, Powell BR, Murphey WH, Westall C, Panton C, Becker LE et al. (1993) Dystrophin expression in the human retina is required for normal function as defined by electroretinography. Nat Genet 4:82-86

Pillers DM, Weleber RG, Woodward WR, Green DG, Chapman VM, Ray PN (1995) mdxCv3 mouse is a model for electroretinography of Duchenne/Becker muscular dystrophy. Invest Ophthalmol Visual Sci 36:462-466

Pittack C, Grunwald GB, Reh TA (1997) Fibroblast growth factors are necessary for neural retina but not pigmented epithelium differentiation in chick embryos. Development 124:805-816

Pulkkinen L, Christiano AM, Airenne T, Haakana H, Tryggvason K, Uitto J (1994) Mutations in the gamma 2 chain gene (LAMC2) of kalinin/laminin 5 in the junctional forms of epidermolysis bullosa. Nat Genet 6:293-297

Pulkkinen L, Christiano AM, Airenne T, Haakana H, Tryggvason K, Uitto J (1994) Mutations in the γ2 chain gene (LAMC2) of kalinin/laminin 5 in the junctional forms of epidermolysis bullosa. Nat Genet 6:293-297

Pulkkinen L, Christiano AM, Gerecke D, Wagman DW, Burgeson RE, Pittelkow MR, Uitto, J. (1994) A homozygous nonsense mutation in the β3 chain gene of laminin 5 (LAMB3) in Herlitz junctional epidermolysis bullosa. Genomics 24:357-360

Pulkkinen L, Christiano AM, Gerecke D, Wagman DW, Burgeson RE, Pittelkow MR, Uitto J (1994) A homozygous nonsense mutation in the beta 3 chain gene of laminin 5 (LAMB3) in Herlitz junctional epidermolysis bullosa. Genomics 24:357-360

Reh TA (1992) Cellular interactions determine neuronal phenotypes in rodent retinal cultures. J Neurobiol 23:1067-1083

Rousselle P, Lunstrum GP, Keene DR, Burgeson RE (1991) Kalinin: an epithelium-specific basement membrane adhesion molecule that is a component of anchoring filaments. J Cell Biol 114:567-576

Ryan MC, Lee K, Miyashita Y, Carter WG (1999) Targeted disruption of the LAMA3 gene in mice reveals abnormalities in survival and late stage differentiation of epithelial cells. J Cell Biol 145:1309-1324

Sarthy PV, Fu M (1990) Localization of laminin B1 mRNA in retinal ganglion cells by in situ hybridization. J Cell Biol 110:2099-2108

Schmitz F, Drenckhahn D (1997) Localization of dystrophin and beta-dystroglycan in bovine retinal photoreceptor processes extending into the postsynaptic dendritic complex. Histochem Cell Biol 108:249-255

Schmitz F, Holbach M, Drenckhahn D (1993) Colocalization of retinal dystrophin and actin in postsynaptic dendrites of rod and cone photoreceptor synapses. Histochemistry 100:473-479

Schmidt MF, Kater SB (1995) Depolarization and laminin independently enable bFGF to promote neuronal survival through different second messenger pathways. Dev Biol 168:235-246

Smyth N, Vatansever HS, Murray P, Meyer M, Frie C, Paulsson M, Edgar D (1999) Absence of basement membranes after targeting the LAMC1 gene results in embryonic lethality due to failure of endoderm differentiation. J Cell Biol 144:151-160

Sunada Y, Campbell P (1996) Dystroglycan: a novel laminin receptor and its involvement in the pathogenesis of muscular dystrophy. In: Ekblom P, Timpl R (eds) The laminins. Harwood Academic Publishers, Amsterdam, pp 291-316

Terranova VP, Rao CN, Kalebic T, Margulies IM, Liotta LA (1983) Laminin receptor on human breast carcinoma cells. Proc Natl Acad Sci USA 80:444-448

Tien L, Rayborn ME, Hollyfield JG (1992) Characterization of the interphotoreceptor matrix surrounding rod photoreceptors in the human retina. Exp Eye Res 55:297-306

Timpl R, Johansson S, van Delden V, Oberbaumer I, Hook M (1983) Characterization of protease-resistant fragments of laminin mediating attachment and spreading of rat hepatocytes. J Biol Chem 258:8922-8927

Timpl R, Rohde H, Robey PG, Rennard SI, Foidart JM, Martin GR (1979) Laminin - a glycoprotein from basement membranes. J Biol Chem 254:9933-9937

Tombran Tink J, Li A, Johnson MA, Johnson LV, Chader GJ (1992) Neurotrophic activity of interphotoreceptor matrix on human Y79 retinoblastoma cells. J Comp Neurol 317:175-186

Tsukamoto Y, Taira E, Yamate J, Nakane T, Kajimura K, Tsudzuki M, Kiso Y, Kotani T, Miki N, Sakuma S (1997) Gicerin, a cell adhesion molecule, participates in the histogenesis of retina. J Neurobiol 33:769-780

Turner DL, Cepko CL (1987) A common progenitor for neurons and glia persists in rat retina late in development. Nature 328:131-136

Turner DL, Snyder EY, Cepko CL (1990) Lineage-independent determination of cell type in the embryonic mouse retina. Neuron 4:833-845

Ueda H, Gohdo T, Ohno S (1998) Beta-dystroglycan localization in the photoreceptor and Müller cells in the rat retina revealed by immunoelectron microscopy. J Histochem Cytochem 46:185-191

Ueda H, Kato Y, Baba T, Terada N, Fujii Y, Tsukahara S, Ohno S (1997) Immunocytochemical study of dystrophin localization in cone cells of mouse retinas. Invest Ophthalmol Visual Sci 38:1627-1630

Ueda H, Kobayashi T, Mitsui K, Tsurugi K, Tsukahara S, Ohno S (1995) Dystrophin localization at presynapse in rat retina revealed by immunoelectron microscopy (see comments). Invest Ophthalmol Visual Sci 36:2318-2322

Uteza Y, Rouillot JS, Kobetz A, Marchant D, Pecqueur S, Arnaud E, Prats H, Honiger J, Dufier JL, Abitbol M, Neuner-Jehle M (1999) Intravitreous transplantation of encapsulated fibroblasts secreting the human fibroblast growth factor 2 delays photoreceptor cell degeneration in Royal College of Surgeons rats. Proc Natl Acad Sci USA 96:3126-3131

Varner HH, Rayborn ME, Osterfeld AM, Hollyfield JG (1987) Localization of proteoglycan within the extracellular matrix sheath of cone photoreceptors. Exp Eye Res 44:633-642

Vidal F, Baudoin C, Miquel C, Galliano MF, Christiano AM, Uitto J, Ortonne JP, Meneguzzi G (1995) Cloning of the laminin alpha 3 chain gene (LAMA3) and identification of a homozygous deletion in a patient with Herlitz junctional epidermolysis bullosa. Genomics 30:273-280

Weidman TA, Kuwabara T (1969) Development of the rat retina. Invest Ophthalmol 8:60-69

Weidman TA, Kuwabara T (1968) Postnatal development of the rat retina. An electron microscopic study. Arch Ophthalmol 79:470-484

Wetts R, Fraser SE (1988) Multipotent precursors can give rise to all major cell types of the frog retina. Science 239:1142-1145

Wetts R, Serbedzija GN, Fraser SE (1989) Cell lineage analysis reveals multipotent precursors in the ciliary margin of the frog retina. Dev Biol 136:254–263

Xu H, Christmas P, Wu XR, Wewer UM, Engvall E (1994) Defective muscle basement membrane and lack of M-laminin in the dystrophic dy/dy mouse. Proc Natl Acad Sci USA 91:5572–5576

Xu H, Wu XR, Wewer UM, Engvall E (1994) Murine muscular dystrophy caused by a mutation in the laminin alpha 2 (Lama2) gene. Nat Genet 8:297–302

Yao XY, Hageman GS, Marmor MF (1992) Recovery of retinal adhesion after enzymatic perturbation of the interphotoreceptor matrix. Invest Ophthalmol Visual Sci 33:498–503

Yao XY, Hageman GS, Marmor MF (1994) Retinal adhesiveness in the monkey. Invest Ophthalmol Visual Sci 35:744–748

Ying SY, Li S, Ishikawa M, Johnson LV (1995) Immunohistochemical localization of inhibin in the retinal interphotoreceptor matrix [letter]. Exp Eye Res 60:585–590

Yurchenco PD, Cheng YS, Colognato H (1992) Laminin forms an independent network in basement membranes. J Cell Biol 117:1119–1133

Zhao S, Barnstable CJ (1996) Differential effects of bFGF on development of the rat retina. Brain Res 723:169–176

Zrenner E, Stett A, Weiss S, Aramant RB, Guenther E, Kohler K, Miliczek KD, Seiler MJ, Haemmerle H (1999) Can subretinal microphotodiodes successfully replace degenerated photoreceptors? Vision Res 39:2555–2567

Adhesive Events in Retinal Development and Function: The Role of Integrin Receptors

Dennis O. Clegg, Linda H. Mullick, Kevin L. Wingerd, Hai Lin,
Jason W. Atienza, Amy D. Bradshaw, Dennis B. Gervin,
and Gordon M. Cann[1]

Summary: Cells in the developing retina contact a vast array of molecular cues in their microenvironment that are thought to guide their development. Many of these cues are embedded in the surface of neighboring cells or deposited within the extracellular matrix (ECM). Evidence has accumulated that cell–cell and cell-ECM interactions are essential in many phases of neural development, including neuroblast migration, determination of cell fate, axon outgrowth and synapse formation. In this chapter, we examine the developmental and functional roles fulfilled by integrins, a family of receptors for ECM molecules and cell adhesion molecules (CAMs). We have approached this problem by addressing a series of three questions: (1) which integrins are expressed in developing retina? (2) when and where are they expressed? and, (3) what functions do they carry out? Integrins have previously been implicated in axon extension, but new evidence suggests that they are also involved in earlier developmental events in preaxonal neuroblasts. High levels of expression of at least eight integrin subunits have been documented in these young retinal cells, and integrins containing the $\beta 1$ subunit have been implicated in migration of adolescent retinal ganglion cells. Integrin expression persists through adulthood, both in the retina and in the neighboring layer of the retinal pigment epithelium (RPE). The integrin $\alpha v\beta 5$ has been shown to reside on the apical surface of the RPE and has been implicated in the phagocytosis of shed photoreceptor outer segments.

1
Introduction

The finding about 15 years ago that conditioned media containing ECM molecules would stimulate neurite outgrowth from cultured neurons gave rise to higher-order questions: how do neurons bind to ECM and extend processes? What are the neuronal receptors for ECM components like the laminins, basal lamina components that have potent neurite promoting activity? Inquiry in this area has led to the integrins, a family of ECM and CAM receptors found on almost all cell types.

[1] Neuroscience Research Institute, Department of Molecular, Cellular and Developmental Biology, University of California, Santa Barbara, California 93106, USA

Results and Problems in Cell Differentiation, Vol. 31
M. E. Fini (Ed.): Vertebrate Eye Development
© Springer-Verlag Berlin Heidelberg 2000

1.1
The Integrin Family of Receptors

Integrins are heterodimeric receptors consisting of an alpha and a beta subunit, both of which span the lipid bilayer and participate in ligand binding (reviewed in Hynes 1992; Sonnenberg 1993; Schwartz et al. 1995; Danen et al. 1998). The family consists of 16 alpha subunits and 8 beta subunits, and different combinations of α/β pairing give rise to different ligand specificity of the receptor. While not all possible pairings have been observed, many subunits can pair with more than one partner, and over 20 different heterodimers have been described. Many integrin receptors are capable of binding multiple ligands, and some can "cooperate" with other receptors in a cis fashion (Hemler 1998). There also appears to be a redundancy in integrin function, as multiple receptors have been identified for a single ligand. For example, at least seven laminin receptors and five fibronectin receptors have been reported. Upon ligand binding, integrins transduce signals to the interior of the cell that can alter cell shape and induce cell movement, prevent apoptosis, stimulate proliferation, and alter gene expression. Signal transduction is accomplished in part by integrin interaction with the actin cytoskeleton, via talin and alpha actinin, and by stimulation of focal adhesion kinase (FAK), a tyrosine kinase localized to focal contacts in cultured cells. Integrin receptor activity can be regulated by inside-out signaling events that alter ligand binding affinity. Genetically engineered mice that lack integrins have severe, often embryonic-lethal phenotypes (Hynes 1996). By way of analogy, integrins can be thought of as the "molecular feet" of the cell, sensing and responding to the environment to bring about adhesion, migration, and survival.

In the eye, integrins are thought to play a number of important roles (Elner and Elner 1996). They are expressed throughout the eye, at all stages of development. Different arrays of integrins have been documented in the retina, the RPE, the cornea, and the trabecular meshwork. Although many in vitro activities and expression patterns have been described, the in vivo functions carried out by integrins are not yet clear.

2
Integrins in the Developing Retina

2.1
Which Integrins are Expressed?

To gain an understanding of integrin function in the developing retina, we used reverse transcription coupled with polymerase chain reaction to identify alpha and beta subunits expressed in embryonic day-6 chick retina (Cann et al. 1996; Gervin et al. 1996). At this developmental time point, retinal ganglion cells RGC's are sending axons to the tectum, and other retinal neuroblasts are engaged in earlier developmental events which may involve integrins. Degenerate primers were designed to recognize alpha and beta subunit conserved sequences and thus amplify many subunit cDNAs. To obtain cDNAs encoding less abundant integrins, the most prevalent cDNAs were removed from the PCR product by digestion with restriction enzymes, and the remaining products were reamplified (Fig. 1).

Three beta and five alpha subunit cDNAs encoding chick homologues of $\alpha2$, $\alpha4$, $\alpha6$, $\alpha8$, αv, $\beta1$, $\beta3$, and $\beta5$ were identified. Other integrin subunits not recognized well by the primers or of very low abundance may also be present. Indeed, other means have been used to document $\alpha1$ and $\alpha3$ subunit expression in developing chick retina (de Curtis et al. 1991; Duband et al. 1992). In addition, because the chick retina is avascular, other integrins may be found in mammalian retina associated with the vasculature.

The $\beta1$ family of heterodimers was found to be the most widely expressed: $\alpha1\beta1$, $\alpha2\beta1$, $\alpha3\beta1$, $\alpha4\beta1$, $\alpha6\beta1$, and $\alpha8\beta1$ have now been documented in retina using various methods to detect protein expression. These appear to function as receptors for laminins, fibronectins, vitronectin, and vascular cell adhesion molecule (VCAM-1; see below). Surprisingly, the $\alpha v\beta5$ heterodimer, a vitronectin receptor, was the most abundant single integrin in developing chick retina. A second vitronectin receptor, $\alpha v\beta3$, was also detected in lesser amounts (Neugebauer et al. 1991; Gervin et al. 1996).

2.2
When and Where are Integrins Expressed?

We next analyzed the temporal expression pattern of integrins to determine if particular developmental periods were characterized by increased integrin expression. Interestingly, $\beta1$ integrins are not coordinately regulated. When plotted as a percent of total RNA, levels of $\alpha2$, $\alpha6$, and $\beta1$ message decreased during development (Fig. 2A). However, on a per retina basis, levels of $\alpha2$, $\alpha6$, and $\beta1$ increased from E6–E9 and then gradually returned to the E6 level (Fig. 3A). The decrease in Fig. 2A may be due to a dilution effect as other nonintegrin messages greatly increase while these integrin messages hover around the same value. The peak observed at E9 may be related to the ten-fold increase in cell number that occurs in the retina from E6–E9 (Dutting et al. 1983; Morris and Cowan 1984).

Expression of the $\beta1$ subunit was further characterized by in situ hybridization (Cann et al. 1996). Whereas previous reports had detected mostly RGC expression of the $\beta1$ subunit (Cohen et al. 1986), we found widespread expression throughout the neuroblast and RGC layers of developing retina. Staining of dissociated E6 cells showed that 64 % were immunopositive for $\beta1$.

The expression of the α2 subunit protein has been examined in greater detail (Bradshaw et al. 1995). Before E6, expression was widespread throughout the neuroblast layer, but RGCs did not express detectable α2 immunoreactivity. The only antibody available did not stain later aged retinal sections, but immunoreactivity was apparent on dissociated cells, suggesting that the epitope was masked in older sections. The number of α2-positive dissociated cells declined from 45 to 20 % by E12, but approximately equal amounts of α2 were detected by Western blotting at E6 and E12. These data are consistent with changes observed in message levels, and show that expression persists through development. Cell types that express

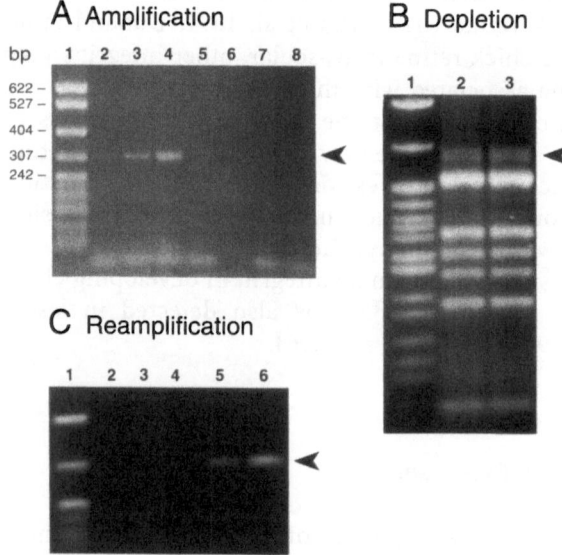

Fig. 1A–C. RT-PCR analysis of E6 chick retinal RNA. PCR products were separated on agarose gels and stained with ethidium bromide. **A** Amplified integrin alpha subunit cDNAs migrate as a single band of about 300 bp (*arrowhead*). *Lane 1* MW markers; *lanes 2–4* products amplified from 15, 30, and 65 ng of cDNA. **B** The most abundant PCR products were depleted by restriction digest, revealing less abundant, undigested products (*arrowhead*). *Lane 1* MW markers; *lanes 2 and 3* PCR products digested with Xbal, Pstl, and Sacl. **C** The undigested band in B was reamplified using standard PCR conditions. This band contained less abundant integrin subunit cDNAs. *Lane 1* MW markers; *lanes 2–6* increasing amounts of reamplified products

Fig. 2A,B. Integrin mRNA per total mRNA in developing chick retina. mRNA levels were quantified by scanning densitometry of Northern blots. Integrin mRNA levels are expressed as the percent of the value at embryonic day 6. **A** Levels of β1 family subunit mRNAs. **B** Levels of vitronectin receptor subunit mRNAs. *Error bars* indicate the standard error of the mean. Cann et al. (1996) and Gervin et al. (1996) with permission of Academic Press and the Association for Research in Vision and Ophthalmology

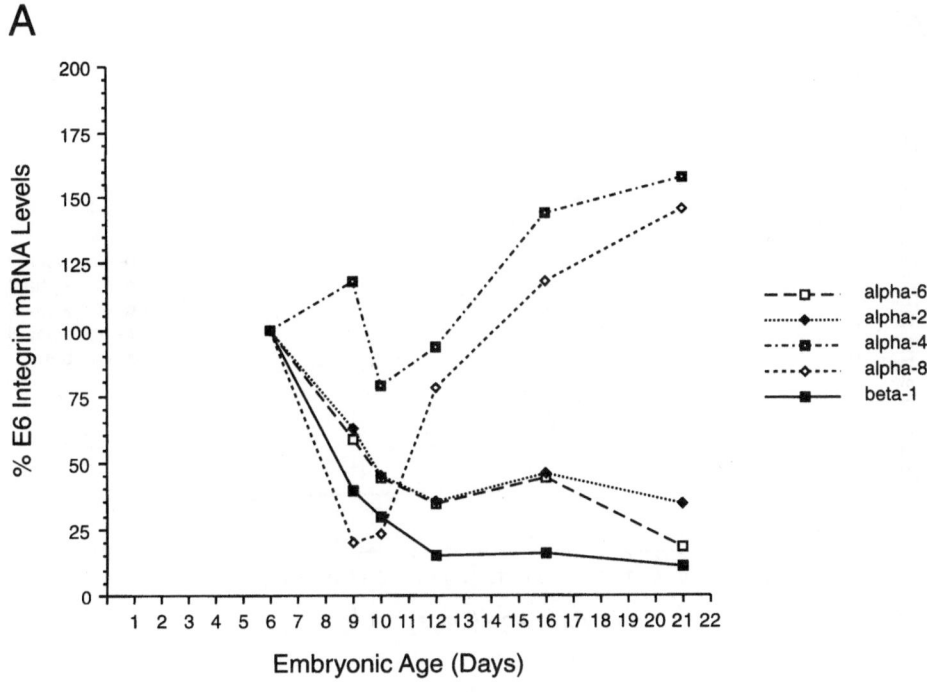

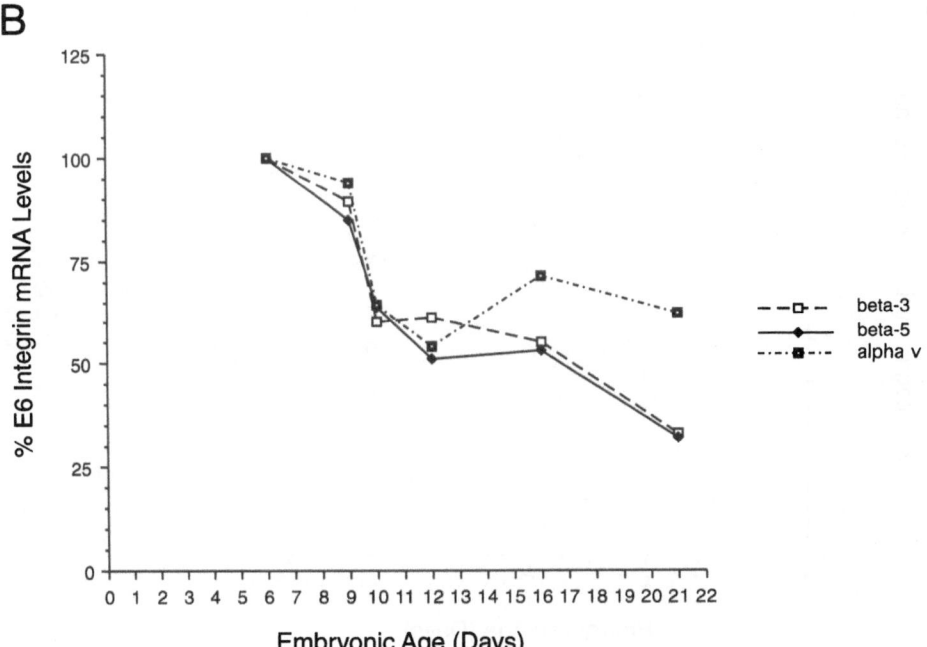

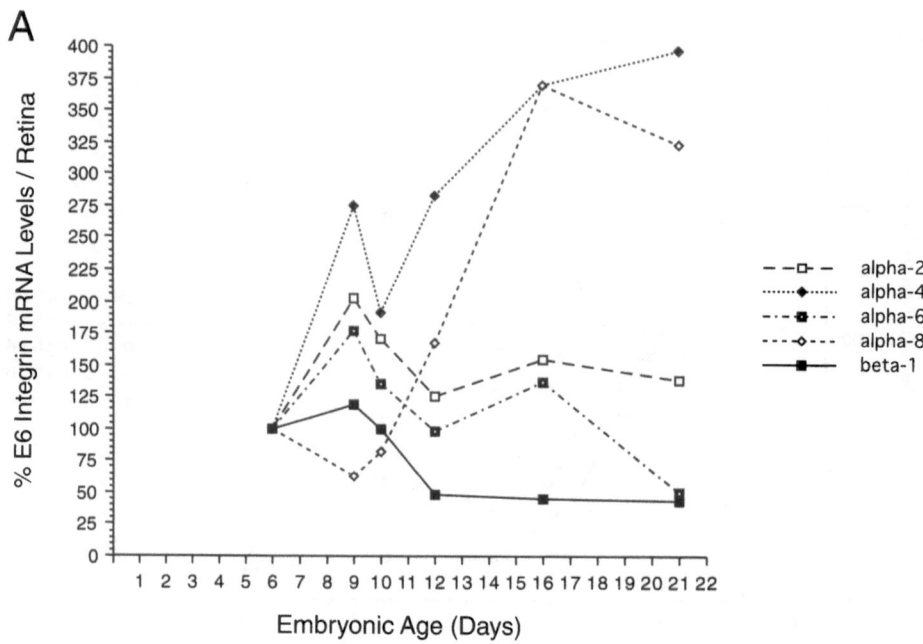

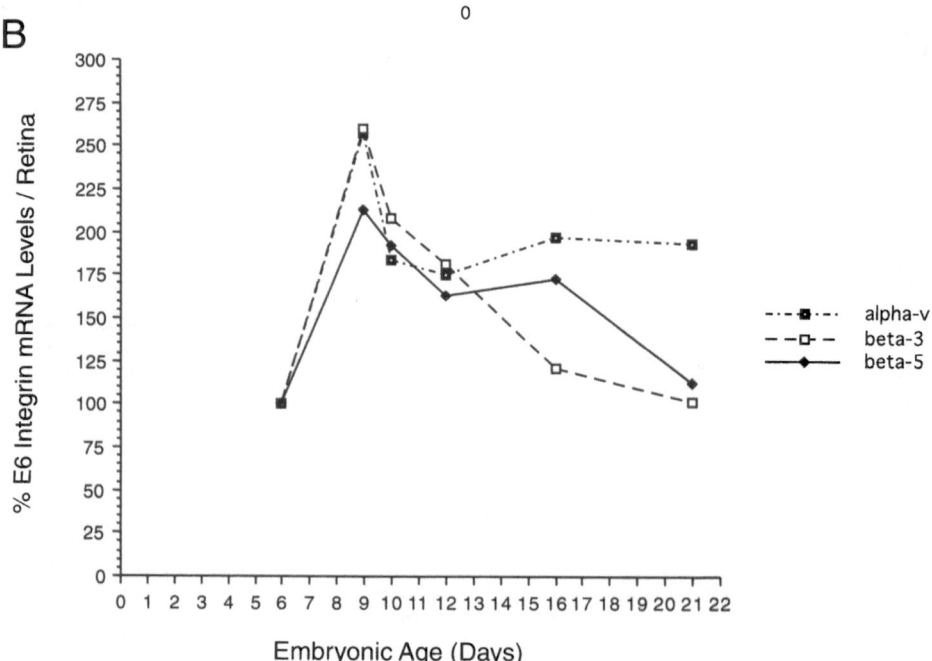

$\alpha 2$ at later times remain to be identified, but in E12 cultures, flat cells resembling Muller glial cells were immunopositive, suggesting a role for this integrin in glial adhesion (Bradshaw et al. 1995). Other investigators have noted $\alpha 2$ immunoreactivity in adult human retina (Duguid et al. 1992; Brem et al. 1994).

The $\alpha 6$ subunit is expressed throughout the retina at E6, with strong expression in RGCs (de Curtis et al. 1991). At E12, expression is limited to RGCs. Interestingly, an $\alpha 6$ splice variant with an alternate cytoplasmic domain ($\alpha 6A$) is selectively expressed in areas of the retina adjacent to the optic nerve (de Curtis and Reichardt 1993). RGCs appear to be well endowed with integrin laminin receptors, as they also express $\alpha 1$ and $\alpha 3$ subunits.

The $\alpha 4$ and $\alpha 8$ subunits showed a markedly different temporal expression pattern, as both increased during development, regardless of how the data are plotted (Figs. 2A, 3A). Increases coincided with the timing of synapse formation in the developing retina. In the case of $\alpha 8$, immunoreactivity was found associated with RGC axons at E6 and E12, and at E12 it was also detected in the inner plexiform layer (Bossy et al. 1991). $\alpha 4$ immunoreactivity has been observed in mouse retina in undifferentiated neuroblasts (Sheppard et al. 1994; G.M. Cann and D.O. Clegg, unpubl.). As was the case with the antibody against $\alpha 2$, the epitope for the $\alpha 4$ antibody used appeared to be masked in sections, as neuroblast immunoreactivity was widespread when analyzed in dissociated cells. Mouse RGCs also express the $\alpha 4$ subunit, as shown from double label immunofluorescence of dissociated cells (G.M. Cann and D.O. Clegg, unpubl.). Cells that account for the increase in $\alpha 4$ message in chick at later developmental times remain to be identified.

In contrast to $\beta 1$ integrins, the vitronectin receptor subunit mRNAs (αv, $\beta 3$, and $\beta 5$), are coordinately expressed in developing retina (Figs. 2B, 3B). Levels declined when measured as a percent of total RNA, but on a per retina basis they peaked at E9 and then returned to E6 levels. In situ hybridization revealed that both $\beta 3$ and $\beta 5$ were expressed throughout the developing retina, especially in undifferentiated neuroblasts (Gervin et al. 1996). As development proceeded, $\beta 5$ expression was highest in the inner nuclear layer, which may correspond to Muller glial cell expression.

A surprising outcome from the analysis of expression patterns was the finding that integrin levels are consistently elevated in neuroblasts. This suggests that these receptors are important for developmental events in ad-

---

Fig. 3A, B. Integrin mRNA per retina in developing chick retina. mRNA levels were quantified by scanning densitometry of Northern blots. Integrin mRNA levels are expressed as the percent of the value at embryonic day 6. A Levels of $\beta 1$ family subunit mRNAs. B Levels of vitronectin receptor subunit mRNAs. *Error bars* indicate the standard error of the mean. Cann et al. (1996), and Gervin et al. (1996) with permission of Academic Press and the Association for Research in Vision and Ophthalmology

dition to axon elongation. Retinoblasts must proliferate, migrate, and along the way decide what cell type to assume, and integrins may be involved in any or all of these processes.

2.3
What Functions do Integrins Fulfill?

This third question is proving to be the most difficult to answer. However, a combination of in vitro and in vivo approaches using a variety of perturbing agents is starting to yield answers. The best defined role for integrins in the retina is in mediating adhesion and axon/dendrite outgrowth. More recent experiments, described below, implicate β1 integrins in migration of RGCs.

β1 integrins were shown to be required for retinal cell neurite outgrowth and adhesion on a number of ECM components more than 10 years ago (Cohen et al. 1986, 1989; Hall et al. 1987). Their importance for RGC axon and dendrite extension has been shown in *Xenopus*, where expression of the chick β1 subunit had dominant negative effects that dramatically decreased axon and dendrite elongation (Lilienbaum et al. 1995). Expression of the β1 subunit before axon outgrowth prompted us to examine the role of these integrins in neuroblast migration. E4 eye cups were cultured in the presence of the blocking β1 monoclonal antibody CSAT, and migration of newborn RGCs from the ventricular to vitreal retina was followed by staining of sections with anti-neurofilament. RGCs were scored as ventricular, migrating, or vitreal. (Table 1) The anti-β1 antibody delayed migration, resulting in fewer vitreal cells and more ventricular cells. Neuroblasts may require β1 integrins to interact with matrix as they extend a process through the retina and contact the inner limiting membrane basal lamina. Adhesion to this matrix may be required before the RGC soma is translocated to the vitreal border and axon outgrowth commences (Prada et al. 1981).

A role for β1 integrins in retinal migration is also suggested by studies where anti-β1 antibodies disrupted retinal morphology in developing chick and *Xenopus* embryos (Svenevik and Linser 1992; Leonard and Sakaguchi 1995). β1 integrins have also been shown to be important in myoblast migration, glial precursor migration, neural crest cell migration and neuroblast migration in the CNS (along radial glia as well as in chain migration) (Jaffredo et al. 1988; Milner et al. 1996; Bronner-Fraser 1985; Galileo et al. 1992; Jacques et al. 1998). While some perturbations of the β1 subunit have not inhibited migratory events (Lilienbaum et al. 1995; Fassler and Meyer 1995), the majority of evidence supports an important integrin role in many types of cell migration. More recently, the α4β1 integrin has been implicated in neural crest migration (Kil et al. 1998).

In addition to the radial migration, some retinal neuroblasts also undergo a tangential displacement during development (Fekete et al. 1994; Reese et al.

Table 1. Inhibition of RGC migration by anti-$\beta1$ antibody. (Cann et al. 1996)

Expt. no.	Ventricular	Migrating	Vitreal
Anti-β_1			
1	39.7 ± 7.7	36.0 ± 7.2	12.0 ± 4.5
2	60.2 ± 4.8	29.9 ± 4.2	9.8 ± 3.4
3	77.0 ± 15	10.0 ± 10.0	13.0 ± 8.3
4	45.8 ± 5.9	30.4 ± 5.5	23.9 ± 4.4
5	14.6 ± 3.6	58.7 ± 3.4	26.5 ± 2.5
6	48.7 ± 5.2	35.0 ± 6.2	16.2 ± 2.6
7	30.4 ± 6.6	65.1 ± 7.1	5.54 ± 1.9
8	48.4 ± 5.7	39.0 ± 4.6	12.2 ± 2.8
9	16.8 ± 3.3	46.7 ± 4.4	36.5 ± 3.4
Average	42.4 ± 6.6	39.0 ± 5.5	17.3 ± 3.3
IgG Control			
1	19.3 ± 4.0	60.0 ± 8.7	16.8 ± 4.7
2	15.2 ± 3.8	44.0 ± 6.0	39.6 ± 7.5
3	25.8 ± 7.3	51.8 ± 8.3	21.2 ± 5.7
4	16.4 ± 4.3	40.8 ± 4.0	42.8 ± 5.2
5	22.3 ± 4.9	57.2 ± 8.2	21.8 ± 3.0
6	39.9 ± 7.9	35.9 ± 7.2	24.0 ± 7.8
7	8.9 ± 1.8	58.8 ± 3.5	31.9 ± 3.4
8	21.6 ± 4.6	55.9 ± 4.3	22.5 ± 3.5
9	29.0 ± 7.7	31.2 ± 6.7	39.7 ± 5.8
Average	22.0 ± 3.0	48.4 ± 3.6	28.9 ± 3.2

Results from nine eye cup culture experiments are presented. Retinas incubated in the presence of anti-β_1 antibodies are indicated in the top panel with the corresponding control eyes incubated in the presence of nonspecific mouse IgG in the bottom panel. Each value is presented plus and minus the standard error of the mean.

1995). Recent studies implicate $\beta1$ integrin function in the dispersion of clonally related cells in the chick retina (Skeith et al. 1999). Future studies will allow identification of specific integrin heterodimers that are involved in these diverse migratory movements.

The alpha subunits that pair with $\beta1$ and mediate RGC migration have not been identified, but functions are beginning to emerge for some alpha subunits. The $\alpha2\beta1$ integrin was shown to be necessary for embryonic retina cell adhesion and process outgrowth on collagens I and IV, but not on laminin (Bradshaw et al. 1995). Interestingly, this activity is down regulated with age. (Fig. 4). E12 retinal cells were unable to adhere to collagen, even though the $\alpha2$ subunit was still expressed in levels comparable to E6. This posttranslational downregulation of integrins in retinal development has been observed for other $\beta1$ integrins as well (Reichardt and Tomaselli 1991). The mechanism for this regulation is unknown, but it may be related to stable conformational changes in the $\beta1$ subunit, since an anti-$\beta1$ activating antibody (TASC) can reverse the inhibition (Neugebauer and Reichardt, 1991). The decrease in activity of other $\beta1$ integrins may be important for RGC axonal pathfinding,

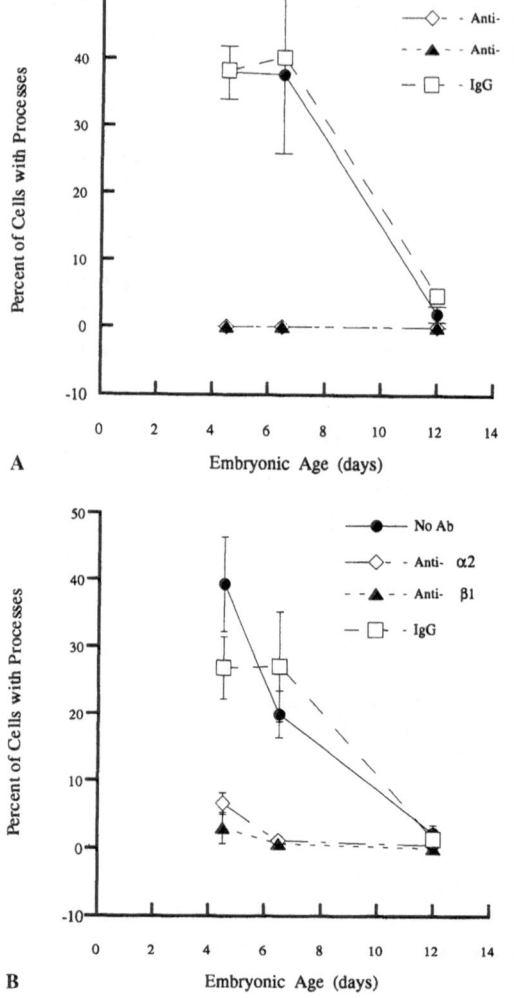

Fig. 4A, B. Inhibition of retinal process outgrowth on A collagen I and B collagen IV by anti-α2 antibodies. Cells from progressively older retinas were dissociated and plated on collagen-coated plastic overnight. No antibody *closed circles*; anti-α2 (10 μg/ml Mep-17) *open diamonds*; anti-β1 (10 μg/ml CSAT) *closed triangles*; mouse IgG control (10 μg/ml) *open squares*. *Error bars* indicate the standard error of the mean. Bradshaw et al. (1995), with permission of The Company of Biologists Limited

leading to a switch in preference from matrix substrates in the retina to cellular substrates found in the optic nerve and tectum (Hynes and Lander 1992). α2 integrins have not been detected on chick RGCs, and may instead play a role in adhesion to collagen-containing ECM by neuroblasts and Muller glial cells. Loss of activity may be related to migration or differentiation.

Similar approaches have defined activities for other β1 integrins. The α6β1 heterodimer is implicated in neurite outgrowth on laminins, and the αv-containing integrins mediate attachment and outgrowth on vitronectin (de Curtis et al. 1991; Neugebauer et al. 1991; de Curtis and Reichardt 1993).

α3β1 is thought to be a secondary laminin receptor. The role of α8 integrins in retina has not been thoroughly examined, but this integrin mediates responses to tenascin-c and fibronectin in sensory neurons (Muller et al. 1995; Varnum-Finney et al. 1995).

We have shown that the α4β1 heterodimer mediates adhesion and outgrowth on both fibronectin and VCAM-1 (L.H. Mullick, G.M. Cann and D.O. Clegg, unpubl.). However, only low levels of these substrates can be detected in the developing retina thus far, suggesting that α4β1 may recognize some other ligand there.

Genetic approaches in mice have been for the most part unhelpful in determining integrin functions in the retina. Because integrins play important roles on other cells, gene knockouts have often resulted in lethality before the initiation of retinal development. For example, mice lacking α4 die at embryonic day 10 due to lack of heart formation (Yang et al. 1995), and embryonic mice lacking β1 die at E4.5 due to inner cell mass failure (Stevens et al. 1995). The one exception to this experience has been mice lacking the α6 integrin subunit (Georges-Labouesse et al. 1998), which show a defective retinal phenotype. These mice die at birth with severe skin blisters, and ectopic neuroblast-like outgrowths were observed in the brain and retina. The ectopic cell clusters in the retina were found in the nerve fiber layer and in the vitreous. The latter cells had apparently migrated out of the retina through the basal lamina of the inner limiting membrane. Laminin persists longer in the mutant brains, suggesting that altered laminin deposition on radial glia may contribute to the failure to arrest neuroblast migration. The RGC axonal organization, which might be expected to be perturbed, has not yet been examined.

Interestingly, antisense inhibition of integrin α6 expression had almost the opposite effect in the chick tectum (Zhang and Galileo 1998). Retrovirally marked clones expressing antisense α6 showed a decrease in migration to upper laminae. The difference in results may be due to the timing or nature of the perturbation.

3
Integrins in the Mature Retina

Integrin expression persists in the adult retina, as well as other parts of the eye (Brem et al. 1994; Elner and Elner 1996). In the RPE, strong immunoreactivity to the αv subunit was reported on the apical surface, and it was hypothesized that integrins may be important in retinal attachment to the RPE (Anderson et al. 1995). Because αv integrins had been implicated in phagocytosis by monocytes, we decided to test a different hypothesis that αv integrins were involved in RPE phagocytosis of shed photoreceptor outer segments.

First, we identified the integrin αvβ5 as the αv-containing integrin on the apical surface of human RPE. To test the function of this integrin in phagocytosis, a blocking monoclonal antibody and RGD peptides were tested for inhibition of phagocytosis of purified bovine rod outer segments by human RPE cells. Both agents inhibited binding and ingestion, with a greater effect seen on outer segment binding (Lin and Clegg 1998). Two additional groups have obtained similar results (Finnemann et al. 1997; Miceli et al. 1997), but a third could not replicate the RGD peptide inhibition using a different assay for phagocytosis (Hall et al. 1997). Studies in this area are difficult, since it is not clear that in vitro assays recapitulate the in vivo event. For example, serum is required for phagocytosis in vitro, but is not present in vivo. Furthermore, the different species used and assays employed have often yielded different experimental outcomes.

The partial inhibition achieved in these experiments suggests that other receptors are also involved in phagocytosis, and the αvβ5 integrin is one of at least four receptors that have been implicated. Other studies have identified a mannose receptor, the CD-36 scavenger receptor, and a 55-kDa glycoprotein related to prolylcarboxy peptidase as receptors in this process (Boyle et al. 1991; Ryeom et al. 1996; Hall et al. 1998). Such a multiplicity of receptors is indeed possible, considering the complex molecular events that must occur to ingest and degrade a large entity like an outer segment.

4
Future Directions

Integrins may also play roles in other aspects of neuroblast life, such as proliferation and determination of cell fate. Integrin receptors are known to transduce proliferative signals in non-neuronal cells, and may do the same in retinoblasts. For example, expression of β1 integrins in basal epidermal cells is downregulated as cells migrate, differentiate, and cease division. If integrins are turned on again in differentiated cells in transgenic mice, proliferation recommences, giving rise to a state that mimics psoriasis (Carroll et al. 1995; Hotchin et al. 1995). A proliferation signaling element has been localized to the cytoplasmic domain of the β1 subunit, which may stimulate entry into the S or G1 phase of the cell cycle (Zhu et al. 1996). CNS precursor cells require αvβ1 and α5β1 for proliferation in cultured neurospheres (Jacques et al. 1998). Furthermore, integrin mutants in *Drosophila* have fewer retinal rhabdomeres and a disrupted retinal morphology (Longley and Ready 1995). We assayed retinoblast proliferation in explanted eyecups by measuring incorporation of 3H-thymidine, and tested the effects of the blocking antibody CSAT. Surprisingly, CSAT did not decrease neuroblast proliferation as measured by DNA replication (data not shown). We speculate that the more abundant αvβ5 integrin may be able to fill in for the missing β1 integrins in this function. Further perturbations eliminating both β1 and β5 integrins should test this hypothesis.

Integrins may also function in retinal synapse formation. Recently, a *Drosophila* mutation called Volado (Chilean slang for absent-mindedness) was traced to an integrin, and integrin perturbing agents can inhibit long term potentiation (Jones 1996; Grotewiel et al. 1998). A clear synaptic localization has been observed for the αvβ8 heterodimer in rat brain (Nishimura et al. 1998), and the α8 subunit has been observed on dendritic spines and postsynaptic densities in the hippocampus (Einheber et al. 1996). Eight α and 2 β mRNAs have been detected throughout the brain (Pinkstaff et al. 1999). α4 and α8 subunit mRNAs increase in retina at the time of maximal synaptic elaboration. Further experiments will be necessary to test what function integrins carry out in this process.

References

Anderson DH, Johnson LV, Hageman GS (1995) Vitronectin receptor expression and distribution at the photoreceptor-retinal pigment epithelial interface. J Comp Neurol 360:1-16

Bossy B, Bossy-Wetzel E, Reichardt LF (1991) Characterization of the integrin α8 subunit: a new integrin β1-associated subunit which is prominently expressed on axons and on cells in contact with basal laminae in chick embryos. EMBO J 10:2375-2385

Boyle D, Tien L, Shepherd V, McLaughlin B (1991) A mannose receptor is involved in retinal phagocytosis. Invest Ophthalmol Visual Sci 32:1464-1470

Bradshaw AD, McNagny KM, Gervin DB, Cann GM, Graff T, Clegg DO (1995) Integrin α2β1 mediates interactions between developing embryonic retinal cells and collagen. Development 121:3593-3602

Brem RB, Robbins SG, Wilson DJ, O'Rourke LM, Mixon RN, Robertson JE, Planck SR, Rosenbaum JT (1994) Immunolocalization of integrins in the human retina. Invest Ophthalmol Visual Sci 35:3466-3474

Bronner-Fraser M (1985) Alterations in neural crest migration by a monoclonal antibody that affects cell adhesion. J Cell Biol 101:610-617

Bronner-Fraser M (1986) An antibody to a receptor for fibronectin and laminin perturbs cranial neural crest development in vivo. Dev Biol 117:528-536

Cann GM, Bradshaw AD, Gervin DB, Hunter AW, Clegg DO (1996) Widespread expression of β1 integrins in the developing chick retina: evidence for a role in migration of retinal ganglion cells. Dev Biol 180:82-96

Carroll JM, Romero MR, Watt FM (1995) Suprabasal integrin expression in the epidermis of transgenic mice results in developmental defects and a phenotype resembling psoriasis. Cell 83:957-68

Cohen J, Burne JF, Winter J, Burne J (1986) Retinal ganglion cells lose their response to laminin with maturation. Nature 322:465-467

Cohen J, Nurcombe V, Jeffrey P, Edgar D (1989) Developmental loss of functional laminin receptors on retinal ganglion cells is regulated by their target tissue, the optic tectum. Development 107:381-387

Danen EH, Lafrenie RM, Miyamoto S, Yamada KM (1998) Integrin signaling: cytoskeletal complexes, MAP kinase activation, and regulation of gene expression. Cell Adhes Com 6:217-224

de Curtis I, Reichardt LF (1993) Function and spatial distribution in developing chick retina of the laminin receptor α6β1 and its isoforms. Development 118:377-388 de Curtis I, Quaranta V, Tamura RN, Reichardt LF (1991) Laminin receptors in the retina: sequence analysis of the chick integrin α6 subunit. J Cell Biol 113:405–416

Duband J-L, Belkin AM, Syfrig J, Thiery JP, Koteliansky VE (1992) Expression of α1 integrin, a laminin-collagen receptor during myogenesis and neurogenesis in the avian embryo. Development 116:585–600

Duguid IGM, Boyd AW, Mandel TE (1992) Adhesion molecules are expressed in the human retina and choroid. Curr Eye Res Supp 11:153–159

Dutting D, Gierer A, Hansmann G (1983) Self-renewal of stem cells and differentiation of nerve cells in the developing chick retina. Dev Brain Res 10:21–32

Einheber S, Schnapp LM, Salzer JL, Cappiello ZB, Milner TA (1996) Regional and ultrastructural distribution of the alpha 8 integrin subunit in developing and adult rat brain suggests a role in synaptic function. J Comp Neurol 370:105–134

Elner SG, Elner VM (1996) The integrin superfamily and the eye. Invest Ophthalmol Visual Sci 37:696–701

Fassler R, Meyer M (1995) Consequences of lack of β1 integrin gene expression in mice. Genes Dev 9:1896–1908

Fekete DM, Perez-Miguelsanz J, Ryder EF, Cepko CL (1994) Clonal analysis in the chicken retina reveals tangential dispersion of clonally related cells. Dev Biol 166:666–682

Finnemann SC, Bonilha VL, Marmorstein AD, Boulan ER (1997) Phagocytosis of rod outer segments by retinal pigment epithelial cells requires αvβ5 integrin for binding but not for internalization. Proc Natl Acad Sci USA 94:12932–12937

Galileo DS, Majors J, Horwitz AF, Sanes JR (1992) Retrovirally introduced antisense integrin RNA inhibits neuroblast migration in vivo. Neuron 9:1117–1131

Georges-Labouesse E, Mark M, Messaddeq N, Gansmuler A (1998) Esential role of α6 integrins in cortical and retinal lamination. Curr Biol 8:983–986

Gervin DB, Cann GM, Clegg DO (1996) Temporal and spatial regulation of integrin vitronectin receptor mRNAs in the embryonic chick retina. Invest Ophthalmol Visual Sci 37:1084–1096

Grotewiel MS, Beck CD, Wu KH, Zhu XR, Davis RL (1998) Integrin-mediated short-term memory in Drosophila. Nature 391:455–460

Hall DE, Neugebauer KM, Reichardt LF (1987) Embryonic neural retinal cell response to extracellular matrix proteins: developmental changes and effects of the cell substratum attachment antibody (CSAT). J Cell Biol 104:623–634

Hall MO, AbramsTA, Burgess BL, Ershov AV (1997) Further studies on the phagocytosis of photoreceptor outer segments by rat retinal pigment epithelial cells. In: LaVail MM, Hollyfield JG, Anderson RE (eds) Degenerative retinal diseases. Plenum Press, New York, pp 385–397

Hall MO, Burgess, BL, Abrams TA (1998) Molecular studies of a candidate receptor for outer segment phagocytosis. Exp Eye Res 67:S163

Hemler ME (1998) Integrin associated proteins. Curr Op in Cell Biol 10:578–585

Hotchin NA, Gandarillas A, Watt FM (1995) Regulation of cell surface beta 1 integrin levels during keratinocyte terminal differentiation. J Cell Biol 128:1209–1219

Hynes RO (1992) Integrins: versatility, modulation, and signaling in cell adhesion. Cell 69:11–25

Hynes RO (1996) Targeted mutations in cell adhesion genes: what have we learned from them? Dev Biol 180:402–412

Hynes RO, Lander AD (1992) Contact and adhesive specificities in the associations, migrations, and targeting of cells and axons. Cell 68:303–322

Jacques TS, Relvas JB, Nishimura S, Pytela R, Edwards GM, Streuli CH, Ffrench-Constant C (1998) Neural precursor cell chain migration and division are regulated through different beta1 integrins. Development 125:3167–3177

Jaffredo T, Horwitz AF, Buck CA, Rong PM, Dieterlen-Lievre F (1988) Myoblast migration specifically inhibited in the chick embryo by grafted CSAT hybridoma cells secreting an anti-integrin antibody. Development 103:431–446

Jones LS (1996) Integrins: possible functions in the adult CNS. Trends Neurosci 19:68–72

Kil SH, Krull CE, Cann GM, Clegg DO Bronner-Fraser M (1998) The integrin α4 subunit is important for neural crest cell migration. Dev Biol 202:29–42

Leonard J, Sakaguchi DS (1995) Effects of anti-β1 integrin antibodies on the histogenesis and differentiation of the *Xenopus* retina in vivo. Soc Neurosci Abstr 611.2

Lilienbaum A, Reszka AA, Horwitz AF, Holt CE (1995) Chimeric integrins expressed in retinal ganglion cells impair process outgrowth in vivo. Mol Cell Neurosci 6:139–152

Lin H, Clegg DO (1998) Integrin αvβ5 participates in the phagocytosis of photoreceptor rod outer segments by cultured human retinal pigment epithelium. Invest Ophthalmol Visual Sci 39:1703–1712

Longley RL, Ready DF (1995) Integrins and the development of three-dimensional structure in the *Drosophila* compound eye. Dev Biol 171:415–433

Miceli MV, Newsome DA, Tate DJ (1997) Vitronectin is responsible for serum-stimulated uptake of rod outer segments by cultured retinal pigment epithelial cells. Invest Ophthalmol Visual Sci 38:1588–1597

Milner R, Edwards G, Streuli C, Ffrench-Constant C (1996) A role in migration for the alpha v beta 1 integrin expressed on oligodendrocyte precursors. J Neurosci 16:7240–7252

Morris VB, Cowan R (1984) A growth curve of cell numbers in the neural retina of embryonic chicks. Cell Tissue Kinet 17:199–208

Muller U, Bossy B, Venstrom K, Reichardt LF (1995) Integrin α8β1 promotes attachment, cell spreading, and neurite outgrowth on fibronectin. Mol Biol Cell 6:433–448

Neugebauer KM, Reichardt LF (1991) Cell surface regulation of β1-integrin activity on developing retinal neurons. Nature 350:68–71

Neugebauer, KM, Emmet CJ, Venstrom KA, Reichardt LF (1991) Vitronectin and thrombospondin promote retinal neurite outgrowth: developmental regulation and role of integrins. Neuron 6:345–358

Nishimura SL, Boylen KP, Einheber S, Milner TA, Ramos DM, Pytela R (1998) Synaptic and glial localization fo the integrin alpha v beta 8 in mouse and rat brain. Brain Res 791:271–282

Pinkstaff JK, Detterich J, Lynch G, Gall C (1999) Integrin subunit expression is regionally differentiated in adult brain. J Neurosci 19:1541–1556

Prada C, Puelles L, Genis-Galvez JM (1981) A Golgi study on the early sequence of differentiation of ganglion cells in the chick embryo retina. Anat Embryol 161:305–317

Reese BE, Harvey AR, Tan SS (1995) Radial and tangential dispersion patterns in the mouse retina are cell-class specific. Proc Natl Acad Sci USA 92:2494–2498

Reichardt LF, Tomaselli KJ (1991) Extracellular matrix molecules and their receptors: functions in neural development. Annu Rev Neurosci 14:531–570

Ryeom SW, Sparrow JR, Silverstein RL (1996) CD36 participates in the phagocytosis of rod outer segments by retinal pigment epithelium. J Cell Sci 109:387–395

Schwartz MA, Schaller MD, Ginsberg MH (1995) Integrins: emerging paradigms of signal transduction. Annu Rev Cell Dev Biol 11:549–599

Sheppard AM, Onken, MD, Rosen GD, Noakes PG, Dean DC (1994) Expanding roles of α4 integrin and its ligands in development. Cell Adhes Comm 2:27–43

Skeith A, Dunlap L, Galileo DS, Linser P (1999) Inhibition of β1 integrin expression reduces clone size during early retinogenesis. Dev Brain Res 116:123–126

Sonnenberg A (1993) Integrins and their ligands. Curr Top Microbiol Immunol 184:7–35

Stevens LE, Sutherland AE, Klimanskaya IV, Andrieux A, Meneses J, Pedersen RA, Damsky CH (1995) Deletion of beta 1 integrins in mice results in inner cell mass failure and peri-implantation lethality. Genes Dev 9:1883–1895

Svennevik E, Linser PJ (1992) The inhibitory effects of integrin antibodies and the RGD tripeptide on early eye development. Invest Ophthalmol Visual Sci 35:1774–1784

Varnum-Finney B, Venstrom K, Muller U, Kypta R, Backus C, Chiquet M, Reichardt L (1995) The integrin receptor α8β1 mediates interactions of embryonic chick motor and sensory neurons with tenascin-C. Neuron 14:1213–1222

Yang JT, Rayburn H, Hynes RO (1995) Cell adhesion events mediated by α4 integrins are essential in placental and cardiac development. Development 121:549–560

Zhang Z, Galileo DS (1998) Retroviral transfer of antisense integrin alpha 6 or alpha 8 sequences results in laminar redistribution or clonal cell death in developing brain. J Neurosci 18 : 6928–6938

Zhu X, Ohtsubo M, Bohmer RM, Roberts JM, Assoian RK (1996) Adhesion-dependent cell cycle progression linked to the expression of cyclin D1, activation of cyclin E-cdk2, and phosphorylation of the retinoblastoma protein. J Cell Biol 133 : 391–403

Connecting the Eye with the Brain:
The Formation of the Retinotectal Pathway

Karl G. Johnson and William A. Harris[1]

1
Introduction

The retino-tectal system has served as a model for the growth and navigation of axons in the developing brain for over 50 years, when the first
anatomical studies of embryos revealed details of the ontogeny of this
pathway (Herrick 1941). Such observations led to various enthusiastic
speculations concerning axon growth, including ideas that growth cones
were towed to their targets, or directed there by electric fields. These hypotheses were succeeded by others, slightly less fantastic, which had retinal
axons following emerging cracks and channels in the brain that blazed the
trail for their long journey (reviewed in Jacobson, 1991). In the 1960s, Sperry
made what at first seemed like an equally wild hypothesis, that retinal axons
were led to their specific targets in the tectum by cytochemical tags (Sperry
1963). For more than 30 years, Sperry's hypothesis gained force as attempts
to disprove or replace it with less biochemical hypotheses failed. Though
success in the search for the molecular basis of retinal axon navigation, and
the vindication of Sperry, had to wait until the 1990s and the second coming
of age of molecular biology, the attraction of the retino-tectal system for
investigating these fundamental issues has never diminished. As a result,
this is one of the most carefully studied and well-characterized developing
axon pathways known.

The pathway retinal ganglion cell (RGC) axons must follow from the retina
to the tectum is not a simple one. It consists of a variety of different envi-
ronments through which an RGC growth cone must navigate (Holt 1989).
Axons originate on the vitreal surface of ganglion cells and grow toward the
optic disk. After reaching the optic disk, RGC axons dive through the deeper
layers of the retina and exit from the back of the eye in the optic nerve. These
axons grow along the optic nerve to the optic chiasm, where axons representing the right and left visual fields are segregated. After leaving the chiasm,
axons grow dorsally along the optic tract, where they diverge to several

---

[1] University of Cambridge, Department of Anatomy, Downing St. Cambridge, CB2 3DY, UK

Results and Problems in Cell Differentiation, Vol. 31
M. E. Fini (Ed.): Vertebrate Eye Development
© Springer-Verlag Berlin Heidelberg 2000

targets including the optic tectum, the dorsal lateral geniculate nucleus, the ventral lateral geniculate nucleus, the hypothalamus, the pretectal area, and the accessory optic nuclei. In the mid-diencephalon, RGC axons turn posteriorly, spread out, and separate into dorsal and ventral brachia, and then cross the tectal border and innervate the optic tectum, the target of 90 % of RGC axons in most vertebrates. Within the optic tectum there is a topographic representation of the retina such that neighboring RGCs project to neighboring tectal targets. In the following section, we will follow these retinal axons along their entire pathway (Fig. 1), from retina to tectum, in an attempt

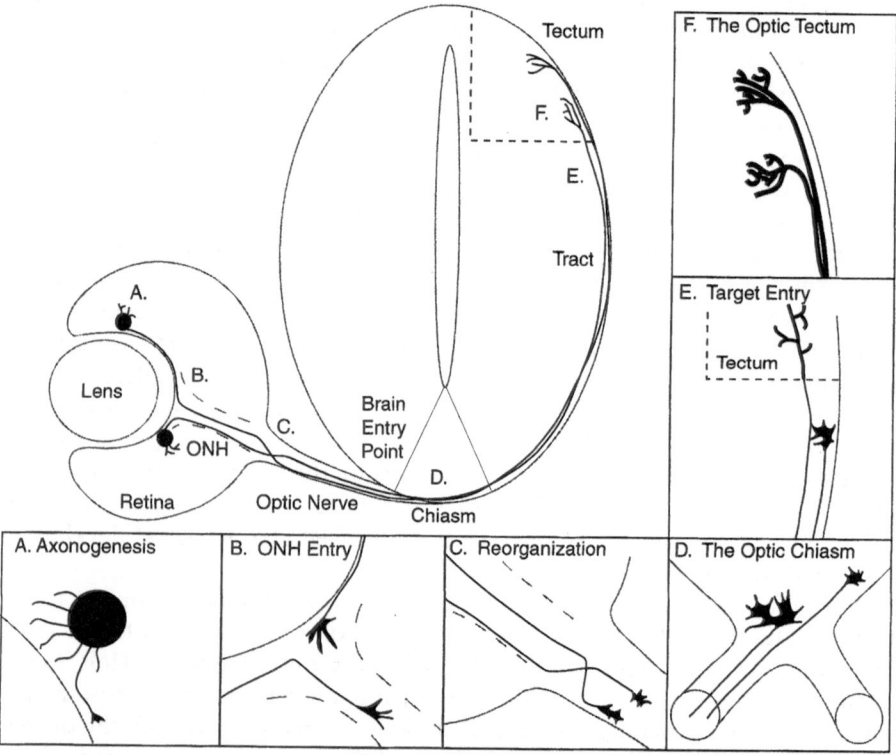

Fig. 1. Schematic diagram of the development of the retinotectal connection. *Boxed diagrams* depict growth cones at different positions along the optic pathway which are discussed in more detail in the text. Shortly after differentiation, RGCs send out transient processes, the most central of which becomes the axon *(A)*. These axons navigate along the vitreal surface of the retina to the ONH where the fibers from younger RGCs enter more centrally *(B)*. Fibers extend out from the back of the retina to the ZOR, where peripheral fibers move from their central position to more peripheral positions in the optic nerve *(C)*. As these axons reach the optic chiasm, they must decide whether to cross to the contralateral side of the brain, or to remain on the ipsilateral side *(D)*. As RGC axons climb the optic tract and reach the tectal border, there is a dramatic change in growth cone morphology *(E)*. Finally, RGC axons from branches and establish synaptic connections with tectal cells while maintaining a topographic map *(F)*

to understand anatomically, experimentally, and molecularly how they find their way in developing brain.

2
The Retinotectal Pathway

2.1
Navigation to the Optic Nerve Head

Retinal ganglion cells are the first retinal cells to be born. These cells exit the cell cycle and migrate to the vitreal surface of the retina. The first RGCs to differentiate are located centrally, in close proximity to the presumptive optic nerve head. Later RGCs differentiate in progressively peripheral waves, extending to the edge of the retina. One of the first tasks for a newly developed RGC is the initiation of process outgrowth. Prior to axonogenesis, RGCs extend multiple transient processes from their vitreal surfaces, exploring the optic fiber layer (Fig. 1A). These processes are not able to elongate because at this stage the retinal neuroepithelium is nonpermissive to RGC axon outgrowth (Halfter 1996). RGC axon initiation is dependent on normal N-cadherin and β1 integrin function, as process outgrowth can be blocked by expressing dominant negative forms of these constructs in retinal ganglion cells (Lilienbaum et al. 1995; Riehl et al. 1996). The nonpermissive substrate eventually becomes a permissive substrate when a chondroitin sulfate (CS)-containing matrix, which inhibits axon growth, dissipates from the optic fiber layer in a central to peripheral gradient (Fig. 2A). As this CS matrix dissipates, the transient processes directed away from the higher concentrations of CS and towards the optic nerve head are maintained and becomes RGC axons (Brittis et al. 1995). Thus, axonogenesis appears to require cell autonomous both N-cadherin and β1 integrin function and occurs on the side of the RGC nearest the center of the retina.

These newly formed RGC axons extend away from the cell body and grow along the glial endfeet in the optic fiber layer, adjacent to the pial surface and in direct contact with the basal lamina of the inner limiting membrane (Easter et al. 1984). In vitro experiments have shown that the basal lamina is an excellent substrate for RGC axon outgrowth, and many cell adhesion molecules (such as L1, NCAM, N-cadherin, NgCAM, axonin-1 and laminin) are expressed in the optic fiber layer of the developing retina. Antibodies generated against the basal lamina severely reduce RGC axon outgrowth from retinal explants (Halfter 1989), while rotated retinal grafts have shown that the basal lamina has an inherent polarity that permits axon growth toward and away from optic disk, but not perpendicular to it (Halfter 1996). The molecular identification of these radial directional cues remains elusive.

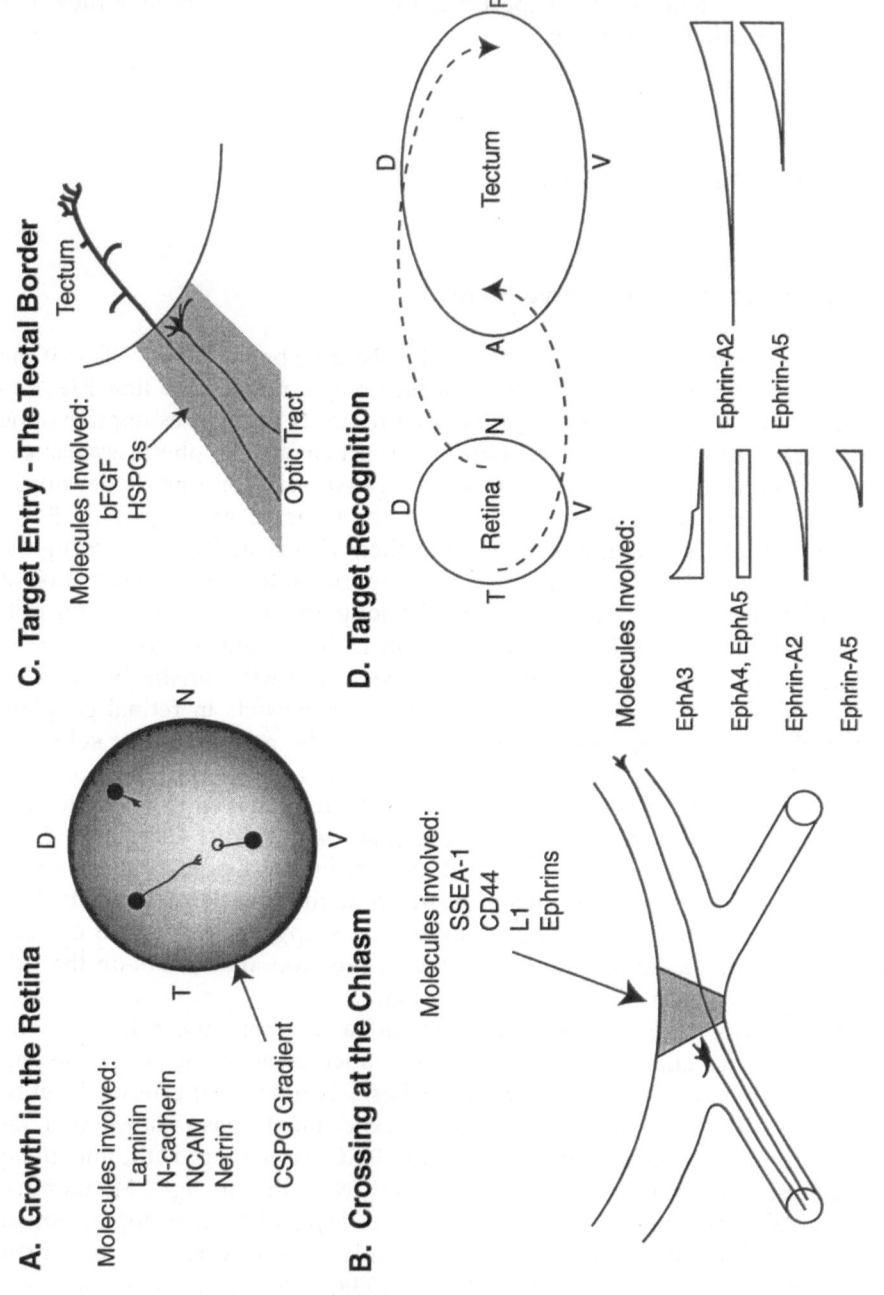

A. Growth in the Retina

Molecules involved:
Laminin
N-cadherin
NCAM
Netrin

CSPG Gradient

D
N
T
V

B. Crossing at the Chiasm

Molecules involved:
SSEA-1
CD44
L1
Ephrins

C. Target Entry - The Tectal Border

Tectum

Molecules Involved:
bFGF
HSPGs

Optic Tract

D. Target Recognition

Retina
Tectum
D
N
V
T
A
P
D
V

Molecules Involved:

EphA3
EphA4, EphA5
Ephrin-A2
Ephrin-A5

Ephrin-A2
Ephrin-A5

Three immunoglobulin superfamily (IgSF) molecules seem likely to be involved in intraretinal axon guidance and fasciculation. The first is L1, a cell adhesion molecule expressed on RGC axons. Exposure of retinae to L1 antibodies in vitro causes defects in RGC axon fasciculation but has no effect on guidance (Brittis et al. 1995). The second IgSF molecule thought to be involved in RGC axon guidance to the optic disk is neural cell adhesion molecule (N-CAM). While in vitro studies have failed to show an axon guidance defect (Brittis and Silver 1995; Halfter and Deiss 1986), in ovo studies have shown that direct injections of N-CAM antisera result in abnormal intraretinal RGC axon projections especially surrounding the optic disk (Thanos et al. 1984). Finally, neurolin, the goldfish homologue of DM-GRASP, has been shown to be involved in directing RGC axons toward the optic disk. Injections of polyclonal antibodies to neurolin cause misrouting of RGC axons in the retina, and a decrease in fasciculation (Ott et al. 1998). Further studies have shown that it is the second immunoglobulin domain that is involved in proper guidance to the optic disk, while the first and third Ig domains mediate RGC fasciculation (Leppert et al. 1999). Growth-promoting cues present within the retina, such as N-cadherin and L1, are thought to mediate their outgrowth-promoting effect via activation of the fibroblast growth factor receptor (FGFR) (Doherty and Walsh 1996). Function blocking FGFR antibodies are able to inhibit the orderly establishment of axon fascicles in the retinal periphery and can cause the defasciculation of axons around the optic nerve head (ONH) (Brittis et al. 1996).

As RGCs differentiate at peripheral positions in the retina and send their axons toward the optic nerve head, they fasciculate with more centrally positioned RGC axons which have already begun to elongate. In many lower vertebrates, the young axons tend to grow superficially in these fascicles, while the older axons occupy deeper positions and appear to be ensheathed in glial processes (Easter et al. 1984). In ferrets, young and old fibers are not segregated in the optic fiber layer, but instead intermingle while growing

---

Fig. 2A–D. Some of the molecules involved in guiding RGC axons at different points in the optic pathway. Axons may initially be directed to the ONH by an increasing central to peripheral gradient of CSPG. Within the retina, RGCs grow along a rich substrate containing laminin, N-cadherin and NCAM, before turning into the netrin-rich optic nerve head (**A**). As these RGC axons reach the chiasm, the encounter a set of neurons expressing SSEA-1, CD44, and L1. Ipsilaterally projecting fibers never traverse this set of neurons, but, instead, make a sharp turn to enter the ipsilateral optic tract. Growth cones are highly complex near the chiasmatic midline (**B**). When RGC axons reach the tectal border, they must change their growth from a substrate rich in bFGF and HSPGs, to a substrate nearly devoid of these. Concurrent with crossing into the tectum is a change in growth cone morphology and the initiation of back-branching (**C**). RGC axons establish a topographic map in the tectum such that more temporal axons project to the anterior tectum, while more nasal axons project to the posterior tectum. The N-T coordinate system is established in part by the graded expression of Eph receptors and ephrins in the retina and tectum (**D**). *D* Dorsal; *N* nasal; *V* ventral; *T* temporal; *A* anterior or rostral; *P* posterior or caudal

toward the ONH (Fitzgibbon and Reese 1992). This species-specific difference in RGC axon organization within the retina predicates further differences in axon organization throughout the retinotectal projection (see below).

2.2
Exiting the Eye at the Optic Nerve Head

Upon reaching the optic nerve head, RGC axons exhibit a dramatic change in behavior and morphology (Fig. 1B). Axons leave the glial endfeet at the vitreal surface of the retina, turn sharply into the ONH, and dive through the deeper layers of the retina to emerge in the optic nerve, which protrudes from the back of the eye. Upon reaching the optic nerve head, RGC growth cones exhibit highly complex structures with extended filopodia (Holt 1989) suggesting that they are examining the surrounding environment in preparation for a change in behavior. Since the ONH has no detectable long-range neurotrophic activity (Halfter 1996), one might wonder how axons converging on this structure know to change their direction of growth. What signals are involved in attracting growth cones into the ONH, or repelling them away from the vitreal surface at this particular location?

Netrin-1 is expressed by cells in the optic stalk prior to RGC axon invasion, while its receptor, DCC, is expressed by RGCs. The Netrin protein can cause chemoattractive turning of RGC axons in vitro, mediated by DCC. Knockout mice deficient in DCC or *Netrin-1* show a reduction in the number of RGC axons exiting the retina at the optic nerve head. Axon guidance defects in these mutant mice show that although RGC axons navigate properly to the optic disk, these axons often bypass the normal turning point and thereafter grow aberrantly in the retina (Deiner et al. 1997). This strongly suggests that Netrin is functioning as a short-range guidance cue at the optic disk, promoting RGC turning and extension through the optic nerve head (Figs. 1B, 2A).

Growth cones turn up concentration gradients of cAMP (Lohof et al. 1992). Recent in vitro work has shown that netrin can function as a chemorepulsive molecule to RGCs when levels of cAMP in the growth cone are low. cAMP is also reduced when RGC axons grow on laminin, and in this situation, netrin is repulsive rather than attractive (Hopker et al. 1999). In the retina, there are high levels of expression of laminin in the optic fiber layer, where RGC axons are growing towards the ONH, while at the ONH, laminin is limited to the vitreal surface. Since laminin expression is limited to the vitreal side of the ONH, while netrin is expressed throughout, it is thought that a gradient of cAMP is established across the growth cone, causing a turn toward the side of the growth cone with the highest levels of cAMP (Lohof et al. 1992). Retinal application of a soluable laminin-1 peptide causes axon misrouting at the ONH, supporting the idea that it is the lack of laminin and the presence of netrin in the ONH that causes proper RGC axon turning at the ONH (Hopker et al. 1999).

Recent studies have also shown that the chick lens functions as the source of a diffusible RGC axon growth repellent. This powerful chemorepulsive activity is easily demonstrated in coculture (Ohta et al. 1999) but the activity emanating from the lens has yet to be identified.

2.3
Fiber Organization in the Optic Nerve

RGC axons entering the optic nerve head are organized roughly according to their position of origin in the retina (Fig. 1C). Different organisms, however, have dramatically different mechanisms for organizing RGC axons along the optic nerve.

The organization of the optic nerve in cichlid fish is surprisingly simple. Cichlid fish have a ribbon-shaped optic nerve folded up within a cylindrical sheath. Retinal fibers are organized in this ribbon according to their radial and circumferential position of origin within the retina. At the optic nerve head, fibres are organized such that the oldest axons occupy the dorsal end of the optic nerve while the youngest enter the ventral surface of the nerve. This establishes a strict dorsal to ventral age-related gradient in the ribbon optic nerve. Fibers are also organized according to their circumferential position in the retina. Nasal axons enter the nasal half of the optic nerve, while temporal axons enter the temporal half. Ventral axons are situated on the lateral fringes of the optic nerve, while dorsal ones project to the medial nerve. Therefore, from the temporal edge of the optic nerve to the nasal edge, axons are arranged in the following order; ventrotemporal, temporal, dorsotemporal, dorsal, dorsonasal, nasal, ventronasal (Scholes 1979). This retinotopic order is maintained until these axons reach the position of the optic chiasm.

In *Xenopus*, RGC fibers entering the ONH are arranged in fascicles that correspond to the circumferential position of origin within the retina, while maintaining a limited age-related or radial organization (i.e., the "hour" position of the RGC determines the fasciculation partners of its axon, and the distance from the ONH plays a modest role) (Taylor 1987). Peripheral RGCs send their axons toward the optic nerve head along the vitreal surface of preexisting fascicles. As these fascicles dive down into the optic nerve head, there is a general trend for peripheral axons to occupy a more central position in the ONH while axons from the central retina are forced peripherally in the nerve head. The major organizational theme at the ONH in frogs, however, is the preservation of circumferential position information. Fibers from a given position in the retina enter the identical position in the ONH; i.e., ventral axons enter the ventral ONH, dorsal fibers enter the dorsal ONH. This pattern is maintained until the zone of reorganization (ZOR) just behind the retina (Taylor 1987).

As *Xenopus* RGC axons enter the ZOR (Fig. 1C), there is a major reorganization of retinal fibers according to their radial position of origin. Fibers emanating from cells in the central retina (the oldest fibers), which were growing along the peripheral edges of the optic nerve, shift to the central region of the optic nerve. Similarly, axons from peripheral RGCs lose their central position and grow along the periphery of the nerve. This radial fiber organization is maintained as these axons grow through the rest of the optic nerve, chiasm, and optic tracts (Taylor 1987).

In the ferret, which seems typical of mammals, the ONH is also organized with respect to the circumferential position of origin of the RGC axon. The axons arising in the four retinal quadrants travel toward the ONH in radial bundles and enter the ONH according to their respective quadrant. Unlike the *Xenopus* ONH, however, there is no radial organization of retinal fibers. Peripheral fibers travel toward the ONH throughout the depth of the optic fiber layer, and enter the optic nerve interspersed with more central fibers (Fitzgibbon and Reese 1996). This pattern may reflect the fact that in ferrets there is not the dramatic central to peripheral temporal pattern of differentiation that there is in fish or frogs. This suggests that in ferrets, the retinotopic order present in the ONH is simply a passive consequence of combining merging fascicles in the optic fiber layer. As these axons travel along optic nerve, there is a gradual loss of circumferential organization; as these axons approach the optic chiasm, there is no longer any detectable order of retinal axons (Fitzgibbon and Reese 1996).

2.4
Crossing at the Optic Chiasm

Upon reaching the optic chiasm, RGC axons must decide whether they will cross to the opposite side of the brain or stay on the ipsilateral side (Figs. 1D, 2B). While this segregation is relatively simple in lower vertebrates, where RGCs project almost exclusively to the contralateral tectum, higher vertebrates must segregate axons at the chiasm based on the position in the retina from which the axons originate.

In order to understand how axons navigate through the optic chiasm, we must first consider the structure of the chiasm, which lies at the surface of the ventral diencephalon. Developing astrocytes form the predominant cell type in this region, but their morphology varies, such that more stellate astrocytes occupy positions in the lateral region of the optic chiasm, while radial glia form a dense network immediately surrounding the midline where future RGC axon divergence will occur (Marcus et al. 1995). Differentiating hypothalamic neurons are also present within the optic chiasm prior to RGC invasion. These neurons are some of the earliest to differentiate in the CNS, and are arrayed in an inverted V shape (Mason and Sretavan 1997). These cells express CD44 (a cell adhesion receptor which inhibits RGC axon growth), L1 (a cell adhesion molecule which promotes it) and stage-specific embryonic antigen 1 (SSEA-1). Elimination of these cells by complement killing with antibodies to CD44 causes RGC axons to stop at the chiasm and fail to project to any CNS structures (Sretavan et al. 1995). Non-crossing axons never traverse this population of cells, but instead project to the ipsilateral side (Marcus et al. 1995; Fig. 2B).

As RGC axons approach the optic chiasm, they exhibit saltatory growth, characterized by periods of rapid elongation followed by periods of stasis (Godement et al. 1994). During these periods of stasis, RGC growth cones often assume a hypercomplex conformation, characterized by a Y-shaped or tri-partite growth cone unique to the chiasmatic midline (Mason and Wang 1997). These growth cones use extensive filopodial projections to examine the extracellular environment for directional cues. The occurrence of this exploratory behavior at the medial chiasm suggests that, as at the optic nerve head, RGC axons are encountering a new growth substrate or new guidance cues (Mason and Wang 1997; Fig. 1D).

In vitro studies have confirmed that a subset of ventrotemporal RGC axons, which do not normally cross at the chiasm, avoided chiasm explants when grown in coculture, while crossing RGC axons readily grow over such explants (Wang et al. 1995). This strongly suggests that ipsilaterally projecting RGC axons are more sensitive to an inhibitory signal at the chiasm than are the contralaterally projecting RGC axons. Perhaps differential sensitivity to these growth-promoting and growth-inhibiting cues at the chiasm is responsible for segregating crossing and non crossing fibers. Defects in crossing at the chiasm are observed in a variety of albino animals (Chan et al. 1993; Guillery 1971; Guillery et al. 1975, 1984, 1987, 1999; Guillery and Updyke 1976). The albino mutation is actually a mutation in the tyrosinase gene, an enzyme involved in melanin synthesis. In this mutation, there is a 50 % reduction in the number of axons in the ipsilateral projection. Cocultures of mutant and normal retinal and chiasmatic cells have suggested that tyrosinase's site of action is in the retina, as the albino chiasm appears indistinguishable from the normal chiasm in chimeric explants (Mason et al. 1996). Although albino animals have defects in the spatiotemporal pattern of cellular differentiation and lamination in the retina (Jeffery and Kinsella 1992; Webster and Rowe 1991), the correlation between this observation and the presence of a reduced ipsilateral projection in albino mutants is still unknown.

It is important to note that the position of RGCs within the retina determines the behavior of the growth cone at the chiasm. In organisms with predominantly overlapping visual fields, such as primates, axons from the nasal half of the retina cross at the chiasm to form the contralateral projection while axons from the temporal half of the retina project to the ipsilateral tectum. While a variety of molecules have been identified in the chiasm that may contribute to the separation of ipsilaterally projecting axons from contralaterally projecting axons, the molecular characterization of differences between the nasal and temporal axons has been elusive. An appealing possibility has been raised based on the fact that ligands for Eph receptor tyrosine kinases (RTKs) are expressed at the chiasmatic midline (Marcus et al. 1995). Nasally originating RGC axons express fewer receptors for the Eph ligands than do temporally originating axons, and nasal axons cross at the chiasm while temporal axons do not. Perhaps the decision to cross at the chiasm is mediated in part by the increased sensitivity of temporal axons to repulsive Eph ligands found in the optic chiasm.

Prior to metamorphosis in *Xenopus*, all RGC axons cross at the chiasm and project to contralateral structures in the brain. During metamorphosis, a small subset of RGCs concentrated in the ventronasal periphery begin to send their axons to ipsilateral structures. The development of this ipsilateral pathway occurs concurrently with the increase in levels of the hormone thyroxine during metamorphosis. In fact, unilateral retinal application of thyroxine can induce an eye-specific ipsilateral retinothalamic projection in premetamorphic tadpoles (Hoskins and Grobstein 1984), while blocking thyroid hormone production can prevent the formation of this ipsilateral projection (Hoskins 1986). This strongly suggests that thyroxine expression during metamorphosis induces a change in the retinal ganglion cells that affects RGC axon decisions at the chiasm. The molecular nature of this change is unknown.

The decision of what to do at the chiasm also appears to be mediated by an intracellular growth cone protein called GAP-43. GAP-43 is a growth associated protein that appears to play a major role in transducing intracellular and extracellular signals into reorganizations of the cytoskeleton (reviewed by Benowitz and Routtenberg, 1997). In GAP-43 deficient mice, RGC axons reach the chiasm but are not able to enter the appropriate optic tract. RGC axons wander extensively around the chiasm, often recrossing the midline, and eventually decide at random to enter the contralateral or the ipsilateral optic tract (Sretavan and Kruger 1998). This suggests that the cues involved in guiding growth cones into the appropriate optic tract appear to mediate their effects through GAP43.

Growth cones on axons originating in the temporal retina grow preferentially along other axons from the temporal retina, while nasal axons grow equally well on nasal or temporal axons (Bonhoeffer and Huf 1985). It has been hypothesized that growth cones on temporal axons, but not nasal axons, are sensitive to an inhibitory factor present on the surface of nasal axons. Supporting this hypothesis is the fact that temporal growth cones collapse when they come in contact with nasal axons (Raper and Grunewald 1990). However, when one eye is removed early in development, ipsilaterally projecting (temporal) RGC axons stall at the midline, suggesting that temporal axons actually need the nasal axons from the opposing eye to grow into the ipsilateral optic tract (Godement et al. 1990). It would appear that as nasal axons cross the chiasm, they lose a cue on their cell surface that repels temporal axons and instead become a permissive, if not supportive, substrate for temporal axon outgrowth. The exact nature of these interactions remains unknown.

The chiasm also represents a point of major RGC fiber reorganization in most species. At the chiasm of cichlid fish, the optic nerve subdivides along its dorso-ventral axis into cross-laminae containing a few hundred axons (each representing a particular retinal radius). The position of dorsally originating RGC axons changes from a stripe in the centre of the optic nerve, to a patch of axons at the lateral surface of a series of laminae. This somewhat complex rearrangement has the net effect of arranging the axons into a ventronasal, ventral, ventrotemporal, temporal, dorsotemporal, dorsal, dorsonasal order as they reach the marginal fiber brachia of the contralateral tectum (Scholes 1979). The optic chiasm also marks the boundary between two distinct astroglial territories. Along the optic nerve, RGC axons are in contact with astroglia expressing the fish equivalent of vimentin. As they cross the midline, they enter a domain of GFAP-expressing astroglia. This change in the astroglial environment coincides with a dramatic decrease in the rate of outgrowth, possibly allowing young axons to fasciculate with older ones originating from a similar circumferential retinal position (Maggs and Scholes 1986).

In *Xenopus*, as fibers approach the optic chiasm, young fibers originating from the peripheral retina migrate from their peripheral positions in the optic nerve to accumulate in a crescent in the ventral part of the nerve, while older fibers from the central retina accumulate on the dorsal edge of the nerve (Cima and Grant 1982). This ventral crescent, containing the youngest RGC axons, projects to the most superficial layers of the optic tract as the optic nerve bends around the optic chiasm. As these fibers pass through the chiasm, they reestablish their circumferential organization. Axons originating from dorsally positioned RGCs are directed into the posterior region of the optic tract while nasal axons enter the anterior optic tract (Fawcett et al. 1984).

At the ferret optic chiasm, there is a dramatic rearrangement of RGC axons segregating dorsal from ventral fibers. As axons enter the optic chiasm, ventral fibers cross to the contralateral tract in a more rostral position and dorsal fibres cross more caudally. While there is some overlap in these axons, this pattern has the net effect of segregating nasal fibers from dorsal fibers in the caudal chiasm. These axons are further segregated in the forming optic tract as the ventral axons occupy more lateral positions than dorsal axons (Reese and Baker 1993).

2.5
Climbing the Optic Tract toward the Tectum

Fluorescently labeled retinal axons were first visualized in the optic tract in vivo with time-lapse movies of axons growing in the brain (Harris et al. 1987). These studies showed that within the optic tract, axons grow at a consistently rapid rate and do not hesitate or wander off course (Harris et al. 1987). The first insight into axon guidance along the optic tract, however, came from transplant studies. In one set of experiments, optic vesicles were transplanted so that the optic nerve entered the brain in ectopic locations. RGC axons in such experiments were still able to navigate directly to the optic tectum (Harris 1982). Such a result suggests the presence of a diffusible attractant released from the tectum, but other experiments showed that the removal of the primordial tectum did not affect the ability of optic axons to navigate appropriately within the tract (Taylor 1991). This indicates that guidance information is also present locally within the tract, a suggestion that confirmed experiments in which the optic tract was removed, rotated 90°, and reimplanted into the developing diencephalon (Harris 1989). RGC axons entering the rotated optic tract deflected either clockwise or counterclockwise according to the direction of rotation of the tract. Upon leaving the rotated piece, the axons corrected their routes and found their way to the tectum.

In the normal optic tract, axons growth is supported by N-cadherin and β1-integrin function (Stone and Sakaguchi 1996). Retinal axons grow rapidly along the pial surface of the diencephalon, then turn posteriorly and grow toward the tectum positioned at the border between the midbrain and the diencephalon. Although RGC axons pass over the hypothalamus after crossing at the chiasm, RGC axons appear to actively avoid the hypothalamus and the epithalamus and prefer to grow over the thalamus en route to the tectum. Indeed, in vitro experiments have shown that both the hypothalamus and the epithalamus express neurocan and tenascin, which are known to be inhibitory to RGC axons (Tuttle et al. 1998). The application of cytochalasin B to a living optic tract results in the loss of RGC growth cone filopodia and the subsequent inability of RGC axons to make the posterior turn from the dorsal diencephalon toward the tectum. In such cases, RGC axons continue growing straight and extend over the epithalamus toward the dorsal midline of the diencephalon (Chien et al. 1993). Clearly, the posterior turn of axons in the optic tract requires either filopodial recognition of a substrate bound molecule or the ability of filopodia to effect a change in direction of growth in response to such a signal. Whether the growth cone is guided by a growth-promoting cue within the optic tract, or avoiding growth-inhibitory cues that flank the tract is not currently known.

The optic tract closely follows a preexisting longitudinal tract in the embryonic brain, known as the tract of the postoptic commissure (TPOC). This tract is well formed prior to RGC axon arrival, suggesting that it may be involved in guiding RGC axons from the chiasm to the tectum. Experimental evidence, however, suggests that this is not the case. Heterochronic transplants of eyes into very young embryos has shown that in *Xenopus*, RGCs are able to navigate properly from the chiasm to the tectum in the absence of any other existing CNS pathways including the TPOC. This suggests that RGC axons do not simply fasciculate onto the TPOC, but rather that they are sensitive to cues present in an axonless substrate significantly before RGC axons normally enter the brain (Cornel and Holt 1992).

The boundaries of homeobox gene expression have been shown to coincide with the tracts of early projecting neurons in the brain (Boncinelli 1994;

Figdor and Stern 1993; Wilson et al. 1993). The developing forebrain into which RGC axons grow is a prime example of patterning by homeobox genes. RGC axons grow along a pathway of cells expressing nkx2.2 from the chiasm to the turn in the optic tract. This pathway is flanked anteriorly and posteriorly by cells expressing dll-3. RGC axons leave this band of nkx2.2 as they turn away from the dorsal diencephalon which expresses pax-6 and grow into an emx-1/emx-2 expressing region. Pax6 mutant mouse embryos show striking axonal navigation errors in this part of the brain (Mastick et al. 1997). Axons progress through this emx-1/emx-2 region into the tectum (reviewed by Retaux and Harris, 1996).

Evidence for a guidance cue present along the developing optic tract came from studies examing the role of FGFRs in RGC axon guidance. RGC axons express high levels of FGFRs along their axons and in their growth cones, and basic FGF (bFGF), a ligand for the FGFRs which promotes RGC axon elongation in vitro, is expressed at high levels throughout the retinotectal pathway from the optic nerve head to the tectal border (McFarlane et al. 1995). This small diffusible molecule is held in place on the optic tract by heparan sulfate proteoglycans (HSPGs) (Walz et al. 1997) resulting in a high local concentration of bFGF along the optic tract. Expression of dominant negative FGFRs blocks bFGF stimulated axon outgrowth in vitro (McFarlane et al. 1996), suggesting that RGC axon growth along the optic tract is mediated, at least in part, by the affinity of RGC growth cones to the optic tract's HSPG-bound bFGF.

Just before retinal axons arrive in the tectum they start to separate into dorsal and ventral brachia, based on their topographic arrangement in the tract. Axons in the dorsal brachium arise from ganglion cells of the ventral retina, while the ventral brachium consists of axons from the dorsal retina. This segregation of axons is based on cues that appear to be distinct from the topographic guidance molecules in the tectum. Heterochronic transplants in *Xenopus* show preserved tract order yet disorder in the tectum (Chien et al. 1995). Similarly, fish mutants exist in which the tract topography is preserved while tectal topography is compromised and vice versa (Karlstrom et al. 1996; Trowe et al. 1996). Transplants in which the ventral half of a *Xenopus* eye is exchanged with a dorsal half (or a dorsal half is replaced with a ventral) have been used to create double dorsal and double ventral eyes. RGC axons emanating from such eyes project entirely through the ventral or dorsal brachium respectively (Fawcett and Gaze 1982). These data suggest that there is an active sorting mechanism that segregates dorsal and ventral fibers into the appropriate brachium.

2.6
Target Recognition

RGC axons follow the FGF-rich tract until they reach the tectal border (Figs. 1E, 2C). At the tectal border, axons must cross into a substrate that lacks bFGF (McFarlane et al. 1995). By blocking normal FGFR function in RGCs, either by expressing a dominant negative form of the receptor or by applying swamping levels of bFGF exogenously, RGC axons fail to innervate the optic tectum, and, instead, bypass the tectum on either the anterior or ventral side (McFarlane et al. 1996). This phenotype can also be caused by exogenous application of heparan sulfate, an essential cofactor required for FGF receptor signaling (Walz et al. 1997). RGC axons must sense this drop in bFGF in order to penetrate the anterior boundary of the tectum.

Time-lapse movies of retinal axons entering the tectum show that they slow down by about a factor of 4 as they cross the tectal border (Harris et al. 1987), consistent with the loss of FGF axon growth-promoting activity. The retinal growth cones also change their dynamic morphology dramatically from a standard cone shaped structure that projects filopodia forward to a complex spindly structure with thin back-branches growing and retracting off the main shaft. While progress in the tract is steady, movement in the tectum is saltatory. The axon tip appears to be encountering a partially repulsive or collapsing environment, which indeed appears to be the case (see below).

Once inside the tectum, RGC axons must also select the appropriate lamina of the optic tectum to form their terminals. The choice of the proper lamina is quite complex, as over 20 cadherin superfamily genes are expressed within the different laminae of the tectum during synapse formation (Miskevich et al. 1998); however, recent research has begun to address this complexity. In the chick, retinal axons enter in the stratum opticum, the most superficial layer of the tectum which expresses L1 and axonin-1 (Yamagata et al. 1995). Synaptic connections, however, are confined to 3 of 15 laminae, those that express N-cadherin, neuropilin, polysialyated N-CAM and a plant lectin-binding glycoconjugate (VVA) (Miskevich et al. 1998). This lamina-specific selectivity can be disrupted in the presence of function-blocking N-cadherin antibodies or VVA (Inoue and Sanes 1997). The other molecules involved in choosing the proper tectal lamina are unknown.

2.7
Finding the Proper Tectal Target – Topographic Mapping

Each RGC axon terminates in a position within the optic tectum that correlates precisely to the position of its ganglion cell body in the retina relative to its neighboring cells (Fig. 1F). In this way, axons originating in the most nasal retina project to the most posterior optic tectum, while axons originating more temporally terminate progressively more anteriorly in the optic tectum (Fig. 2D). In addition, axons from the dorsal retina innervate the ventral tectum and axons from the ventral retina innervate the medial tectum. Thus, the entire retinal image is topographically mapped, albeit inverted, onto the tectum.

Sperry detached and rotated frog's eyes by 180°, and analyzed the effects on their behavior. After regeneration, the frogs behaved as if their visual maps had been inverted; leaping backwards for lures presented ahead of them, and diving down for lures presented above them. This suggested to Sperry that a given location in the retina connected to a given location in the tectum by matching gradients of "cytochemical tags" (Sperry 1963). A stimulus presented above the frog will stimulate ventral RGCs. However, since the eye has been inverted, these ventral RGCs now bear the cyto-chemical tag of a dorsal RGC, and connect to the brain as a dorsal RGC would. Thus, the frog interprets a stimulus above as a stimulus below and dives in the wrong direction. In order for each RGC in the retina to have its own unique cytochemical tag, there must be at least two graded patterns of expression across its surface. A combination of a dorsal to ventral gradient with a nasal to temporal gradient would provide each RGC with a unique identity, thought to be necessary for maintaining neighbor relations in the tectum (Fig. 1F).

Within the retina, two members of the Eph receptor tyrosine kinases family are expressed. EphA3 is expressed in a high temporal to low nasal gradient, while EphA4 and EphA5 are expressed uniformly across the retina (Cheng et al. 1995; Connor et al. 1998). Also expressed in the retina are ligands for these Eph receptors, named ephrin-A2 and ephrin-A5. These ligands are expressed in an increasing temporal to nasal gradient, in the opposite direction of the gradient of EphA3. These A ephrins have been shown to phosphorylate and therefore inactivate the retinal Eph receptors (Hornberger et al. 1999). Thus, it is the graded expression of the EphA receptors and the ephrin A ligands that forms a high nasal to low temporal gradient of functional EphA receptors established in the retina.

Ephrin-A2 and ephrin-A5 are also found in the tectum in an increasing rostral to caudal gradient (Cheng et al. 1995; Monschau et al. 1997). These ephrins are repulsive to RGC axon outgrowth and are thought to play a role in establishing a topographic map along the nasal-temporal axis of the retina. As RGC axons enter from the rostral end of the tectum, temporal axons (expressing high levels of Eph receptors) are repelled by the low levels of ephrins in the rostral tectum. Axons expressing lower levels of Eph receptors penetrate further into the tectum, while the most nasal axons, which express the lowest levels of functional Eph receptors, penetrate into the most caudal tectum (Fig. 2D).

The molecules responsible for establishing the dorsal-ventral axis in the retina remain somewhat more elusive. Recently, EphB2 and EphB3 have been shown to be expressed in a high ventral to low dorsal gradient in the developing chick retina (Braisted et al. 1997). Ephrin-B1, a ligand for these receptors, is expressed in a high dorsal to low ventral gradient in the tectum. Since ventral axons form synaptic connections with dorsal tectal cells, it appears that axons expressing high levels of receptor form synapses where there are high levels of ligand, just the opposite of the pattern for nasal and temporal axons. This would suggest that ephrin-B1 can function as an attractive cue, but experimental evidence supporting this theory remains to be generated.

The first topographic maps that retinal axons make in the tectum have various degrees of precision, depending on the species examined, with fish and frogs showing a great deal of initial precision and birds and mammals showing respectively less (reviewed in Roskies et al. 1995). The refinement of the retinotectal map is beyond the scope of this chapter. We shall only say here that there is an activity-independent stage, during which rough corrections are made, followed by an activity-dependent stage, in which normal visual experience helps fine tune the map (reviewed in Holt and Harris, 1993).

2.8
Preventing Retinal Axon Escape

The most posterior edge of the tectum is situated next to the midbrain-hindbrain boundary organizer (MHB organizer). Cells in the MHB organizer can induce midbrain differentiation in the forebrain resulting in abnormal RGC innervation and the induction of engrailed expression via upregulation of FGF8. These abnormalities can also be induced by inserting beads soaked in of FGF8 (or FGF4) protein (Crossley et al. 1996; Shamim et al. 1999) which is normally expressed in the MHB organizer. The zebrafish mutant *acerebellar* is a loss-of-function mutation of FGF8 (Reifers et al. 1998) that lacks the normal MHB organizer. In such fish, there is distinct lack of anterior-posterior patterning in the tectum, resulting in the loss of ephrin-A5b expression and the loss of the graded expression of ephrin-A2 and ephrin-A5a (Picker et al. 1999). This mutant also shows defects in anterior-posterior and dorso-ventral retinotectal mapping, a smaller tectum, and the presence of axons that grow past the tectum, exiting from the ventral most aspect (Picker et al. 1999). Ephrin-A5 knockout mice show the same overshooting phenotype, with retinal axons extending into the inferior colliculus (Frisen et al. 1998). These results suggest that the MHB, through the action of ephrin-A5b, represents a strongly repulsive environment for RGC axons that prevents overshooting of the tectum.

3
Conclusion

The development of the retinotectal connection is possibly the best understood case of axonal navigation, both anatomically and molecularly, from origin to final target. Along this pathway, there are several places where axons change their behavior in response to local cues: the initiation of an axon; the turning of axons toward the optic nerve head; the entry of axons into the optic nerve head; the fiber reorganizations along the optic nerve; the pathway choice at the chiasm; fiber reorganizations at the chiasm; target recognition at the entry point into the tectum; and the identification within the tectum of the appropriate target. Within the past 20 years, molecules have been identified at nearly all of these choice points that help guide axons along the appropriate route.

There are still many unanswered questions with regard to how RGC axons reach their appropriate synaptic targets. What molecules are responsible for organizing and reorganizing optic fibers at the optic nerve head, the optic nerve, the optic chiasm, and the tectal brachia? Why are so many reorganizations necessary? What molecules are responsible for segregating ispilaterally projecting axons from contralaterally projecting axons at the optic chiasm? What transcription factors are involved in establishing the pathway that becomes the optic tract, and what are the local positional cues that these transcription factors regulate? What establishes the dorsal-ventral gradient in the optic tectum? How is this gradient established in the retina? The screen for zebrafish retinotectal mutants conducted in the early 1990s provided a diverse population of mutants that will undoubtedly provide answers to some of these questions (Baier et al. 1996; Karlstrom et al. 1996; Trowe et al. 1996). As these mutants are characterized genetically, a more detailed picture will emerge, giving us a better understanding of how, exactly, the growth cone of a ganglion cell unerringly traverses the vast and diverse pathway from its origin in the retina to its target in the tectum.

References

Baier H, Klostermann S, Trowe T, Karlstrom RO, Nusslein-Volhard C, Bonhoeffer F (1996) Genetic dissection of the retinotectal projection. Development 123:415–425

Benowitz LI, Routtenberg A (1997) GAP-43: an intrinsic determinant of neuronal development and plasticity. Trends Neurosci 20:84–91

Boncinelli E (1994) Early CNS development: distal-less related genes and forebrain development. Curr Opin Neurobiol 4:29–36

Bonhoeffer F, Huf J (1985) Position-dependent properties of retinal axons and their growth cones. Nature 315:409–410

Braisted JE, McLaughlin T, Wang HU, Friedman GC, Anderson DJ, O'Leary DD (1997) Graded and lamina-specific distributions of ligands of EphB receptor tyrosine kinases in the developing retinotectal system. Dev Biol 191:14–28

Brittis PA, Lemmon V, Rutishauser U, Silver J (1995) Unique changes of ganglion cell growth cone behavior following cell adhesion molecule perturbations: a time-lapse study of the living retina. Mol Cell Neurosci 6:433–449

Brittis PA, Silver J (1995) Multiple factors govern intraretinal axon guidance: a time-lapse study. Mol Cell Neurosci 6:413–432

Brittis PA, Silver J, Walsh FS, Doherty P (1996) Fibroblast growth factor receptor function is required for the orderly projection of ganglion cell axons in the developing mammalian retina. Mol Cell Neurosci 8:120–128

Chan SO, Baker GE, Guillery RW (1993) Differential action of the albino mutation on two components of the rat's uncrossed retinofugal pathway. J Comp Neurol 336:362-377

Cheng HJ, Nakamoto M, Bergemann AD, Flanagan JG (1995) Complementary gradients in expression and binding of ELF-1 and Mek4 in development of the topographic retinotectal projection map. Cell 82:371-381

Chien CB, Cornel EM, Holt CE (1995) Absence of topography in precociously innervated tecta. Development 121:2621-2631

Chien CB, Rosenthal DE, Harris WA, Holt CE (1993) Navigational errors made by growth cones without filopodia in the embryonic Xenopus brain. Neuron 11:237-251

Cima C, Grant P (1982) Development of the optic nerve in Xenopus laevis. I. Early development and organization. J Embryol Exp Morphol 72:225-249

Connor RJ, Menzel P, Pasquale EB (1998) Expression and tyrosine phosphorylation of Eph receptors suggest multiple mechanisms in patterning of the visual system. Dev Biol 193:21-35

Cornel E, Holt C (1992) Precocious pathfinding: retinal axons can navigate in an axonless brain. Neuron 9:1001-1011

Crossley PH, Martinez S, Martin GR (1996) Midbrain development induced by FGF8 in the chick embryo. Nature 380:66-68

Deiner MS, Kennedy TE, Fazeli A, Serafini T, Tessier-Lavigne M, Sretavan DW (1997) Netrin-1 and DCC mediate axon guidance locally at the optic disk: loss of function leads to optic nerve hypoplasia. Neuron 19:575-589

Doherty P, Walsh FS (1996) CAM-FGF receptor interactions: a model for axonal growth. Mol Cell Neurosci 8:99-111

Easter SS Jr, Bratton B, Scherer SS (1984) Growth-related order of the retinal fiber layer in goldfish. J Neurosci 4:2173-2190

Fawcett JW, Gaze RM (1982) The retinotectal fibre pathways from normal and compound eyes in Xenopus. J Embryol Exp Morphol 72:19-37

Fawcett JW, Taylor JS, Gaze RM, Grant P, Hirst E (1984) Fibre order in the normal Xenopus optic tract, near the chiasma. J Embryol Exp Morphol 83:1-14

Figdor MC, Stern CD (1993) Segmental organization of embryonic diencephalon. Nature 363:630-634

Fitzgibbon T, Reese BE (1992) Position of growth cones within the retinal nerve fibre layer of fetal ferrets. J Comp Neurol 323:153-166

Fitzgibbon T, Reese BE (1996) Organization of retinal ganglion cell axons in the optic fiber layer and nerve of fetal ferrets. Vis Neurosci 13:847-861

Frisen J, Yates PA, McLaughlin T, Friedman GC, O'Leary DD, Barbacid M (1998) Ephrin-A5 (AL-1/RAGS) is essential for proper retinal axon guidance and topographic mapping in the mammalian visual system. Neuron 20:235-243

Godement P, Salaun J, Mason CA (1990) Retinal axon pathfinding in the optic chiasm: divergence of crossed and uncrossed fibers. Neuron 5:173-186

Godement P, Wang LC, Mason CA (1994) Retinal axon divergence in the optic chiasm: dynamics of growth cone behavior at the midline [published erratum appears in J Neurosci 1995 Mar; 15(3 Pt 1): following table of contents]. J Neurosci 14:7024-7039

Guillery RW (1971) An abnormal retinogeniculate projection in the albino ferret (Mustela furo). Brain Res 33:482-485

Guillery RW, Hickey TL, Kaas JH, Felleman DJ, Debruyn EJ, Sparks DL (1984) Abnormal central visual pathways in the brain of an albino green monkey (Cercopithecus aethiops). J Comp Neurol 226:165-183

Guillery RW, Jeffery G, Cattanach BM (1987) Abnormally high variability in the uncrossed retinofugal pathway of mice with albino mosaicism. Development 101:857-867

Guillery RW, Jeffery G, Saunders N (1999) Visual abnormalities in albino wallabies: a brief note. J Comp Neurol 403:33-38

Guillery RW, Okoro AN, Witkop CJ, Jr (1975) Abnormal visual pathways in the brain of a human albino. Brain Res 96:373-377

Guillery RW, Updyke BV (1976) Retinofugal pathways in normal and albino axolotls. Brain Res 109:235-244

Halfter W (1989) Antisera to basal lamina and glial endfeet disturb the normal extension of axons on retina and pigment epithelium basal laminae. Development 107:281-297

Halfter W (1996) Intraretinal grafting reveals growth requirements and guidance cues for optic axons in the developing avian retina. Dev Biol 177:160-177

Halfter W, Deiss S (1986) Axonal pathfinding in organ-cultured embryonic avian retinae. Dev Biol 114:296-310

Harris WA (1982) The transplantation of eyes to genetically eyeless salamanders: visual projections and somatosensory interactions. J Neurosci 2:339-353

Harris WA (1989) Local positional cues in the neuroepithelium guide retinal axons in embryonic Xenopus brain. Nature 339:218-221

Harris WA, Holt CE, Bonhoeffer F (1987) Retinal axons with and without their somata, growing to and arborizing in the tectum of Xenopus embryos: a time-lapse video study of single fibres in vivo. Development 101:123-133

Herrick CJ (1941) Development of the optic nerves of Amblystoma. J Comp Neurol 74:473-534

Holt CE (1989) A single-cell analysis of early retinal ganglion cell differentiation in Xenopus: from soma to axon tip. J Neurosci 9:3123-3145

Holt CE, Harris WA (1993) Position, guidance, and mapping in the developing visual system. J Neurobiol 24:1400-1422

Hopker VH, Shewan D, Tessier-Lavigne M, Poo M, Holt C (1999) Growth-cone attraction to netrin-1 is converted to repulsion by laminin-1. Nature 401:69-73

Hornberger MR, Dutting D, Ciossek T, Yamada T, Handwerker C, Lang S, Weth F, Huf J, Wessel R, Logan C, Tanaka H, Drescher U (1999) Modulation of EphA receptor function by co-expressed ephrinA ligands on retinal ganglion cell axons. Neuron 22:731-742

Hoskins SG (1986) Control of the development of the ipsilateral retinothalamic projection in Xenopus laevis by thyroxine: results and speculation. J Neurobiol 17:203-229

Hoskins SG, Grobstein P (1984) Induction of the ipsilateral retinothalamic projection in Xenopus laevis by thyroxine. Nature 307:730-733

Inoue A, Sanes JR (1997) Lamina-specific connectivity in the brain: regulation by N-cadherin, neurotrophins, and glycoconjugates. Science 276:1428-1431

Jacobson M (1991) Developmental Neurobiology, pp. 776. New York: Plenum

Jeffery G, Kinsella B (1992) Translaminar deficits in the retinae of albinos. J Comp Neurol 326:637-644

Karlstrom RO, Trowe T, Klostermann S, Baier H, Brand M, Crawford AD, Grunewald B, Haffter P, Hoffmann H, Meyer SU, Muller BK, Richter S, van Eeden FJ, Nusslein-Volhard C, Bonhoeffer F (1996) Zebrafish mutations affecting retinotectal axon pathfinding. Development 123:427-438

Leppert CA, Diekmann H, Paul C, Laessing U, Marx M, Bastmeyer M, Stuermer CA (1999) Neurolin Ig domain 2 participates in retinal axon guidance and Ig domains 1 and 3 in fasciculation. J Cell Biol 144:339-349

Lilienbaum A, Reszka AA, Horwitz AF, Holt CE (1995) Chimeric integrins expressed in retinal ganglion cells impair process outgrowth in vivo. Mol Cell Neurosci 6:139-152

Lohof AM, Quillan M, Dan Y, Poo MM (1992) Asymmetric modulation of cytosolic cAMP activity induces growth cone turning. J Neurosci 12:1253-1261

Maggs A, Scholes J (1986) Glial domains and nerve fiber patterns in the fish retinotectal pathway. J Neurosci 6:424-438

Marcus RC, Blazeski R, Godement P, Mason CA (1995) Retinal axon divergence in the optic chiasm: uncrossed axons diverge from crossed axons within a midline glial specialization. J Neurosci 15:3716-3729

Mason CA, Marcus RC, Wang LC (1996) Retinal axon divergence in the optic chiasm: growth cone behaviors and signalling cells. Prog Brain Res 108:95-107

Mason CA, Sretavan DW (1997) Glia, neurons, and axon pathfinding during optic chiasm development. Curr Opin Neurobiol 7:647-653

Mason CA, Wang LC (1997) Growth cone form is behavior-specific and, consequently, position-specific along the retinal axon pathway. J Neurosci 17:1086-1100

Mastick GS, Davis NM, Andrew GL, Easter SS, Jr. (1997) Pax-6 functions in boundary formation and axon guidance in the embryonic mouse forebrain. Development 124:1985-1997

McFarlane S, Cornel E, Amaya E, Holt CE (1996) Inhibition of FGF receptor activity in retinal ganglion cell axons causes errors in target recognition. Neuron 17:245-254

McFarlane S, McNeill L, Holt CE (1995) FGF signaling and target recognition in the developing Xenopus visual system. Neuron 15:1017-1028

Miskevich F, Zhu Y, Ranscht B, Sanes JR (1998) Expression of multiple cadherins and catenins in the chick optic tectum. Mol Cell Neurosci 12:240-255

Monschau B, Kremoser C, Ohta K, Tanaka H, Kaneko T, Yamada T, Handwerker C, Hornberger MR, Loschinger J, Pasquale EB, Siever DA, Verderame MF, Muller BK, Bonhoeffer F, Drescher U (1997) Shared and distinct functions of RAGS and ELF-1 in guiding retinal axons. Embo J 16:1258-1267

Ohta K, Tannahill D, Yoshida K, Johnson AR, Cook GM, Keynes RJ (1999) Embryonic lens repels retinal ganglion cell axons. Dev Biol 211:124-132

Ott H, Bastmeyer M, Stuermer CA (1998) Neurolin, the goldfish homolog of DM-GRASP, is involved in retinal axon pathfinding to the optic disk. J Neurosci 18:3363-3372

Picker A, Brennan C, Reifers F, Clarke JD, Holder N, Brand M (1999) Requirement for the zebrafish mid-hindbrain boundary in midbrain polarisation, mapping and confinement of the retinotectal projection. Development 126:2967-2978

Raper JA, Grunewald EB (1990) Temporal retinal growth cones collapse on contact with nasal retinal axons. Exp Neurol 109:70-74

Reese BE, Baker GE (1993) The re-establishment of the representation of the dorso-ventral retinal axis in the chiasmatic region of the ferret. Vis Neurosci 10:957-968

Reifers F, Bohli H, Walsh EC, Crossley PH, Stainier DY, Brand M (1998) Fgf8 is mutated in zebrafish acerebellar (ace) mutants and is required for maintenance of midbrain-hindbrain boundary development and somitogenesis. Development 125:2381-2395

Retaux S, Harris WA (1996) Engrailed and retinotectal topography. Trends Neurosci 19:542-546

Riehl R, Johnson K, Bradley R, Grunwald GB, Cornel E, Lilienbaum A, Holt CE (1996) Cadherin function is required for axon outgrowth in retinal ganglion cells in vivo. Neuron 17:837-848

Roskies A, Friedman GC, O'Leary DD (1995) Mechanisms and molecules controlling the development of retinal maps. Perspect Dev Neurobiol 3:63-75

Scholes JH (1979) Nerve fibre topography in the retinal projection to the tectum. Nature 278:620-624

Shamim H, Mahmood R, Logan C, Doherty P, Lumsden A, Mason I (1999) Sequential roles for Fgf4, En1 and Fgf8 in specification and regionalisation of the midbrain. Development 126:945-959

Sperry RW (1963) Chemoaffinity in the orderly growth of nerve fiber patterns and connections. Proc Nat Acad Sci 50:703-710

Sretavan DW, Kruger K (1998) Randomized retinal ganglion cell axon routing at the optic chiasm of GAP-43-deficient mice: association with midline recrossing and lack of normal ipsilateral axon turning. J Neurosci 18:10502-10513

Sretavan DW, Pure E, Siegel MW, Reichardt LF (1995) Disruption of retinal axon ingrowth by ablation of embryonic mouse optic chiasm neurons. Science 269:98-101

Stone KE, Sakaguchi DS (1996) Perturbation of the developing Xenopus retinotectal projection following injections of antibodies against beta1 integrin receptors and N-cadherin. Dev Biol 180:297-310

Taylor JS (1987) Fibre organization and reorganization in the retinotectal projection of Xenopus. Development 99:393-410

Taylor JS (1991) The early development of the frog retinotectal projection. Development Suppl 95-104

Thanos S, Bonhoeffer F, Rutishauser U (1984) Fiber-fiber interaction and tectal cues influence the development of the chicken retinotectal projection. Proc Natl Acad Sci USA 81:1906-1910

Trowe T, Klostermann S, Baier H, Granato M, Crawford AD, Grunewald B, Hoffmann H, Karlstrom RO, Meyer SU, Muller B, Richter S, Nusslein-Volhard C, Bonhoeffer F (1996) Mutations disrupting the ordering and topographic mapping of axons in the retinotectal projection of the zebrafish, Danio rerio. Development 123:439-450

Tuttle R, Braisted JE, Richards LJ, O'Leary DD (1998) Retinal axon guidance by region-specific cues in diencephalon. Development 125:791–801

Walz A, McFarlane S, Brickman YG, Nurcombe V, Bartlett PF, Holt CE (1997) Essential role of heparan sulfates in axon navigation and targeting in the developing visual system. Development 124:2421–2430

Wang LC, Dani J, Godement P, Marcus RC, Mason CA (1995) Crossed and uncrossed retinal axons respond differently to cells of the optic chiasm midline in vitro. Neuron 15:1349–1364

Webster MJ, Rowe MH (1991) Disruption of developmental timing in the albino rat retina. J Comp Neurol 307:460–474

Wilson SW, Placzek M, Furley AJ (1993) Border disputes: do boundaries play a role in growth-cone guidance? Trends Neurosci 16:316–323

Yamagata M, Herman JP, Sanes JR (1995) Lamina-specific expression of adhesion molecules in developing chick optic tectum. J Neurosci 15:4556–4571

Regeneration of the Lens in Amphibians

Panagiotis A. Tsonis[1]

1
Phylogeny

Some urodele amphibians are the only animals that throughout their life are capable of regenerating their lens following lentectomy (Stone 1967). Other vertebrates, such as freshwater fish, chicken, and frogs can occasionally regenerate a lens, but such an ability is very limited with a short time window during their embryonic development. Chicken can regenerate the lens only during 2–4 days of development. Interestingly, regeneration occurs from the ventral iris (van Deth 1940). In *Xenopus*, regeneration of the lens occurs from the inner layer of the outer cornea only before metamorphosis (Filoni et al. 1997). In mammals, the only report dealing with lens regeneration is of an adult rabbit after removal of the lens and only after implantation of cytolyzing fetal tissue (Steward and Espinasse 1959). Among urodeles the ability is not universal. The axolotl, for example, a salamander with very good regenerative abilities of the limb and tail, is not able to regenerate the lens. Such restrictions pose interesting questions as to why this selection exists.

2
Histological and Cellular Events

A section through a newt eye is presented in Fig. 1 where all the basic tissues are present. In Fig. 1b an electron microscope photograph of the iris is presented. Since the dorsal iris is the place from where lens regeneration can occur, this will familiarize the reader with some important cellular structures. An illustration of the pigment epithelium architecture in the dorsal iris is also presented in Fig. 1c with terminology related to the cells capable of transdifferentiation. Once the lens is removed, the dorsal iris pigment epithelium dedifferentiates (Eguchi 1963; Dumont and Yamada 1977; Yamada 1977). The cells start proliferating after 3 days and become depigmented. This event is followed by

---

[1] Laboratory of Molecular Biology, Department of Biology, University of Dayton, Dayton, Ohio 45469-2320, USA

Results and Problems in Cell Differentiation, Vol. 31
M. E. Fini (Ed.): Vertebrate Eye Development
© Springer-Verlag Berlin Heidelberg 2000

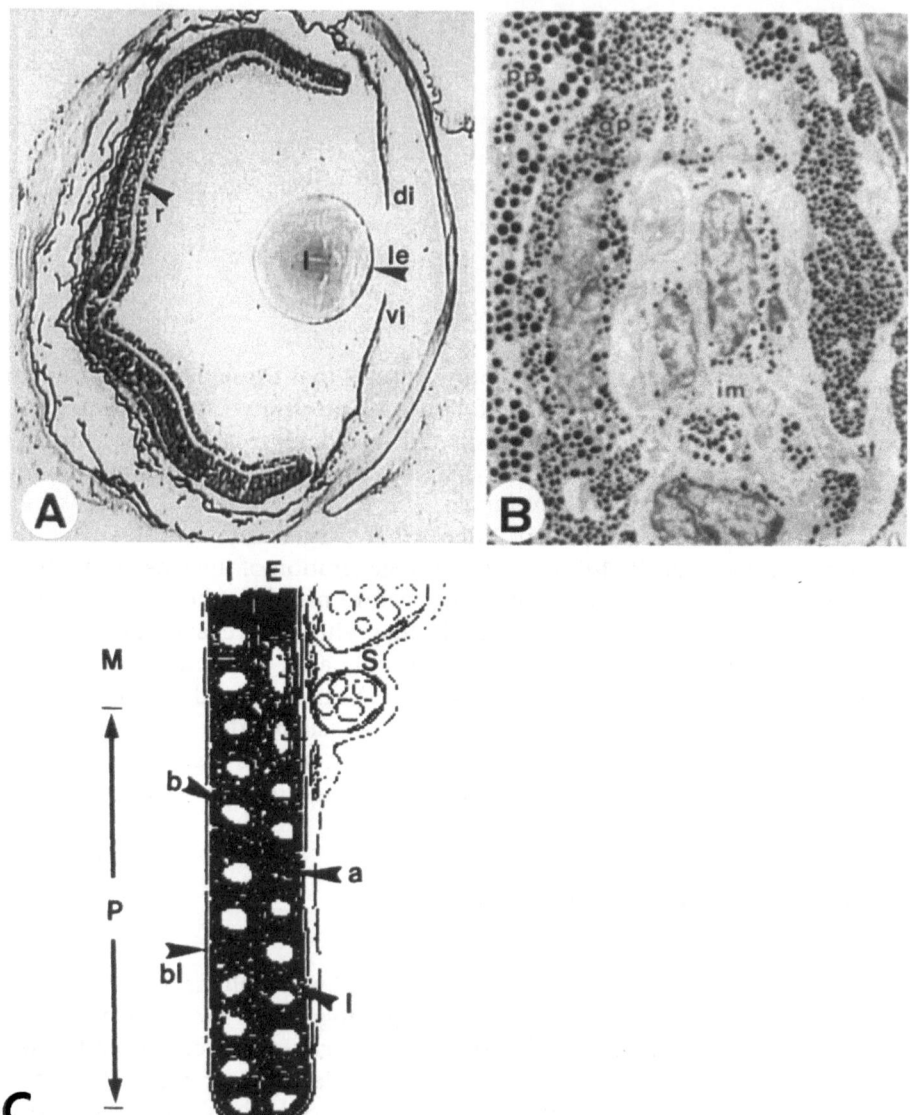

Fig. 1. A A section through an adult newt eye. *r* retina; *l* lens; *le* lens epithelium; *di* dorsal iris; *vi* ventral iris. B An electron micrograph of the dorsal iris. *pp* posterior pigment epithelium; *ap* anterior pigment epithelium; *st* stroma; *im* iris muscle (From Okamoto 1997, with permission). C An illustration of a medial sagittal section of the pupillary region of the dorsal iris showing the important cell population and arrangement. *P* Pupillary region; *M* middle ring; *E* external layer of iris pigment epithelium (anterior). I internal layer of iris epithelium (posterior). *B, l, a* basal, lateral and apical compartments of the two layers respectively. *bl* basal lamina. (After Yang and Zalik 1994)

increase in tyrosinase activity by the depigmented cells (Achazi and Yamada 1972) and by structural changes in the DNA marked by nicks, gaps, and single-stranded material (Collins 1974a, b). In this respect, it is important to note that amplification of ribosomal RNA sequences has been reported in the dorsal iris undergoing dedifferentiation. In fact, these cells contain 60% more rRNA cistrons (Collins 1972), indicating active transcription. Six days later, trans-differentiation of the iris into lens cells begins. The regenerating lens starts as a budding process from the dorsal iris cells. Formation of the lens vesicle by the depigmented progenies of the iris cells is evident between 9 and15 days. Between 12 and 15 days after lentectomy, the internal layer of the lens vesicle thickens and synthesis of β- and γ-crystallins start. Following that period, lens fibers are produced in the internal layer of the vesicle and β-crystallin appears in the external layer. By day 20, the fiber complex grows further and accumulation of α-crystallin is seen in the lens fibers and in the external layer. By day 25, the lens has been completely formed and dividing cells are observed only in the lens epithelium. These events of cellular transdifferentiation are shown in Fig. 2. The regenerated lens cells express readily the specific crystallins, while such expression is not seen in the dedifferentiated pigment epithelial cells (PECs). An important phenomenon that seems to help dedifferentiation is the attraction at the site (dorsal iris) of macrophages, which ingest pigment granules from the pigment epithelial cells of the dorsal iris. Macrophage activity is histologically obvious during the above processes up to 15 days postlentectomy. It has been suggested in the past that the cells of the dorsal iris that are involved in lens regeneration are, in fact, myoepithelial cells (Yang and Zalik 1994). This has been based on experiments in which these cells have been shown to express muscle-specific actin. Also these cells, especially the ones in the pupillary ring have a spindle shape. In other studies it has been suggested that the iris muscle (Fig. 1b) disappears after lentectomy, but appears again some months later perhaps with the contribution of the iris pigment epithelial cells (Okamoto 1997). The ventral iris does not contribute to this phenomenon and all these histological and cellular events are virtually absent.

Work with the developing lens in frogs and in the newt *Notophthalmus viridescens* has shown that synthesis of crystallins parallels the steps seen during lens regeneration. In this sense, similar events might take place during development and regeneration of the lens, even though inductive interactions such as those seen during lens development (surface ectoderm with optic vesicle) are not necessary for lens regeneration (McDevitt et al. 1969; McDevitt and Brahma 1981). In a similar sequence to the regenerating lens, β-crystallin is the first to appear even before the lengthening of the primary fibers during lens development. γ-crystallin appears later in a slightly more advanced stage of the fiber differentiation and α-crystallin appears in a few cells in the area of beginning fiber differentiation. Some differences also exist; α-crystallin is present at the external layer of the regenerating lens, but absent in the developing lens. In regard to crystallin synthesis, therefore, it seems that the embryonic program is regulated very faithfully during transdifferentiation as well.

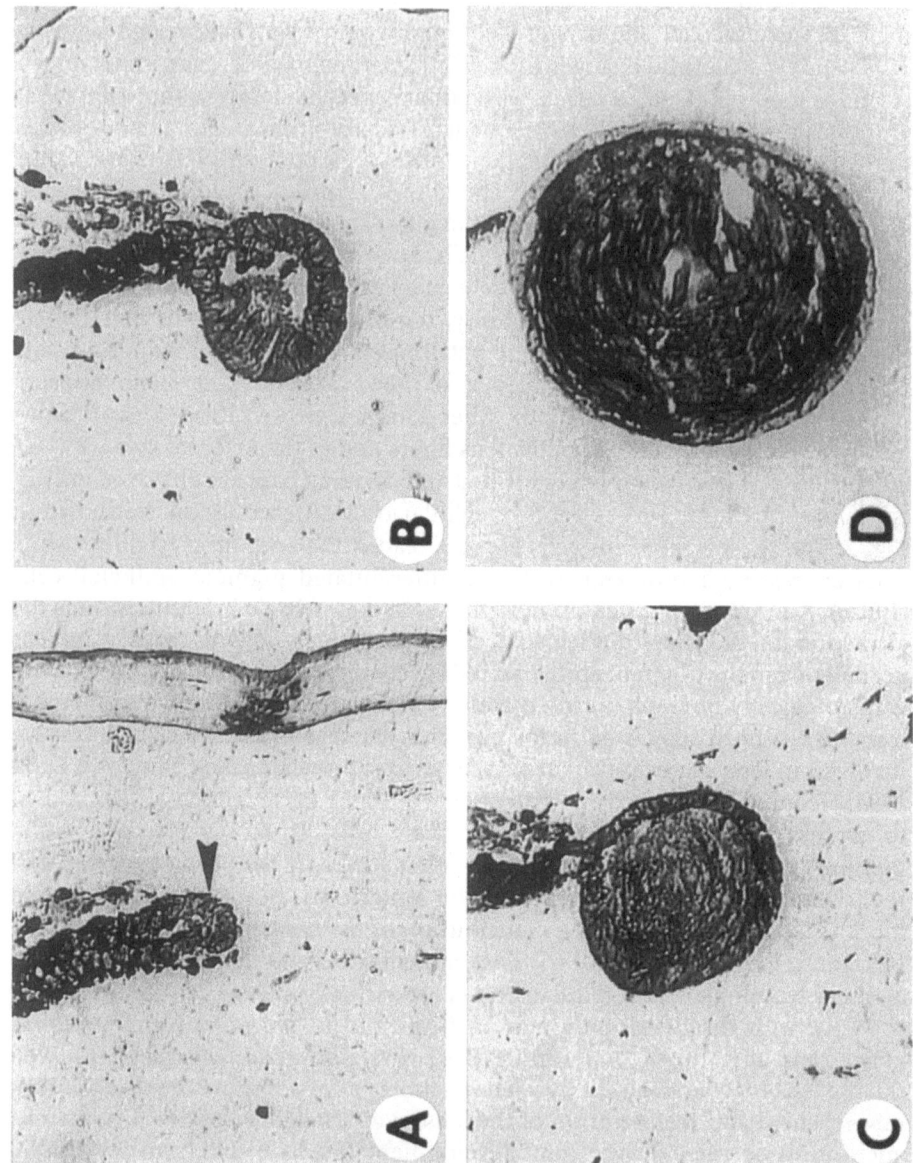

3
Molecular Biology and Gene Regulation

During the past few years information has accumulated indicating specific gene expression along the dorsal-ventral iris. Especially evident are differences in the synthesis of molecules of the extracellular matrix. Eguchi has described a monoclonal antibody that is directed against a cell surface antigen involved in cell adhesion and expressed in the iris PECs among other tissues (Imokawa and Eguchi 1992; Imokawa et al. 1992). This antigen disappears during the process of dedifferentiation after lentectomy. More striking, however, is the fact that ventral iris treated with the antibody in vitro and implanted into a newly lentectomized eye is able to produce a lens (Eguchi 1988). The immediate conclusion from such results is that disappearance of cell adhesion molecule(s) is necessary and sufficient for the dedifferentiation and the subsequent events of regeneration to occur. Several cell surface-associated proteoglycans, for example, seem to sequentially disappear from the dorsal iris after lentectomy (Zalik and Scott 1973). Disappearance of laminin and heparan sulfated proteoglycans has also been reported in studies employing immunofluorescence (Ortiz et al. 1992). This strengthens the association between cell surface alterations and regeneration. Such alterations could implicate molecules such as fibroblast growth factor (FGF). Proteoglycans are bound to FGF and their loss during regeneration might make FGF available to the cells. Indeed, it has been reported that basic FGF is one of the essential factors that enhances and regulates lens transdifferentiation of the pigmented epithelial cells in vitro (Hyuga et al. 1993).

3.1
Expression and Role of FGFs and FGFRs
in Lens Regeneration

Given the fact that FGFs have now been well connected with fiber differentiation during lens formation, these molecules and their receptors were primary candidates for regulation during regeneration. Work in our laboratory has provided interesting data concerning expression of FGF-1, FGFR-1,

◀ ───

Fig. 2A–D. Histological features in sections during lens regeneration. **A** Dorsal iris 10 days after lentectomy. The tip of the dorsal iris (*arrowhead*) has depigmented and dedifferentiated. This is the beginning of the formation of the lens vesicle. **B** Dorsal iris 15 days after lentectomy. The internal layer of the lens vesicle thickens and differentiation of the lens fibers starts. **C** Regenerating lens 20 days after lentectomy. Lens differentiation is almost complete. The lens epithelium is covering the lens at the anterior, while lens fiber differentiation is prominent in the posterior lens. **D** Regenerated lens 25 days after lentectomy showing a definite lens

FGFR-2, and FGFR-3 during lens regeneration using newt probes. FGF-1, FGFR-2 and FGFR-3 were found to be present in the retina and the lens epithelium of the intact eyes. This pattern of expression is consistent with the known presence of FGFs in these tissues and their expression in other animals. Upon lentectomy, expression in the retina was not downregulated but stayed rather at steady levels. However, when the lens vesicle began to differentiate from the dedifferentiated PECs the expression in the lens vesicle was higher than that of the retina, and it stayed high in the lens epithelium and the differentiating fibers (Del Rio-Tsonis et al. 1997). No apparent regulation was observed between the dorsal and the ventral iris, so we believe that while these factors could play a role in lens differentiation, they are not likely linked to the control of the dedifferentiation and regeneration process from the dorsal iris. Expression for FGFR-2 was also studied with an antibody and it was verified that both dorsal and ventral iris PECs expressed this receptor (Fig. 3C, F). However, when we examined expression of FGFR-1 the pattern was different. FGFR-1 product was exclusively found in the dedifferentiated PECs and subsequently in the differentiating lens (Fig. 4A, B). No expression was seen in the ventral iris (Fig. 3A, B, D, E; Del Rio-Tsonis et al. 1998). Obviously, among FGFRs, FGFR-1 is the only one with an expression pattern correlated with the process of dedifferentiation and transdifferentiation from the dorsal iris. This exciting result prompted us to examine the effects of FGFR-1 inhibition on lens regeneration. Recent advances in the field of kinase inhibitors have led to the development of specific FGFR inhibitors. In fact, FGFR-1 specific inhibitors are available. We have obtained such an inhibitor (SU 5402) from SUGEN. SU 5402 has been proven to be a potent FGFR-1 inhibitors with no effects for EGFR, PDGR, or insulin receptor (Mohammadi et al. 1997). Using this inhibitor we observed that regeneration from the dorsal iris was inhibited (Fig. 3G, H). The inhibitor was dissolved in the water where the animals were maintained at a concentration of 20 μM. The first treatment was applied 7 days after lentectomy and was continued every 3 days until day 20 after lentectomy. At day 20, untreated and treated eyes were collected and examined histologically. We examined 22 treated eyes and we found inhibition in 100 % of the cases (Del Rio-Tsonis et al. 1998). These data strongly suggest that FGFR-1 is involved and plays a major role in lens regeneration.

At the same time, the effects of exogenous FGF during lens regeneration were examined. While transgenic technology has not yet been fully developed in newts, application of ectopic FGF in the lentectomized eye could represent an excellent system to study such effects. FGF transgenic mice develop with abnormal lenses characterized by transformation of lens epithelium to lens fibers, abnormal polarity, and formation of cataracts (Robinson et al. 1995). FGFR-1 has also been found to play a major role as a lens fiber-determining factor (Chow et al. 1995). We, therefore, became very interested to examine expression and possible role of these molecules in lens regeneration. After lentectomy, we inserted in the eye ball heparin beads with FGF-1 (newt,

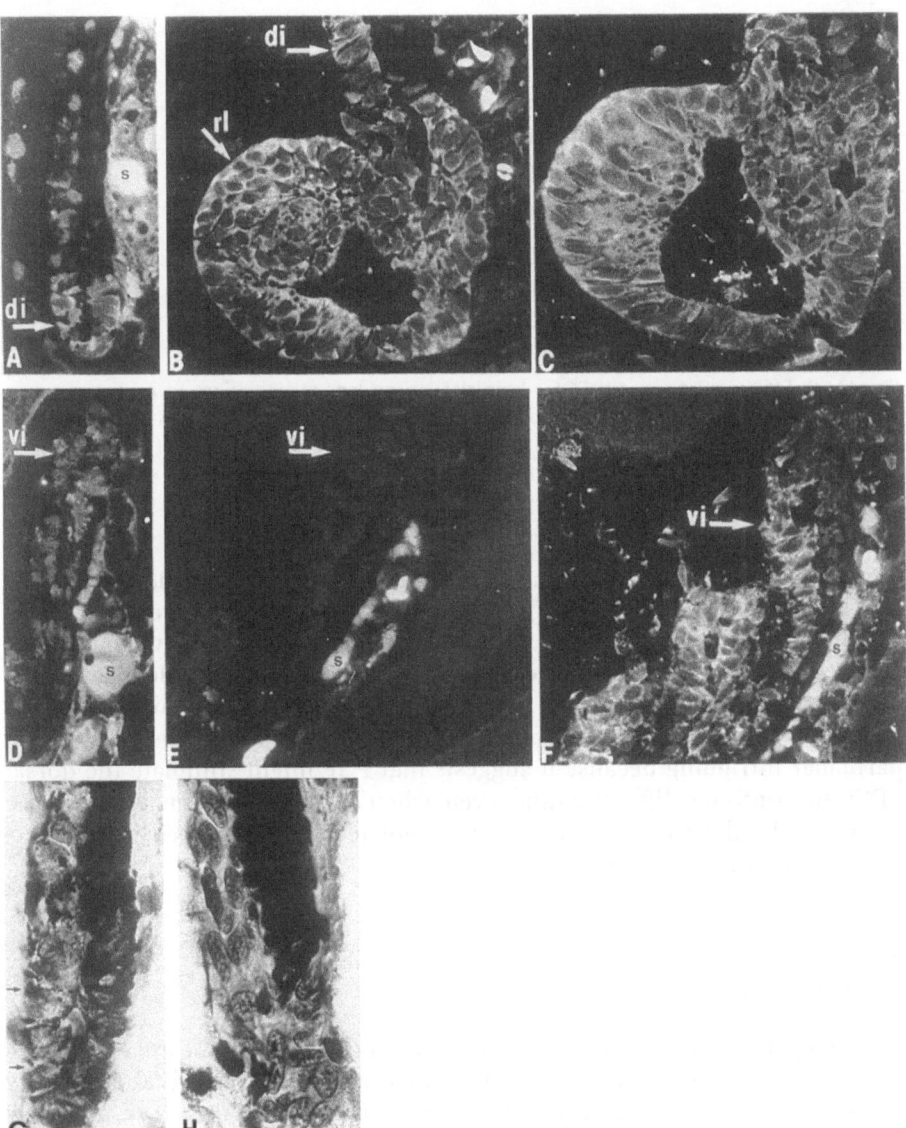

Fig. 3A–H. Expression of FGFR-1 and FGFR-2 during lens regeneration. **A, D** Expression of FGFR-1 in dorsal iris (*di*) and ventral iris (*vi*) respectively, 10 days after lentectomy. Note expression only in the dedifferentiating dorsal iris. **B, E** Expression of FGFR-1 in the dorsal (*di*) and ventral (*vi*) iris 15 days after lentectomy. Expression can be seen only in the regenerating lens (*rl*) from the dorsal iris. **C, F** Expression of FGFR-2 in the dorsal and ventral iris respectively, 15 days after lentectomy. Expression can be seen in both the regenerating lens from the dorsal iris and in the ventral iris. **G, H** Inhibition of transdifferentiation and lens differentiation after treatment of a lentectomized eye with an inhibitor of FGFR-1. In **G**, depigmentation has barely been initiated (*arrows*)

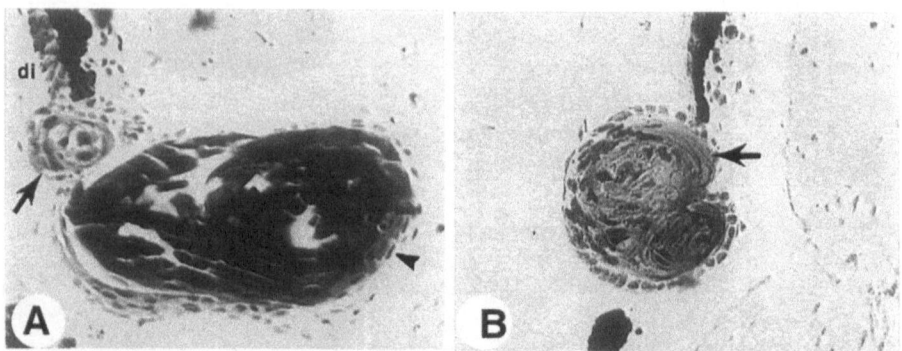

Fig. 4A, B. Effects of exogenous FGF administration on lens regeneration. In **A** we can see an abnormal lens protruding into the anterior chamber with elongated lens epithelial cells (arrowhead). A second lens (arrow) is also regenerating from the dorsal iris (*di*). **B** A case of tow fused lenses with abnormal polarity showing fiber formation in the dorsal/anterior area (*arrow*)

recombinant) or FGF-4 (human, recombinant). These beads have been proven to slowly release FGF. Twenty five days later we collected the eyes for examination. We observed several abnormalities consistent with the abnormalities in transgenic mice. These included transformation of lens epithelial cells to lens fibers, abnormal lens polarity, but the most unique was the formation of double lenses from the dorsal iris (Fig. 4). This result is in particular intriguing because it suggests that FGF might stimulate the dorsal PECs to continue differentiating even when a lens has been regenerated. Interestingly, the additional lens was vacuolated, a feature seen in cataracts (Del Rio-Tsonis et al. 1997).

3.2
Regulatory Factors

For molecules to disappear after lentectomy specifically in one area of the eye (dorsal iris), regulative processes should come into play. Several regulatory factors are known to be involved in eye development and, therefore, are likely to play a role in lens regeneration. Regulatory factors involved in transcription and confined to the ventral retina include the retinoic X receptor α (RXR) and pax-2. Knockout experiment involving RXR and RAR have shown that the ventral retina develops short (Kastner et al. 1994).

Members of the Hox family have been also implicated in eye determination and differentiation. Among these Hox genes two also contain a paired box and belong to the pax family. The pax-2 gene is expressed during development of the mouse eye and its expression is restricted to the ventral optic cap and stalk (Nornes et al. 1990). The *pax-6* gene is expressed in the developing mouse eye, including the lens (Walther and Gruss 1991). In fact, *pax-6* has been directly implicated in lens determination in chick, where it is first expressed in a region of the future head ectoderm close to the anterior margin of the early neural plate, an area where lens induction takes place. *Pax-6* has also been shown to induce ectopic lens formation when injected in *Xenopus* embryos (Altman et al. 1997). Furthermore, the mutation *aniridia* in humans and the equivalent *small eye* of the mouse are induced by a deletion in *pax-6* (Hill et al. 1991; Ton et al. 1991). Most interesting, however, is the fact that the *eyeless* gene (homologous to *pax-6*) is involved in the development of the eye in *Drosophila* despite the different morphology and mode of development (Quiring et al. 1994). *Pax-6* seems to be even involved in the formation of the eye spots in invertebrates (Loosli et al. 1996). *Pax-6* has been found to single handedly induce ectopic eyes in *Drosophila* (Halder et al. 1995). *Pax-6*, therefore, must be considered an eye master gene across species. Among other homeobox-containing genes, *six-3*, the mouse homolog of *Drosophila sine oculis* has also been found to induce ectopic lenses in zebrafish (Oliver et al. 1996). While *pax-6* and *six-3* have been found to have lens-inducing capabilities, the list of homeobox-containing genes being expressed in the developing eye has become longer over the years (Beebe 1994). Among them of particular interest are *msx-1*, which is expressed after formation of the optic cup marking the presumptive ciliary body, *msx-2*, which is expressed in regions corresponding to the future corneal epithelium and neural retina (Monaghan et al. 1991), and the crystallin enhancer binding protein (Funahashi et al. 1993).

3.2.1
Expression of Pax and Hox Genes

To test for these genes, we cloned the newt homolog of *pax-6* and we isolated via PCR several Hox genes that were expressed in the newt eye. The Hox genes were *Hox7* (or *Gbx-2*), *Hox A4*, *Hox B1*, and *Hox X* (named as such because there are no such sequence similarities in the genebank and it may represent a novel homeobox). Later, we added in our list *msx-1*, and the homolog of Xenopus *Xbr1* (Papalopoulou and Kintner 1996).

Expression was studied by in situ hybridization. To examine expression we collected sections from intact newt eye and from lentectomized eye, 1, 5, 10, 15, 20, and 25 days after lens removal. To this collection we added sections from intact axolotl eye as well from lentectomized axolotl eye (5 and 10 days) in order to compare expression in an animal that is unable to regenerate its lens. Such comparison could shed light on specific regulation due to the regenerative process.

Pax-6 is indeed expressed during lens regeneration and appears in the dedifferentiated dorsal PECs. The expression of *pax-6* was also examined in the axolotl, a urodele which is unable to regenerate its lens, and it was found that *pax-6* expression in the adult eye of the axolotl was lower when compared with the newt. This suggested that *pax-6* levels in the adult urodele eye could be important for regenerative abilities (Del Rio-Tsonis et al. 1995).

For the Hox genes we observed three different types of regulation. The first, which we will call it type A involved *Hox A4*, *Xbr1*, *msx-1* and *Hox X*. These genes were expressed in the retina and in the lens epithelium of the intact eye. Upon lentectomy, these genes were downregulated in the retina. This downregulation was most obvious in the photoreceptor and amacrine layers of the retina. Expression of these genes was gradually restored in the retina within a few days and returned to normal levels before the appearance of the lens vesicle (by day 10). We therefore concluded that expression of these genes might be affected because of the removal of the lens, and is unlikely to be correlated with the process of lens differentiation, because their expression levels reached the level of the intact eye before lens formation. However, the possibility exists that they are involved in the early events of the process.

The second type of regulation, which we will call it type B, involved *Hox B1* and *Hox 7* (*Gbx-2*). As in type A these two genes were expressed in the intact retina and lens epithelium. However, upon lentectomy, these genes were completely downregulated in the retina (all layers, including ganglion) and expression was not restored before the appearance of the new lens (Fig. 5). Expression was also prominent in the regenerating lens. Furthermore, in day-5 and day-10 lentectomized axolotl eye, expression was similar to the levels of the intact eye. Therefore, expression of these two genes in the retina was correlated with the differentiation of the lens and with the regenerating ability of the animals. We thus concluded that expression of *Hox 7* and *Hox B1* in the retina must be important for the lens regeneration process. To rule out the possibility that the absence of the lens and not the regeneration process is responsible for such regulation, we examined expression of these genes in eyes where the lens was removed and then displaced back in the eye. Expression of *Hox 7* and *Hox B1* was also downregulated completely in the retina of these eyes, further indicating that this regulation is connected to the process of regeneration and not to injury and/or absence of the lens (Jung et al. 1998).

A third interesting pattern was obtained for *prox 1*, which is the homologue for the *Drosophila prospero* (Tomarev et al. 1996). The product of this gene was preferentially present in the intact and regenerating dorsal iris. This expression pattern, in fact, associates *prox 1* with the lens regeneration competent cells of the dorsal iris (Del Rio-Tsonis et al. 1999).

For all the six homeobox genes that were involved in type A and type B regulation there was no apparent regulation along the dorsal-ventral axis in the iris. However, these expression patterns provided the first molecular evidence of alterations in regulation in retina because of the process of lens regeneration. Retina has been thought in the past to be involved in lens regeneration. Earlier experiments where retina was removed had shown that

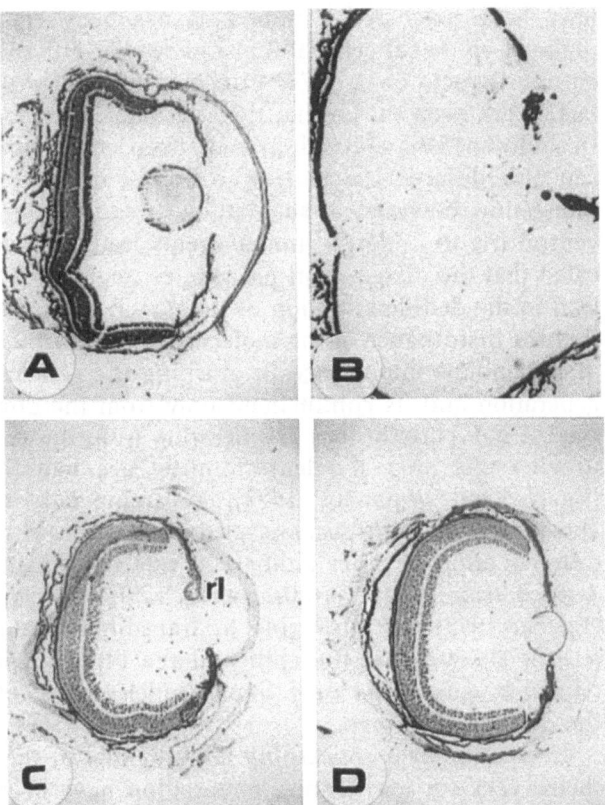

Fig. 5A–D. Regulation of Hox 7 during lens regeneration. A Expression in the adult intact eye is confined in the retina and lens. B Downregulation in the retina 10 days after lentectomy. C, D expression is slowly restored as lens is regenerated (*rl*) 15 and 25 days respectively

lens regeneration does not proceed unless retina restoration by regeneration has been accomplished (Stone 1958). This has led scientists in the past to speculate that retina might produce a factor that is important for the initiation of lens regeneration. Our results thus suggest that these Hox genes might regulate factors in the retina that are important for lens regeneration. These were unexpected results, but provide the first molecular evidence for retina involvement in lens regeneration.

4
In Vitro Systems for Lens Regeneration

The field of lens regeneration has benefited enormously from the ability of eye cells to grow efficiently in culture. During the past 20 years such cultures have been extensively studied. It has been demonstrated that newt iris pigment epithelial cells (PECs) can readily differentiate to lentoid bodies in culture (Eguchi et al. 1974; Yamada and MacDevitt 1974; Yamada 1977). In fact, it has been shown that this can happen in clonal cultures of these cells. In addition, lens epithelium cells from chick embryos (Okada et al. 1971) can also differentiate to lentoid bodies in vitro (Fig. 6). What was most interesting, however, in such studies was the ability of newt PECs from the ventral iris to undergo similar events leading to lens formation. This indicated that the dissociation procedures might provoke the initial signals that lead to the dedifferentiation of the dorsal PECs. This, in turn, could suggest that the disturbance of the molecules of the extracellular matrix could prove of paramount importance in grasping the cellular mechanisms of lens regeneration and its confinement only from the dorsal iris in vivo. The only case of induction of lens regeneration from the ventral iris after lentectomy in vivo was after the lentectomized eye had been treated with MNNG (Eguchi and Watanabe 1973), a carcinogenic substance. As mentioned above, the in vitro systems also extended the role of iris PECs in lens differentiation to other animals. It was subsequently found that these cells derived from other vertebrates, including humans (even of very old age) (Eguchi 1993), are also able to transdifferentiate in vitro to lens. These studies showed that the ability of the PECs to produce lens exists in the animal kingdom, but only some urodeles have retained this capacity from the dorsal iris in vivo.

Interesting data concerning the potential of the ventral iris pigment epithelial cells for lens transdifferentiation have been gained from transplantation studies. When PECs are isolated from dorsal and ventral iris and placed in culture after dissociation they can both undergo transdifferentiation. The potential, though, to form a lens after reaggregated cells are transplanted in the eye cavity after lentectomy depends on the duration of the culture. The ventral iris cells were not able to form any lens after being cultured for 6 days (Okamoto et al. 1998). Obviously, for these cells to form a lens they must be cultured for longer periods. In other experiments, the ability of these cells to form lens was tested after transplantation in a foreign environment, such as the regenerating limb. The results showed that when dissociated dorsal PECs were transplanted in the blastema they were able to depigment and transdifferentiate to form a complete lens with the lens epithelium! The ventral PECs were not able to do so (Fig. 7). However, when the same number of dorsal PECs (capable of giving rise to lens) were mixed with ventral PECs, their ability for lens formation declined (Ito et al. 1999). These spectacular results indicate two very interesting points. First, a com-

----

Fig. 6. A Transdifferentiation of chick pigment epithelial cells into lens cells in vitro. *lb* lentoid body. **B** Differentiation of chick lens epithelial cells into fiber cells in vitro. *lb* lentoid body. **C** Same as B, but stained with an antibody to δ-crystallin

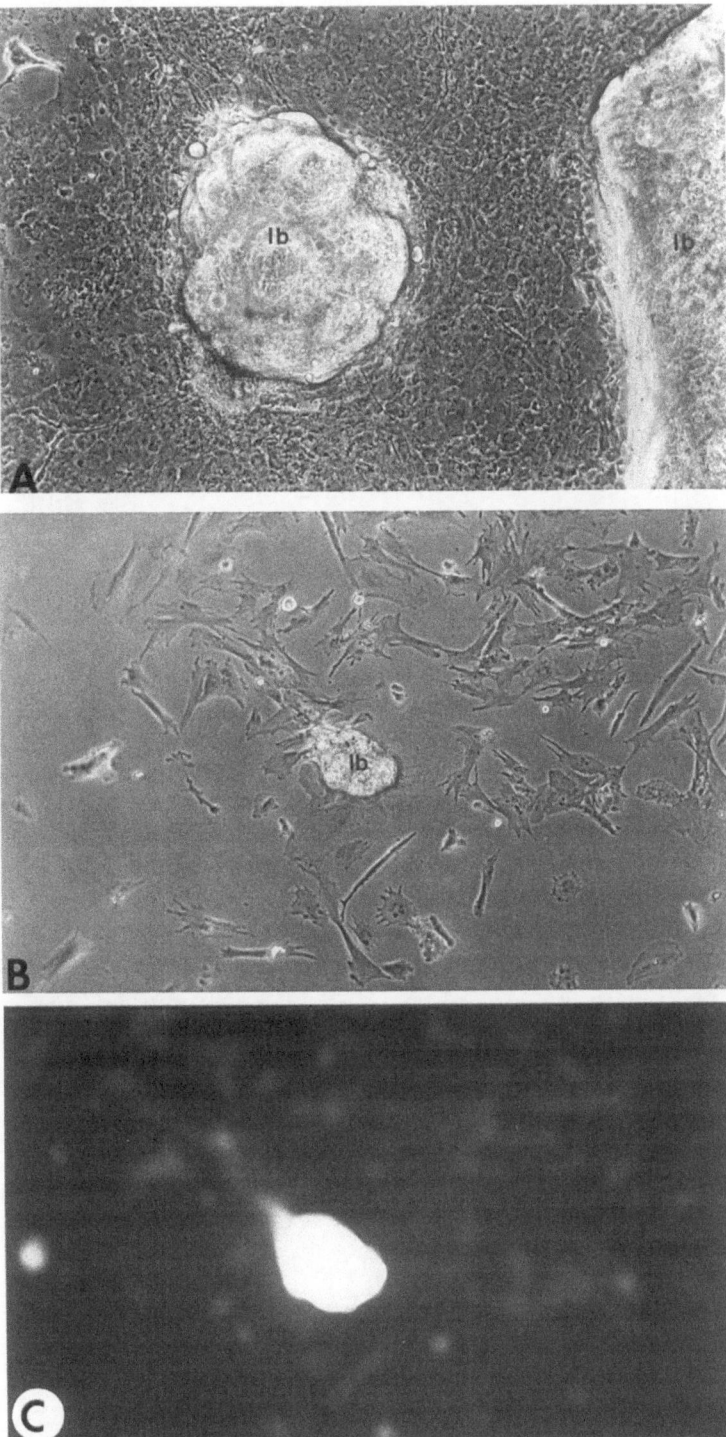

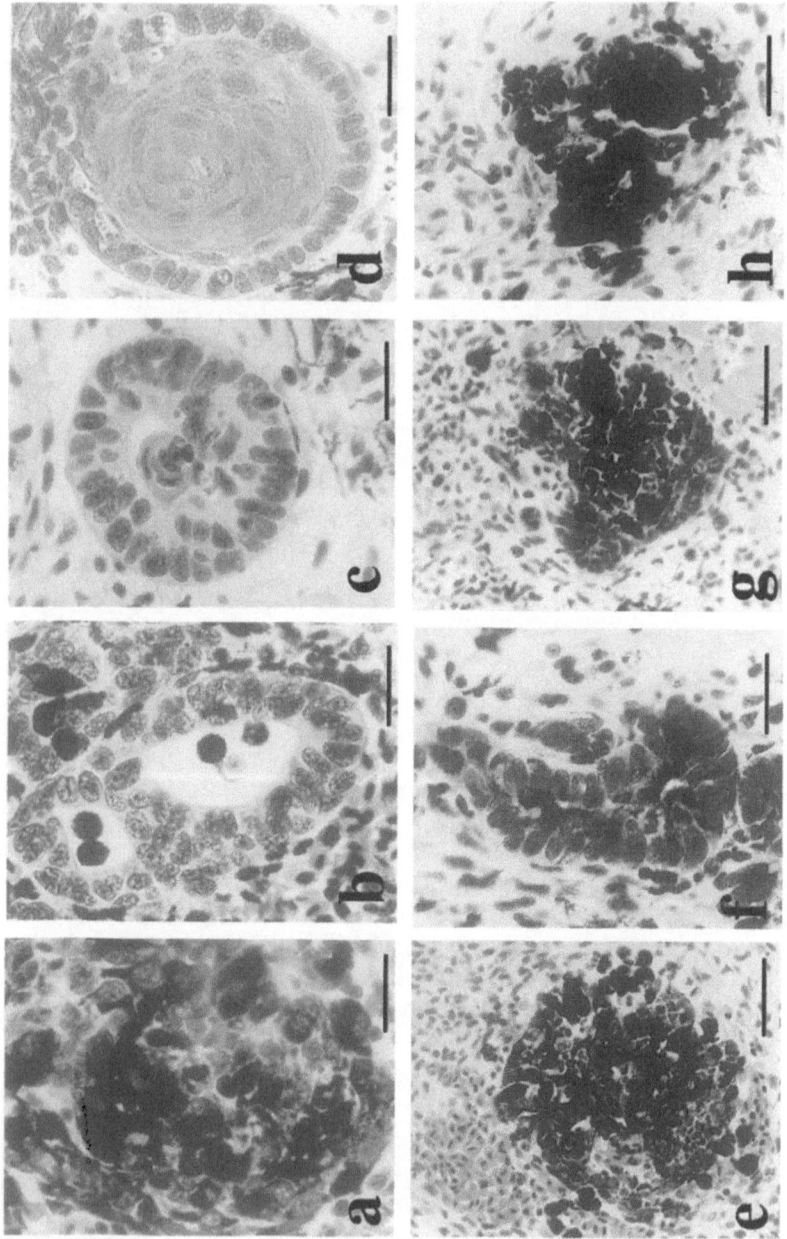

◀───

Fig. 7a–h. Transdifferentiation of pigment epithelial cells into lens after transplantation into a regenerating limb blastema. **a–d** transplantation of reaggregated dorsal PECs showing depigmentation after 9 days (**a**), tubular structures with depigmented cells after 14 days (**b**), lens vesicle-like structure with a few lens fibers inside the vesicle after 19 days (**c**) and induced lens with mature fibers after 30 days (**d**). **e–h** same time series as in **a–d**, but with implantation of ventral reaggregated cells. Only the dorsal cells are capable of inducing lens in the regenerating limb. (courtesy Dr. M. Okamoto)

plete lens can be formed in an environment other than the eye. This probably means that the physical presence of eye tissues is not necessary, but rather that factors (most likely involved in limb regeneration as well) are. This might support the argument that the retina might not be important, but a factor secreted or regulated in the retina. Second, the inability of the ventral PECs to transdifferentiate is probably due to a factor (most likely secreted or on the cell surface) that is inhibitory to lens transdifferentiation; perhaps something that is upregulated in the ventral iris is important for the process.

5
Clinical Applications

Knowing the factors that govern lens transdifferentiation and regeneration would be of enormous advantage in eye research with implications in diseases such as cataracts. Secondary cataracts occur after extracapsular cataract surgery and are characterized by the formation of secondary lens fibers on the posterior capsule. These fibers differentiate from the remaining epithelial cells which were not removed completely after surgery. Such complications affect nearly 50% of cataract patients (Apple et al. 1984; Kappelhof and Vrensen 1992). One way to treat this is by inhibiting the differentiation process by specific inhibitors. Since research, such as ours, deals with differentiation events from the dorsal iris, it is conceivable that identification of dorsal specific factors could lead to the design of inhibitors which can then be used to inhibit formation of such cataracts. Furthermore, if induction of lens regeneration is realized in the future, it could replace synthetic lens. Since the iris pigment epithelial cells possess the ability to transdifferentiate to lens (even in old humans), it could be possible to induce lens regeneration if we knew the mechanism. Obviously, research with newts could provide this much needed information.

Acknowledgements. I wish to thank Katia Del Rio-Tsonis for help with the preparation of the manuscript and Dr. M. Okamoto, Nagoya University for providing figures and unpublished results. This work was supported by NIH grant EY10540.

References

Achazi R, Yamada T (1972) Tyrosinase activity in the Wolffian lens regeneration system. Dev Biol 27:295-306

Apple DJ et al. (1984) Complications of intraocular lenses: a historical and histopathological review. Surv Ophthalmol 101:1-30

Altman CR, Chow RL, Lang RA, Hemmati-Brivanlou A (1997) Lens induction by pax-6 in *Xenopus laevis*. Dev Biol 185:119-123

Beebe DC (1994) Homeobox genes and vertebrate eye development. Invest Ophthalmol Visual Sci 35:2897-2900

Chow RL, Roux GD, Roghani M, Palmer MA, Rifkin DB, Moscatelli DA, Lang RA (1995) FGF suppresses apoptosis and induces differentiation of fibre cells in the mouse lens. Development 121:4383-4393

Collins JM (1972) Amplification of ribosomal ribonucleic acid cistrons in the regenerating lens of *Triturus*. Biochemistry 11:1259-1263

Collins JM (1974a) Structural changes in DNA during early stages of lens regeneration in *Triturus*. J Biol Chem 249:1839-1847

Collins JM (1974b) Template ability of activated DNA from the regenerating lens. Biochem Biophys Res Commun 57:359-364

Del Rio-Tsonis K, Washabaugh CH, Tsonis PA (1995) Expression of pax-6 during eye development and lens regeneration. Proc Natl Acad Sci USA 92:5092-5096

Del Rio-Tsonis K, Jung J-C, Chiu I-M, Tsonis PA (1997) Conservation of fibroblast growth factor function in lens regeneration. Proc Natl Acad Sci USA 94:13701-13706

Del Rio-Tsonis K, Trombley MT, Mcmahon G, Tsonis PA (1998) Regulation of lens regeneration by fibroblast growth factor receptor 1. Dev Dyn 213:140-146

Del Rio-Tsonis K, Tomarev SI, Tsonis PA (1999) Regulation of Prox 1 during lens regeneration. Invest Ophthalmol Vis Sci 40:2039-2045

Dumont JN, Yamada T (1977) Dedifferentiation of iris epithelial cells. Dev Biol 29:385-401

Eguchi G (1963) Electron microscopic studies on lens regeneration. I. Mechanism of depigmentation of the iris. Embryologia 8:45-62

Eguchi G (1988) Cellular and molecular background of Wolffian lens regeneration. In: Eguchi G, Okada TS, Saxen L (eds). Regulatory mechanisms in developmental processes. Elsevier, Amsterdam, pp 147-158

Eguchi G (1993) Lens transdifferentiation in the vertebrate retinal pigmented epithelial cells. Prog Retinal Res 12:205-230

Eguchi G, Watanabe K (1973) Elicitation of lens formation from the ventral iris epithelium of the newt by a carcinogen, N-methyl-N'-nitro-N-nitrosoguanidine. J Embryol Exp Morphol 30:63-71

Eguchi G, Abe S-I, Watanabe K (1974) Differentiation of lens-like structures from newt iris epithelial cells in vitro. Proc Natl Acad Sci USA 71:5052-5056

Funahashi J, Sekido R, Murai K, Kamachi Y, Kondoh H (1993) δ-crystallin enhancer binding protein δEF1 is a zinc finger-homeodomain protein implicated in postgastrulation embryogenesis. Development 119:433-446

Halder G, Callaerts P, Gehring WJ (1995) Induction of ectopic eyes by targeted expression of the eyeless gene in Drosophila. Science 267:1788-1792

Hill RE, Favor J, Hogan BLM, Ton CCT, Saunders GF, Hanson IM, Prosser J, Jordan T, Hastie ND, van Heyningen V (1991) Mouse small eye results from mutations in a paired-like homeobox-containing gene. Nature 354:522-525

Hyuga M, Kodama R, Eguchi G (1993) Basic fibroblast growth factor as one of the essential factors regulating lens transdifferentiation of pigmented epithelial cells. Int J Dev Biol 37:319-326

Imokawa Y, Eguchi G (1992) Expression and distribution of regeneration-responsive molecule during normal development of the newt, Cynops pyrrhogaster. Int J Dev Biol 36:407–412

Imokawa Y, Ono S-I, Takeichi T, Eguchi G (1992) Analysis of a unique molecule responsible for regeneration and stabilization of differentiated state of tissue cells. Int J Dev Biol 36:399–405

Ito M, Hayashi T, Kuroiwa A, Okamoto M (1999) Lens formation by pigmented epithelial cell reaggregate from dorsal iris implanted in the adult newt. Develop Growth Differ 41:429–440

Jung J-C, Del Rio-Tsonis K, Tsonis PA (1998) Regulation of homeobox-containing genes during lens regeneration. Exp Eye Res 66:361–370

Kappelhof JP, Vrensen GF (1992) The pathology of after-cataract. A mini review. Acta Ophthalmol 205:13–24

Kastner P, Grondona JM, Mark M, Gansmuller A, LeMeur M, Decimo D, Vonesch J-L, Dolle P, Chambon P (1994) Genetic analysis of RXR α developmental function: convergence of RXR and RAR signaling pathways in heart and eye morphogenesis. Cell 78:987–100

Loosli F, Kmita-Cunisse M, Gehring WJ (1996) Isolation of a pax-6 homolog from the ribbonworm, Lineus sanguineus. Proc Natl Acad Sci USA 93:2658–2663

McDevitt DS, Brahma SK (1981) Ontogeny and localization of the α, β and γ crystallins in newt eye lens development. Dev Biol 84:449–454

McDevitt DS, Meza I, Yamada T (1969) Immunofluorescence localization of the crystallins in amphibian lens development with special reference to the γ-crystallins. Dev Biol 19:581–607

Mohammadi M, McMahon G, Sun L, Tang C, Hirth P, Yeh BK, Hubbard SR, Schlessinger J (1997) Structure of the tyrosine kinase domain of fibroblast growth factor receptor in complex with inhibitors. Science 276:955–960

Monaghan KP, Davinson DR, Sime C, Graham E, Baldock R, Bhattacharya SS, Hill RE (1991) The Msh-like homeobox genes define domains in the developing vertebrate eye. Development 112:1053–1061

Nornes HO, Dressler GR, Knapik EW, Deutsch U, Gruss P (1990) Spatially and temporally restricted expression of pax-2 during murine neurogenesis. Development 109:797–809

Okada TS, Eguchi G, Takeichi M (1971) The expression of differentiation by chicken lens epithelium in vitro culture. Dev Growth Differ 13:323–336

Okamoto M (1997) Appearance of iris muscle-like cells in the pigmented iris epithelium after disappearance of intact iris muscle cells in the lentectomized newt eye. J Submicrosc Cytol Pathol. 29:435–441

Okamoto M, Ito M, Owaribe K (1998) Difference between dorsal and ventral iris in lens producing potency in normal lens regeneration is maintained after dissociation and reaggregation of cells from the adult newt Cynops pyrrhogaster. Dev Growth Differ 40:11–18

Oliver G, Loosli F, Koster R, Wittbrodt J, Gruss P (1996) Ectopic lens induction in fish in response to the murine homeobox gene six-3. Mech Dev 60:233–239

Ortiz JR, Vigny M, Courtois Y, Jeanny J-C (1992) Immunocytochemical study of extracellular matrix components during lens and neural retina regeneration in the adult newt. Exp Eye Res 54:861–870

Papalopoulou N, Kintner C (1996) A Xenopus gene, Xbr1, defines a novel class of homeobox genes and is expressed in the dorsal ciliary margin of the eye. Dev Biol 174:104–114

Quiring R, Walldorf U, Kloter U, Gehring, WJ (1994) Homology of the eyeless gene of Drosophila to the small eye in mice and aniridia in humans. Science 265:785–789

Robinson ML, Overbeek PA, Verran DJ, Grizzle WE, Stockard CR, Friesel R, Maciaq T, Thompson JA (1995) Extracellular FGF-1 acts as a lens differentiation factor in transgenic mice. Development 121:505–514

Steward DS, Espinasse PG (1959) Regeneration of the lens of the rabbit. Nature 183:1815

Stone LS (1958) Inhibition of lens regeneration in newt eyes by isolating the dorsal iris from the neural retina. Anat Rec 131:151–169

Stone LS (1967) An investigation recording all salamanders which can and cannot regenerate a lens from the dorsal iris. J Exp Zool 164:87–104

Tomarev SI, Sundin O, Banerjee-Basu S, Dunkan MK, Yang J-M, Piatigorsky J (1996) Chicken
    homeobox gene Prox 1 related to *Drosophila prospero* is expressed in the developing lens
    and retina. Dev Dyn 206:354–367
Ton CCT, et al (1991) Positional cloning and characterization of a paired box- and homeobox-
    containing gene from the *aniridia* region. Cell 67:1059–1074
Walther C, Gruss P (1991) Pax-6, a murine paired box gene, is expressed in the developing CNS.
    Development 113:1435–1449
Yamada T (1977) Control mechanisms in cell-type conversion in newt lens regeneration.
    Monographs in Dev Biol 13. Karger, Basel
Yamada T, McDevitt DS (1974) Direct evidence for transformation of differentiated iris epi-
    thelial cells into lens cells. Dev Biol 38:104–118
Yang Y, Zalik SE (1994) The cells of the dorsal iris involved in lens regeneration are myoepi-
    thelial cells whose cytoskeleton changes during cell conversion. Anat Embryol 189:475–487
Zalik SE, Scott V (1973) Sequential disappearance of cell surface components during lens
    dedifferentiation in lens regeneration. Nat New Biol 244:212–214

How the Neural Retina Regenerates

Pamela A. Raymond[1] and Peter F. Hitchcock[1,2]

1
Neurogenesis and Neuronal Stem Cells

The rules that govern cellular behavior during development and regeneration of tissues are complex and enigmatic, but substantial progress is being made toward understanding the molecular basis of proliferation and differentiation. Although most of the recent mechanistic insights have been gained from studies of embryonic development, the capacity of differentiated tissues and organs to regenerate is also an intriguing and important question (Lewis 1991; Brockes 1997; Ferrari et al. 1998). The greatest challenge to understanding how damaged tissues are repaired is to identify the stem cells and to understand the molecular factors that regulate their proliferation and differentiation.

Until recently, it was generally believed that neurogenesis does not occur in the nervous system of adult mammals and birds, except in certain populations of specialized nerve cells, such as sensory receptors of the olfactory epithelium (Graziadei and Monti Graziadei 1978). Despite long-standing evidence of postnatal neurogenesis in the central nervous system of adult rodents (Altman 1970; Kaplan and Hinds 1981; Kaplan et al. 1985), skepticism prevailed until Nottebohm and colleagues showed that new neurons are generated seasonally in the brains of adult song birds (Goldman and Nottebohm 1983; Nottebohm et al. 1990; Alvarez-Buylla and Kirn 1997). Subsequent reexaminations of the capacity and extent of neurogenesis in postnatal mammalian brains confirmed that in rodents (Reynolds et al. 1992; Morshead et al. 1994; Weiss et al. 1996; McKay 1997), primates (Gould et al. 1998), and even in humans (Eriksson et al. 1998), the adult brain contains proliferating stem cells that retain the ability to generate new neurons. Much of the recent excitement generated by these findings is linked to the potential therapeutic uses of neuronal stem cells (McKay 1997).

[1] Department of Cell and Developmental Biology, University of Michigan Medical School, 4610 Medical Science Building II, Ann Arbor, Michigan 48109-0616, USA
[2] Department of Ophthalmology and Visual Science, University of Michigan Medical School, Ann Arbor, Michigan 48109, USA

Results and Problems in Cell Differentiation, Vol. 31
M. E. Fini (Ed.): Vertebrate Eye Development
© Springer-Verlag Berlin Heidelberg 2000

Notwithstanding the recent furor provoked by the realization that neuronal stem cells persist in the adult mammalian brain, the existence of postembryonic neurogenesis in the central nervous system of anamniotic vertebrates (amphibians and fish) has been known for decades (Segaar 1965; Kirsche 1967). This ability to generate new neurons accounts for the profound capacities these animals possess for postembryonic growth and regeneration of neural tissues (Kirsche 1960; Kirsche and Kirsche 1961; Reyer 1977; Richter and Kranz 1981a, b; Easter 1983). Most of the work has focused on one specialized part of the central nervous system: the neural retina (Gaze and Watson 1968; Raymond 1985, 1991; Fernald 1991; Hitchcock and Raymond 1992; Reh and Pittack 1995; Grigorian 1996; Mitashov 1997; Raymond and Hitchcock 1997). The remainder of this chapter deals with regeneration of the neural retina and compares the cellular and molecular events that underlie regeneration in various vertebrate species.

2
Regeneration of the Neural Retina in Adult Urodele Amphibians

2.1
The Classical Model of Retinal Regeneration

The first studies of retinal regeneration were published in the 18th and 19th centuries by investigators who reported that when large portions of the eye of urodele amphibians (salamanders and newts) are removed, the lost parts are regenerated (reviewed by Stone 1950b). In the classical model of retinal regeneration in urodeles, the eye is enucleated and then replaced in the orbit; the temporary loss of ocular vascularization causes the neural retina to degenerate, but after a few weeks the retina is replaced and vision is restored (Stone 1950a,b; Reyer 1977). Subsequent work by many investigators used both immunocytochemistry and electrophysiology to document the biochemical and functional steps, and these studies showed that the regeneration process in urodele amphibians largely mimics normal embryonic development (Lam 1977; Sarthy and Lam 1983; Bugra et al. 1992; Kaneko and Saito 1992; Negishi et al. 1992; Ortiz et al. 1992; Saito et al. 1994; Chiba and Saito 1995; Chiba et al. 1997; Cheon et al. 1998; Chiba 1998). In general, these results are consistent with the postulate that common mechanisms control both the generation and regeneration of retinal neurons.

The major candidates for the source of regenerated retinal neurons in urodeles include the ciliary margin (i.e., the transition zone where neural retina meets iris/ciliary epithelium) and the pigmented retinal epithelium that underlies the neural retina (see Chap 10, this Vol.). The ciliary margin in adult urodeles contains proliferative, multipotent, neuroepithelial cells that continuously generate annuli of new retinal neurons contributing to the sustained growth of the adult eye (Gaze and Watson 1968). It is now generally accepted that following retinal destruction in adult newts, the ciliary germinal zone can restore an annulus of new retina at the anterior margin of the eye, but this is properly interpreted as an exaggeration of the normal growth process (Keefe 1973a, b; Levine 1975, 1977; Stroeva and Mitashov 1983; Grigorian 1996; Mitashov 1996, 1997).

Although initially a controversial idea, differentiated retinal pigmented epithelial (RPE) cells have been unambiguously identified as the source of regenerated retinal neurons in amphibians (Fig. 1A). The process by which RPE cells are transformed into retinal neurons is known as transdifferentiation or metaplasia (Okada 1980), and it involves several rounds of cell division and dedifferentiation prior to the appearance of retinal progenitors (Stone 1950b; Stroeva and Mitashov 1983). With antibody markers to distinguish RPE cells from retinal neurons, a transitional state (characterized by expression of both RPE and neuronal markers of differentiation) has been demonstrated in transdifferentiating RPE cells from adult newts (Klein et al. 1990; Grigorian and Anton 1995; Kajiwara et al. 1999) and from anuran tadpoles, *Rana catesbienna* (Reh and Nagy 1987). The development of in vitro culture systems in which the molecular factors controlling RPE transdifferentiation can be studied has led to progress in understanding this phenomenon (Eguchi and Kodama 1993; Kodama and Eguchi 1995; Reh and Pittack 1995), as described below.

2.2
Molecular Mechanisms of RPE Transdifferentiation

The transdifferentiation of RPE cells into other ocular tissue types, including neural retina and lens, has been demonstrated experimentally in several vertebrate species during embryonic stages, and in anuran amphibians (frogs and toads) up to metamorphosis, but only in certain urodele amphibians is this capacity maintained into adulthood (Okada 1980; Stroeva and Mitashov 1983; Eguchi and Kodama 1993; Opas 1994). In fact, phenotypic switching, i.e., the interchangeability of differentiated cell fates, may be a characteristic of all cells of the ocular primordium that derive from the ectodermal lineage, which includes the optic vesicle and lens vesicle (see Chap 4, this Vol.). The use of carefully controlled in vitro conditions, especially the analysis of clonal cell cultures, has provided definitive evidence that differentiated (i.e., melanin-containing) cells from the RPE or iris, or neuroepithelial cells from the presumptive neural retina, can give rise to any of three differentiated, ocular cell types: pigmented epithelial cells, neuronal cells, or lentoid bodies that express lens crystallins (Okada 1980; Eguchi and Kodama 1993; Kodama and Eguchi 1995; Zhao et al. 1995).

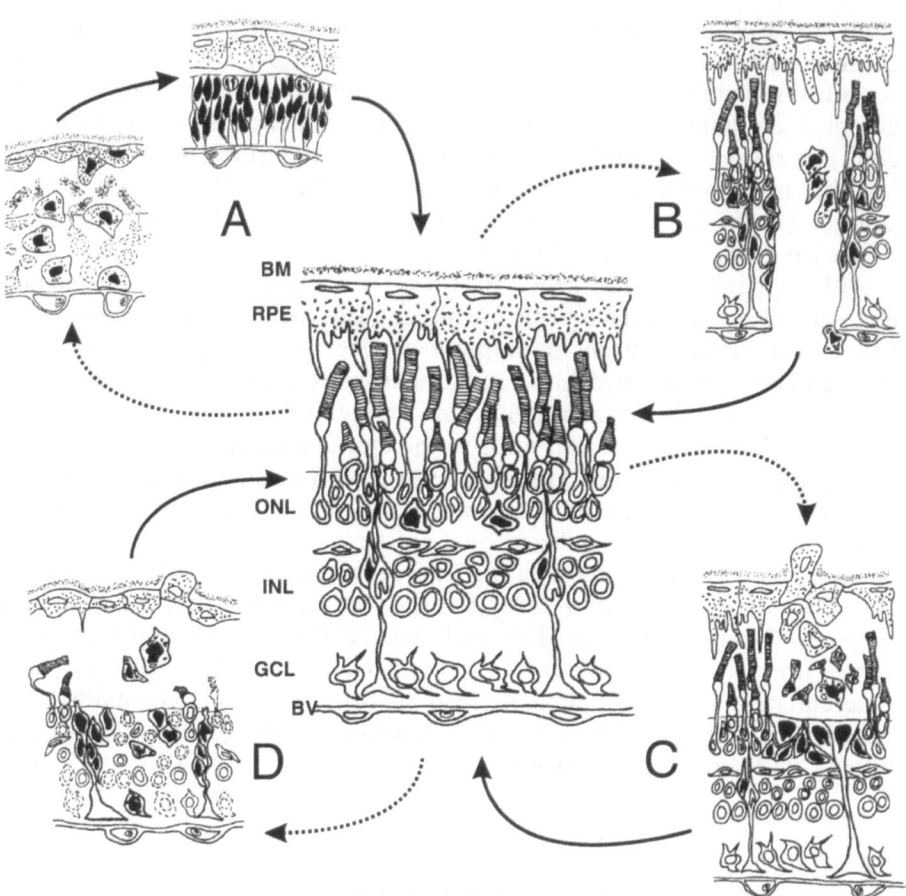

Fig. 1A–D. Schematic diagrams illustrating retinal lesions and regeneration. The *central figure* indicates the intact retina, the *dotted arrows* indicate various lesioning paradigms, and the *solid arrows* represent regeneration. Cells with filled (*black*) nuclei are proliferating. **A** Devascularization of the retina in urodeles. **B** Surgical excision of retina in goldfish. **C** Photocoagulation of photoreceptors in goldfish. **D** Ouabain-induced degeneration of retinal neurons in goldfish. *RPE* Retinal pigmented epithelium; *ONL* outer nuclear layer; *INL* inner nuclear layer; *GCL* ganglion cell layer; *BV* blood vessels

Several studies have emphasized the importance of the cellular environment, such as growth factors and cell–cell and cell–substrate interactions, in regulating differentiation of cells in the ocular primordium (Reh 1991; Eguchi and Kodama 1993; Park and Hollenberg 1993; Opas 1994; Kodama and Eguchi 1995; Reh and Pittack 1995). The earliest indications of the importance of these interactions came from reports (beginning in the 1930s) that cells of the early optic vesicle differentiate in vitro exclusively as neural retina when allowed to aggregate in the absence of contact with mesenchyme, but differentiate as pigmented epithelial cells when allowed to spread as a single layer on a substratum of extracellular matrix (cited in the review by Park and Hollenberg 1993). Interactions between the inner and outer walls of the optic cup were demonstrated by early experiments in amphibia and chick embryos, which showed that separating the neural retina from the RPE during the period when RPE phenotype remained labile caused an ectopic neural retina to form locally within the RPE underlying the area of retinal detachment (cited in Coulombre and Coulombre 1965).[3] In contrast to the regenerated newt retinas described above, the ectopic retinas generated from RPE in these examples have a reversed basal-apical orientation, with photoreceptors oriented toward the inside of the eye. This curious arrangement occurs because RPE cells producing these ectopic retinas transdifferentiate in situ, without detaching from their own basal lamina (Bruch's membrane), as happens in the regenerating newt retina (Fig. 1A). Because the apical-basal polarity of the endogenous RPE is reversed (mirror-image) compared to the endogenous neural retinal (see Fig. 1), and because photoreceptors arise from the apical surface of the neuroepithelium, the polarity of the ectopic retina is reversed. Whether these ectopic retinas can mediate visual behavior has never been examined, although it seems unlikely.

To induce transdifferentiation of the RPE in chick embryos, it is necessary to leave a fragment of neural retina within the eye, although not in contact with the denuded RPE; if the entire neural retina is removed, the RPE does not transdifferentiate (Coulombre and Coulombre 1965; Park and Hollenberg 1993). Because certain other tissues such as embryonic otocyst could be substituted for the fragment of retina, it was suggested that soluble factors might be involved (Coulombre and Coulombre 1965; Lopashov and Sologub 1972; Sologub 1975, 1977; Sakaguchi et al. 1997). Park and Hollenberg showed that following retinectomy, the RPE in the embryonic chick is induced to transdifferentiate into neural retina in situ after exposure to FGF-2 (Park and Hollenberg 1989, 1991, 1993). A more complete explanation of how FGF-2 promotes the formation of neural retina from RPE was provided by culture experiments using cells from embryonic chick (Pittack et al. 1991, 1997; Guillemot and Cepko 1992), or larval amphibians (Rana spp. and Xenopus laevis) (Reh 1991; Sakaguchi et al. 1997), or rat (Zhao et al. 1995). No other growth factors tested, including EGF, NGF, TGF-β, and activin, produced the effects observed with FGF (Pittack et al. 1991; Guillemot and Cepko 1992; Opas and Dziak 1994; Zhao et al. 1995). The changes in gene expression produced in RPE cells by FGF-2 are likely to be both indirect and irreversible, since cultures exposed to FGF-2 for only 24 h or continuously for several days show the same degree of transdifferentiation when examined several days later (Zhao et al. 1995). The order of appearance of markers for specific

---

[3] A curious, and strikingly similar, phenomenon has been described in the homozygous silver plumage mutant of Japanese quail: the RPE at the back (fundus) of the eye lacks melanin, and in this locally nonpigmented area, an ectopic retina develops underneath the normal retina (Fuji and Wakasugi 1993).

retinal cell classes matches their birth order during normal retinal development (Zhao et al. 1997), again suggesting a commonality of mechanisms
underlying normal development and regeneration.

The ability of RPE cells to transdifferentiate into neural retina in vitro
declines with developmental age (Pittack et al. 1991; Zhao et al. 1995), as it
does in vivo (Park and Hollenberg 1993). Since FGF receptors are expressed
by the RPE beyond the age at which it ceases to respond to the effects of FGF,
the decline in responsiveness is presumably not due to loss of receptors
(Zhao et al. 1997). The intracellular signaling cascade triggered by the activation of the FGF receptor is mediated through the G-protein *ras* (Baird
1994), and cell lines derived from adult human RPE transfected with activated
ras express neuronal markers, indicating that oncogenes can overcome the
age-dependent decline in the capacity to transdifferentiate (Dutt et al. 1993).
Although transdifferentiation in vivo is invariably accompanied by proliferation and at least partial dedifferentiation of RPE cells, e.g., loss of melanin
pigment (Stroeva and Mitashov 1983; Reh and Pittack 1995), RPE cells in
vitro can undergo direct conversion to neural retina without cell division
(Guillemot and Cepko 1992).

The phenotypic changes induced in competent RPE cells by exposure to
FGF are presumably mediated by alterations in expression of genes that
determine pathways of differentiation. Candidates include members of the
gene families that encode paired box and/or homeobox motifs that exhibit
regionalized patterns of gene expression in the early optic vesicle prior to any
morphological evidence of differentiation, which occurs at the optic cup
stage. Some of these transcriptional regulators include *rx/rax*, *chx10/vsx-2/
alx*, *otx-1*, *hox-8*, and *dlx-1* (Reh and Pittack 1995; Zhao et al. 1997). The
surface ectoderm expresses high levels of FGF-2 during optic vesicle evagination (Pittack et al. 1997), and receptors for FGF have been localized to the
developing optic vesicles in chick and rat embryos (Wanaka et al. 1991;
Tcheng et al. 1994), suggesting that FGF-2 may be involved in promoting
retinal determination during normal development (Pittack et al. 1997).

Cell culture studies have demonstrated that inducing RPE cells to switch to
a neuronal phenotype requires a combination of elements, including the
competence to transdifferentiate, the presence of a growth factor (probably
FGF-2), and finally, a specific multicellular arrangement imparted by cell–cell
and cell–substrate interactions. For example, dissociated RPE cells from
larval amphibians (*Rana catesbienna* tadpoles) can be induced to transdifferentiate into retinal neurons by the extracellular matrix molecule laminin
(Reh et al. 1987), and transdifferentiation is blocked in vivo by an antibody
that inhibits the interaction of RPE cells with the laminin-heparan sulfate
proteoglycan complex of the retinal basal lamina (Nagy and Reh 1994). These
results may also help to explain the role of FGF in promoting differentiation
of retinal neurons, since growth factors in the FGF family are typically bound
to heparan sulfate proteoglycan molecules in the basal lamina (Baird and
Böhlen 1990). Somewhat surprisingly, Bruch's membrane contains both

laminin and heparan sulfate proteoglycans, and could therefore bind FGF, yet RPE cells attached to it do not normally transdifferentiate. The endogenous factors promoting the establishment and maintenance of the RPE cell phenotype are largely unknown, although it has been suggested that activin, a member of the TGF-β family, which is expressed in developing RPE and the surrounding mesenchyme, may be involved (Reh and Pittack 1995), since activin can block FGF-induced transdifferentiation of chick RPE (Pittack et al. 1997).

3
Regeneration of the Neural Retina in Adult Teleost Fish

3.1
Source of the Regenerated Neurons – Stem Cells

Regeneration of the neural retina in adult goldfish (*Carassius auratus*) was first described over 30 years ago (Lombardo 1968, 1972). These and subsequent investigations have used several methods to destroy the neural retina, including removal by aspiration of all or part of the retina (Lombardo 1968, 1972; Knight and Raymond 1994), excision of a small patch (approximately 1–2 mm^2) of retina (Hitchcock et al. 1992; Cameron and Easter 1995; Cameron et al. 1997), intraocular injection of the metabolic poison ouabain (Maier and Wolburg 1979; Kurz-Isler and Wolburg 1982; Raymond et al. 1988a), intraocular injection of selective neuronal toxins including tunicamycin and 6-hydroxydopamine (Negishi et al. 1988, 1991a, b; Braisted and Raymond 1992, 1993), and photocoagulation with an argon laser (Braisted et al. 1994). Most of these experiments have used goldfish (reviewed by Raymond 1991; Hitchcock and Raymond 1992; Raymond and Hitchcock 1997), although a few other species have been studied, such as rainbow trout (*Salmo gairdneri*, now called *Onchoryncus mykiss*; Kurz-Isler and Wolburg 1982), green sunfish (*Lepomis cyanellus*; Cameron and Easter 1995; Cameron et al. 1997), and zebrafish (*Danio rerio*; D.A. Cameron, L.K. Barthel and P.A. Raymond, unpublished observations).

There is no evidence for transdifferentiation of RPE into neural retina in adult teleost fish, independent of which injury model is used (Raymond 1991; Hitchcock and Raymond 1992; Knight and Raymond 1994; Raymond and Hitchcock 1997). Mitotic activity in the RPE does not increase substantially following retinal destruction (Lombardo 1972; Maier and Wolburg 1979; Kurz-Isler and Wolburg 1982; Hitchcock et al. 1992; Knight and Raymond 1994), and although a few pigmented cells migrate away from Bruch's membrane, and pigmented inclusions are occasionally seen within the lesioned retina, these are not associated with foci of mitotic activity (Raymond et al. 1988b; Hitchcock et al. 1992). Instead, a marked increase in mitotic activity is observed in the circumferential germinal zone at the retinal (cili-ary) margin and within the residual retina (Lombardo 1968, 1972; Maier and Wolburg 1979; Kurz-Isler and Wolburg 1982; Negishi et al. 1988, 1991a, b; Raymond et al. 1988b; Hitchcock et al. 1992; Knight and Raymond 1994). Similar to larval anuran and adult urodele amphibians (discussed above), the germinal zone at the retinal margin in teleosts normally produces annuli of new retina, associated with continued growth of the fish (Raymond Johns 1977). As in the case of amphibian retinal regeneration, it is generally agreed that enhanced neurogenesis in the circumferential germinal zone following retinal damage in teleost fish is an accelerated form of normal retinal growth, and cell proliferation in this region is not responsible for regeneration of damaged central retina.

Within the first few days following excision of a patch of retina (Fig. 1B), a blastema of proliferating neuroepithelial cells appears at the edge of the retinal wound (Lombardo 1972; Hitchcock et al. 1992; Knight and Raymond 1994; Cameron and Easter 1995), and the wound boundary contracts toward the center of the lesion (Cameron and Easter 1995). Over the next several weeks, the wound cavity is filled in by the appositional addition of regen-erated neurons (Hitchcock et al. 1992). The neuroepithelial cells of this wound "blastema" are derived from a local cellular source intrinsic to the differentiated neural retina, and not from the germinal zone at the margin, located up to several hundred micrometers away (Hitchcock et al. 1992). Many of the dividing cells, especially those in the inner nuclear layer (INL), are grouped in clusters that are elongated radially, although the majority of proliferating cells are located in the outer nuclear layer (ONL), which con-tains the nuclei of rod and cone photoreceptor cells (Lombardo 1972; Hitchcock et al. 1992; Cameron and Easter 1995). A comparable pattern of proliferation is seen following laser lesions that destroy only the ONL (Fig. 1C), including increased proliferation in the ONL adjacent to the damaged region, and clusters of dividing cells that appear to originate from the un-damaged INL immediately underlying the damaged ONL (Braisted et al. 1994). Following cytochemical destruction (Fig. 1D), radially elongated clusters of proliferating "neurogenic" cells are seen scattered across the damaged retina (Maier and Wolburg 1979; Kurz-Isler and Wolburg 1982; Negishi et al. 1988; Raymond et al. 1988b; Braisted and Raymond 1992, 1993).

The relative abundance of mitotic activity in the ONL of the damaged retinas seen in all these studies led to the initial suggestion that the regen-erated neurons most likely derive from intrinsic retinal progenitor cells lo-cated in this layer (Lombardo 1972; Negishi et al. 1987, 1991a, b; Raymond et al. 1988b; Raymond 1991; Hitchcock and Raymond 1992; Hitchcock et al. 1992; Raymond and Hitchcock 1997). Other possible cellular sources of re-generated retinal neurons have been considered. For example, following laser lesions in which cell loss is nearly or completely confined to a local region of photoreceptors, Müller glia immediately proximate to the lesion proliferate, their nuclei migrate into the "empty" ONL (Fig. 1C), and they upregulate expression of glial fibrillary acidic protein (GFAP) (Braisted and Raymond 1993; Raymond and Hitchcock 1997). Since neurons and Müller glia in the vertebrate retina derive from a common progenitor (Cepko et al. 1996), and Müller cells have been shown to transdifferentiate into lens cells in culture (Kodama and Eguchi 1995), it is possible that the proliferating Müller cells give rise to neuronal progenitors. However, Müller cells undergo identical behaviors following laser lesions that destroy photoreceptors in the retina of adult rats, in which there is no subsequent neuronal regeneration (Humphrey et al. 1993, 1997).

The bulk of current evidence favors the more recent suggestion that re-generated retinal neurons in the adult teleost retina derive from pluripotent, normally quiescent, stem cells that reside in the INL (Reh and Levine 1998). Many studies have reported mitotic cells in the INL of the teleost retina following trauma or intraocular injection, with or without actual loss of retinal neurons (Lombardo 1972; Negishi et al. 1987, 1988, 1991a, b; Raymond et al. 1988b; Braisted and Raymond 1992, 1993; Hitchcock et al. 1992; Negishi and Shinagawa 1993; Braisted et al. 1994; Cameron and Easter 1995; Julian et al. 1998). Although occasionally seen in undamaged, adult goldfish retina, a recent study by Korenbrot and colleagues provided unequivocal evidence that neuronal progenitors exist in the INL of rapidly growing juvenile and adult rainbow trout (Julian et al. 1998). Cumulative, systemic labeling with BrdU, or PCNA immunocytochemistry, reveals radially elongated clusters of proliferating cells in the INL; their progeny then migrate into the ONL, where they continue to proliferate as rod precursors (Julian et al. 1998), similar to what was shown previously in larval goldfish (Raymond and Rivlin 1987). These clusters of proliferating cells in the INL of intact, undamaged trout retina are identical in histological appearance and similar in relative abun-dance to those described in damaged retinas of goldfish, zebrafish, and green sunfish following a pulse-label with ^{3}H-thymidine or BrdU, suggesting that damage induces mitotic stimulation of a resident population of cells in the INL. The cells in the postembryonic teleost retina meet some or all of the criteria of stem cells (McKay 1997): they are normally relatively quiescent and few in number; they self-renew; they respond to damage by increased pro-liferation; they appear to be capable of replacing all the cell lineages in the retina.

There are also a few reports in the literature that clusters of proliferating cells persist in the INL of larval *Xenopus* retina (Levine 1981; Taylor et al. 1989; Dorsky et al. 1997), and studies by Levine have implicated these cells in the regeneration of the retina in *Xenopus* tadpoles following small surgical lesions (Levine 1981). Recent investigations of retinal regeneration in adult newts have identified proliferating neuroepithelial cells in the INL that are believed to be responsible for restoring damaged photoreceptor cells (Grigorian et al. 1996). Harris and colleagues (Dorsky et al. 1997) described proliferating cells that persist late in development in the INL of larval *Xenopus* retina and that express *Notch* and *Delta*, signaling molecules involved in lateral inhibitory interactions important in regulating cell determination and differentiation (Nye and Kopan 1995; Fleming et al. 1997; Robey 1997). The *xNotch-1* receptor and its ligand *Delta-1* are also expressed by the mitotic neuroepithelial cells and newly postmitotic retinal cells in the germinal zone at the ciliary margin in *Xenopus* retina (Dorsky et al. 1995, 1997; Perron and Harris 1997). The expression of these and other developmental regulatory genes in the regenerating retina, described in the next section, provides additional evidence for the existence of stem cells in the adult teleost retina.

3.2
Injury-Induced Gene Expression

Death of retinal neurons in goldfish induces the reexpression of developmental regulatory genes important in embryonic and postembryonic growth of the retina (Fig. 2). To date, the injury-induced expression of three genes has been described, two of which, *vsx-1* and *pax6* (Levine et al. 1994), are members of the superfamily of transcription factors that encode paired-type homeodomains, and the third, *gNotch-3* (Sullivan et al. 1997), is a member of the family of *Notch* receptor genes. In the retina of adult goldfish, *vsx-1* is expressed in newly postmitotic cells adjacent to the circumferential germinal zone and in presumptive bipolar cells in the outer INL of the mature retina. Following a surgical lesion, *vsx-1* is also expressed adjacent to the regeneration blastema in radial clusters of cells spanning the INL and ONL (Levine et al. 1994). This study was the first to demonstrate that retinal destruction induces the reexpression of developmental regulatory genes, and that the molecular pathways active during retinal development are reactivated during retinal regeneration. Similar to *vsx-1*, the expression of the homeobox gene, *pax6*, is also modulated by retinal injury. Immunostaining with antibodies against recombinant, zebrafish pax6 (Hitchcock et al. 1996) showed that pax6 protein is expressed in mitotically active retinal progenitors in the circumferential germinal zone of goldfish retina (but not in rod precursors in the ONL), and in differentiated amacrine and ganglion cells. Following surgical injury, pax6 immunoreactivity is seen in the regeneration blastema (Hitchcock et al. 1996).

In the brain of embryonic goldfish, *gNotch-3* is expressed widely in proliferating cells, and in the adult brain and retina, *gNotch-3* is expressed by

--→

Fig. 2. Model of retinal injury in fish, amphibians, and mammals. Expression levels of growth factors and transcription factors listed are modulated by injury to the central nervous system, CNS. The proliferation of microglia/macrophages and endothelial cells has not been examined in the urodele model of retinal regeneration, as indicated on the diagram by ??. See text for further discussion

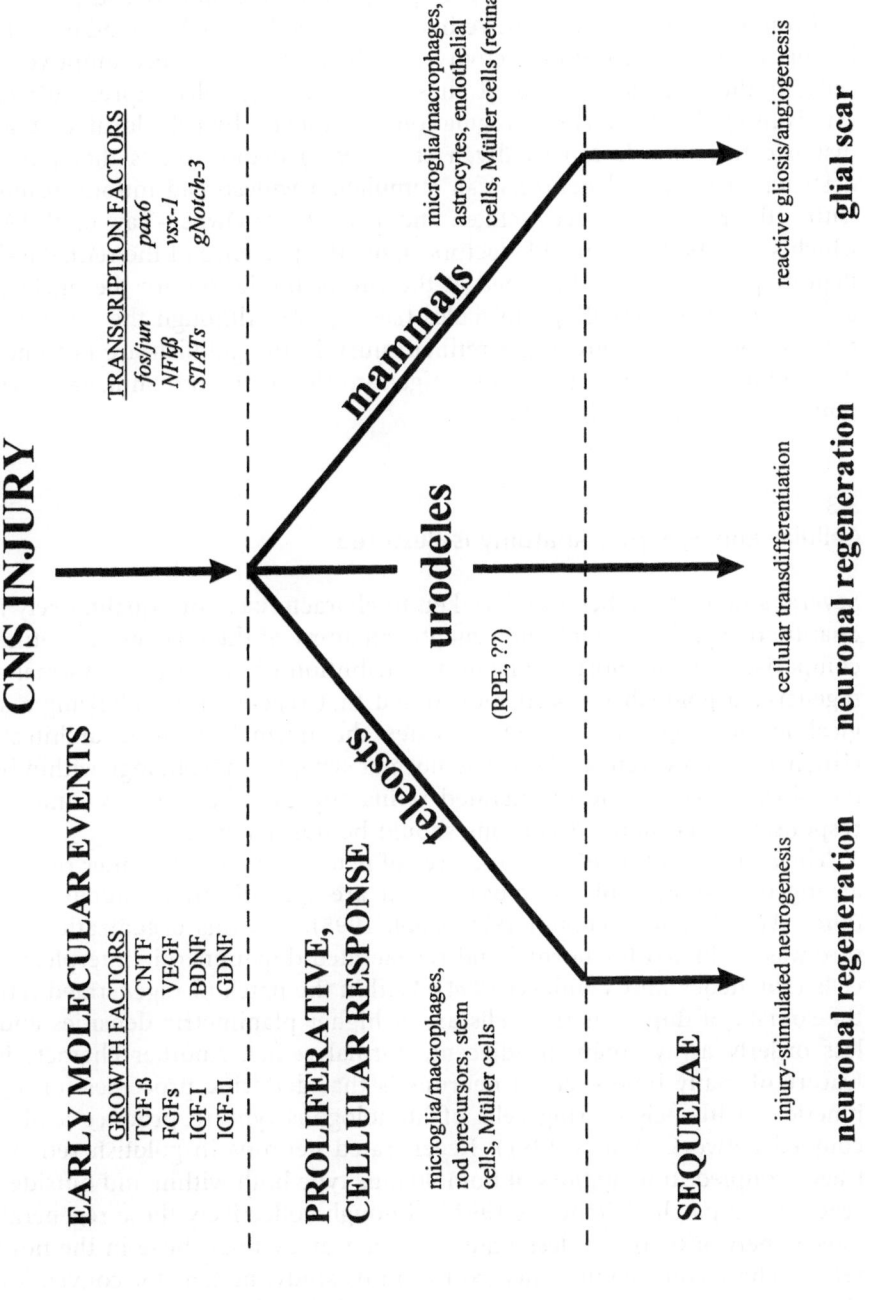

mitotically active neuronal and glial progenitors, but not by rod precursors (Sullivan et al. 1997). When the retina is damaged by a stab wound or by laser lesions to the ONL, mitotic activity is induced locally in presumptive stem cells in the INL, and some of these proliferating cells express *gNotch-3* (Sullivan et al. 1997). The upregulation of *gNotch-3* by cells located at some distance from the lesion (Sullivan et al. 1997) demonstrates that punctate destruction of retinal neurons can stimulate a widespread mitotic response within the retina (also see Henken and Yoon 1989; Owusu-Yaw et al. 1992), which suggests that diffusible factors signal the presence of the retinal lesion. Peptide growth factors liberated at the site of the lesion are the likely candidates for these signaling molecules (see Fig. 2). Although the identities of factors involved in signaling a retinal injury in the goldfish are not known, this subject is under active investigation (Negishi and Shinagawa 1993; Boucher and Hitchcock 1998).

3.3
Cellular and Synaptic Anatomy is Restored

Several studies have been undertaken to characterize, both qualitatively and quantitatively, the cellular and synaptic anatomy of the regenerated retina. A comparison of the morphology and distribution of synapses in normal and regenerated goldfish retina demonstrated that regeneration following a surgical lesion (Fig. 1B) largely recreates the normal synaptic architecture (Hitchcock and Cirenza 1994). The normal synaptic morphology within both plexiform layers of the regenerated retina suggests that visually stimulated responses of regenerated neurons should be normal as well.

The somata and dendritic processes of neurons within the inner retina are arranged in unique planar mosaics that are spatially independent of each other (Wässle and Riemann 1978; Cook 1998), and this organizational feature was evaluated for normal and regenerated dopaminergic interplexiform cells (Hitchcock and Vanderyt 1994). Within the patch of regenerated retina, the somata of dopaminergic cells are at higher planimetric densities and in less orderly arrays than in adjacent, normal retina. Another characteristic feature of many inner–retinal neurons is that dendritic processes form gap junctions with neighboring cells of homologous type to produce a planar, coupled network (Vaney 1991). Regenerated neurons in goldfish retina are tracer-coupled to neighbors of homologous type both within and outside the regenerated patch (Hitchcock 1997), although collectively these regenerated, mosaic networks form a less regular, planar array than those in the normal retina. Three conclusions emerged from this study: first, as for conventional, chemical synapses in regenerated retina (Hitchcock and Cirenza 1994), the circuitry represented by gap junctions in regenerated retina is normal. Second, the boundary between the patch of regenerated retina and the surrounding normal retina is not a barrier, and dendritic processes are able to grow into and out of regenerated retina. Third, regenerated retinal neurons are functionally integrated into the preexisting synaptic circuitry, and thus the functions subserved by the junctional coupling between normal retinal neurons are reestablished between normal and regenerated retinal neurons.

The regeneration (and ectopic generation) of cone photoreceptors in green sunfish has been examined following surgical excision of a patch of retina (Cameron and Easter 1995). The green sunfish lends itself to this analysis, because its retina contains only two types of cones, single and double cones, which form a very precise, rhombic pattern that is largely invariant across the retina (Cameron and Easter 1993). Following a surgical wound, the regenerated cones form a relatively disordered planar array, with anomalous triple and quadruple cones, and a higher than normal planimetric density, largely due to an excess of single cones (Cameron and Easter 1995). The authors suggested that the higher-than-normal density of regenerated cones increases the frequency of contact between adjacent cones, leading to the creation of these anomalous multiple-cone fusions. Surprisingly, within the stretched, but otherwise undamaged, retina adjacent to the wound, new cone photoreceptors are also generated (Cameron and Easter 1995). It is well known that stretch imposed on proliferative tissues is mitogenic (Squier 1980), and there is indirect evidence that similar mechanisms may modulate neurogenesis in the fish retina (Raymond et al. 1988b). Stretch alone, however, cannot explain the generation of new cones in the region surrounding the wound, because the fish retina stretches enormously over its lifetime, albeit slowly, and cone photoreceptors are not normally generated in central retina, despite the continual production of rod photoreceptors (Raymond 1985). Therefore, the ectopic generation of new cones in the undamaged retina following a surgical lesion is most likely related to an accelerated rate of stretch, which may result in the dilution of local inhibitory signals or the disruption of cell–cell interactions that normally prevent cone photoreceptors from being generated.

To further investigate the anomalous cone phenotypes in regenerated retina, microspectrophotometry was used to determine their photopigment content (Cameron et al. 1997). In the normal green sunfish retina, the two cone types each contain a unique photopigment with a characteristic absorption spectrum: the absorption of the single cones peaks at 532 nm (green), and that of double cones peaks at 620 nm (red). Within regenerated retina the absorption spectra for double cones is normal, but regenerated single cones contain either red or green photopigment. The anomalous triple cones contain a photopigment whose spectrum peaks in the red, again matching the double cones. This observation suggests a linkage during both development and regeneration between the mechanisms that regulate photoreceptor fusion and photopigment expression.

3.4
Vision is Restored

Relatively few studies have examined functional recovery of vision following retinal regeneration in teleosts, and all have examined recovery following ouabain-induced retinal degeneration in adult goldfish. The optic tectum becomes reinnervated a few months after destruction of the retina with oubain, although the axonal pathways of the regenerated retinal ganglion cells are abnormal, erratic, and circuitous (Stuermer et al. 1985). Functional recovery has been assessed by observation of a visuomotor reflex (optokinetic nystagmus) which returns between 1 and 3 months after retinal destruction (Kästner and Wolburg 1982). The electroretinogram (ERG) has a characteristic wave form that reflects the amplitude and duration of the electrical activity of various populations of cells and their synaptic interconnections, and, as the retina regenerates, the different components of the ERG appear sequentially (Mensinger and Powers 1999). The fully regenerated retina produces an ERG that is normal with respect to waveform and spectral sensitivity; however, at the survival times examined (up to 210 days), the amplitude of the ERG b-wave remains below normal levels. The area of regenerated retina is not completely restored, and the ouabain causes optical abnormalities (decreased pupil size and cloudy lens), and any of these factors could contribute to the decreased amplitude of the electrical response (Mensinger and Powers 1999). This partial recovery of ERG function is similar to what had been described previously for regenerated newt retina (Lam 1977; Sarthy and Lam 1983). Behavioral recovery in goldfish, measured by a vestibulo-ocular reflex, parallels the electrophysiological improvement (Mensinger and Powers 1999). The level of behavioral recovery is also quantitatively related to the amount of retina regenerated (Powers et al. 1998). Together, these studies indicate that the regenerated retina can recover normal function and reestablish connections to the brain that mediate visual behavior.

4
A Model of Retinal Injury in Fish, Amphibians and Mammals

Figure 2 illustrates a model of retinal regeneration for adult fish and urodele amphibians and the injury response in the mammalian brain and retina.[4]

---

[4] The cellular response of the mammalian brain to a lesion that kills neurons is well characterized, although the molecular details are incomplete. This subject is the topic of numerous recent reviews (Raivich et al. 1996; Streit 1996; Pennypacker 1997; Ridet et al. 1997; Minghetti and Levi 1998).

This model summarizes and organizes the data reviewed in this chapter and highlights a proposed scheme of the salient molecular and cellular events for each class of animal.

We suggest that the injury response can be divided into three phases. During the initial phase, intracellular signaling pathways are activated by the injury-induced release of growth factors and by the binding of these factors to their receptors. Growth factors, acting as both autocine and paracrine factors in pleiotropic networks, bind their cognate receptors and activate intracellular signaling pathways whose final targets are transcription factors. It has been suggested that within the mammalian central nervous system (CNS), TGF-β organizes the injury response (Finch et al. 1993), and the first transcriptional events are initiated by fos/jun dimers in neurons, STAT proteins in microglia, and NF$\kappa\beta$ transcription factors in astrocytes (Pennypacker 1997). We propose that these initial injury-induced molecular events are conserved across the three classes of animals. In particular, for each group of animals, similar growth factors are released by injured and dying neurons, and similar receptors are activated. During the second, proliferative phase, however, these common molecular events lead to divergent cellular responses, perhaps because the growth-factor receptors are expressed by different cells. As discussed above, following a retinal lesion in fish, rod precursors, putative stem cells, Müller cells, and microglia/macrophages proliferate; in urodeles, RPE cells proliferate and transdifferentiate; in mammals, Müller cells, astrocytes, microglia/macrophages, and endothelial cells proliferate. During the third phase, the proliferation of these different cellular elements results in distinctly different outcomes. In fish, retinal progenitors emerge and give rise to postmitotic neurons in situ. In urodeles, the RPE gives rise to a neuroepithelium from which retinal neurons differentiate. In mammals, the proliferation of glial and endothelial cells in vivo leads to the formation of a glial scar and the consequent loss of functions subserved by the destroyed neurons. It is unclear why neuronal stem cells, which we now know exist in the adult mammalian CNS (see above), fail to generate new neurons following CNS injury.

Among these three models, we know the least about the early molecular events that stimulate retinal regeneration in fish and amphibians. The injury models used for these animals are all well characterized, the regeneration response is robust and largely invariant, the cells that give rise to regenerated neurons are identified, and the structural and functional outcomes are known. The goal of future studies is to identify the molecules and signaling pathways that initiate and control spontaneous retinal regeneration that leads to functional recovery in fish and amphibian retinas. The hope is to mimic these events therapeutically for treating injuries to the human central nervous system, including the retina.

References

Altman J (1970) Postnatal neurogenesis and the problem of neural plasticity. In: Himwich WH (ed) Developmental Neurobiology. CC Thomas, Springfield, Illinois, pp 197–230

Alvarez-Buylla A, Kirn JR (1997) Birth, migration, incorporation, and death of vocal control neurons in adult songbirds. J Neurobiol 33:585–601

Baird A (1994) Fibroblast growth factors: activities and significance of non-neurotrophin neurotrophic growth factors. Curr Opin Neurobiol 4:78–86

Baird A, Böhlen P (1990) Fibroblast growth factors. In: Sporn MB, Roberts AB (eds) Peptide growth factors and their receptors. Handbook of experimental pharmacology, vol 95/I. Springer, Berlin Heidelberg New York, pp 369–418

Boucher SE, Hitchcock PF (1998) Insulin-related growth factors stimulate proliferation of retinal progenitors in the goldfish. J Comp Neurol 394:386–394

Braisted JE, Raymond PA (1992) Regeneration of dopaminergic neurons in goldfish retina. Development 114:913–919

Braisted JE, Raymond PA (1993) Continued search for the cellular signals that regulate regeneration of dopaminergic neurons in goldfish retina. Dev Brain Res 76:221–232

Braisted JE, Essman TF, Raymond PA (1994) Selective regeneration of photoreceptors in goldfish retina. Development 120:2409–2419

Brockes JP (1997) Amphibian limb regeneration: rebuilding a complex structure. Science 276:81–87

Bugra K, Jacquemin E, Ortiz JR, Jeanny JC, Hicks D (1992) Analysis of opsin mRNA and protein expression in adult and regenerating newt retina by immunology and hybridization. J Neurocytol 21:171–183

Cameron DA, Easter SS (1993) The cone photoreceptor mosaic of the green sunfish, *Lepomis cyanellus*. Vis Neurosci 10:375–384

Cameron DA, Easter SS (1995) Cone photoreceptor regeneration in adult fish retina: phenotypic determination and mosaic pattern formation. Vis Neurosci 15:2255–2271

Cameron DA, Cornwall MC, MacNichol EF (1997) Visual pigment assignments in regenerated retina. J Neurosci 17:917–923

Cepko CL, Austin CP, Yang X, Alexiades M, Ezzeddine D (1996) Cell fate determination in the vertebrate retina. Proc Natl Acad Sci USA 93:589–595

Cheon EW, Kaneko Y, Saito T (1998) Regeneration of the newt retina: order of appearance of photoreceptors and ganglion cells. J Comp Neurol 396:267–274

Chiba C (1998) Appearance of glutamate-like immunoreactivity during retinal regeneration in the adult newt. Brain Res 785:171–177

Chiba C, Saito T (1995) Development of responses to excitatory and inhibitory amino acids in spiking cells during retinal regeneration in the adult newt. Jpn J Physiol 45:869–887

Chiba C, Matsushima O, Muneoka Y, Saito T (1997) Time course of appearance of GABA and GABA receptors during retinal regeneration in the adult newt. Dev Brain Res 98:204–210

Cook J (1998) Getting to grips with neuronal diversity. In: Chalupa L, Finlay B (eds) Development and organization of the retina. Plenum Press, New York, pp 91–120

Coulombre JL, Coulombre AJ (1965) Regeneration of neural from the pigmented epithelium in the chick embryo. Dev Biol 12:79–92

Dorsky RI, Rapaport DH, Harris WA (1995) *Xotch* inhibits cell differentiation in the *Xenopus* retina. Neuron 14:487–496

Dorsky RI, Chang WS, Rapaport DH, Harris WA (1997) Regulation of neuronal diversity in the *Xenopus* retina by Delta signalling. Nature 385:67–70

Dutt K, Scott M, Sternberg PP, Linser PJ, Srinivasan A (1993) Transdifferentiation of adult human pigment epithelium into retinal cells by transfection with an activated H-ras proto-oncogene. DNA Cell Biol 12:667–673

Easter SS (1983) Postnatal neurogenesis and changing connections. Trends Neurosci 6:53–56

Eguchi G, Kodama R (1993) Transdifferentiation. Curr Opin Cell Biol 5:1023–1028

Eriksson P, Perfilieva E, Bjork-Eriksson T, Alborn A, Nordborg C, Peterson D, Gage F (1998) Neurogenesis in the adult human hippocampus. Nat Med 4:1313–1317

Fernald RD (1991) Teleost vision: seeing while growing. J Exp Zool 5:167–180

Ferrari G, Cusella-De Angelis G, Coletta M, Paolucci E, Stornaiuolo A, Cossu G, Mavilio F (1998) Muscle regeneration by bone marrow derived myogenic progenitors. Science 279: 1528–1530

Finch CE, Laping NJ, Morgan TE, Nichols NR, Pasinetti GM (1993) TGF-β1 is an organizer of responses to neurodegeneration. J Cell Biochem 53:314–322

Fleming RJ, Purcell K, Artavanis-Tsakonas S (1997) The NOTCH receptor and its ligands. Trends Cell Biol 7:437–441

Fuji J, Wakasugi N (1993) Transdifferentiation from retinal pigment epithelium (RPE) into neural retina due to silver plumage mutant gene in Japanese quail. Dev Growth Differ 35:487–493

Gaze R, Watson WE (1968) Cell division and migration in the brain after optic nerve lesions. In: Wolstenholme GEW, O'Connor M (eds) Growth of the nervous system. Ciba Foundation Symposium. Churchill, London, pp 53–67

Goldman SA, Nottebohm F (1983) Neuronal production, migration, and differentiation in a vocal control nucleus of the adult female canary brain. Proc Natl Acad Sci USA 80:2390–2394

Gould E, Tanapat P, McEwen B, Flugge G, Fuchs E (1998) Proliferation of granule cell precursors in the dentate gyrus of adult monkeys is diminished by stress. Proc Natl Acad Sci USA 95:3168–3171

Graziadei PPC, Monti Graziadei GA (1978) The olfactory system: a model for the study of neurogenesis and axon regeneration in mammals. In: Cotman CW (ed) Neuronal plasticity. Raven Press, New York, pp 131–151

Grigorian EN (1996) [The urodelean retina as a model for studying the retinal regeneration potentials of other vertebrates] [in Russian]. Ontogenez 27:173–185

Grigorian EN, Anton GJ (1995) [An analysis of keratin expression in the cells of the retinal pigment epithelium during transdifferentiation in newts] [in Russian]. Ontogenez 26: 310–323

Grigorian EN, Ivanova IP, Poplinskaia VA (1996) [The discovery of new internal sources of neural retinal regeneration after its detachment in newts. Morphological and quantitative research] [in Russian]. Izv Akad Nauk Ser Biol 3:319–332

Guillemot F, Cepko CL (1992) Retinal fate and ganglion cell differentiation are potentiated by acidic FGF in an in vitro assay of early retinal development. Development 114:743–754

Henken DB, Yoon MG (1989) Optic nerve crush modulates proliferation of rod precursor cells in goldfish retina. Brain Res 501:247–259

Hitchcock PF (1997) Tracer coupling among regenerated amacrine cells in the retina of the goldfish. Vis Neurosci 14:463–472

Hitchcock PF, Cirenza P (1994) Synaptic organization of regenerated retina in the goldfish. J Comp Neurol 343:609–616

Hitchcock PF, Raymond PA (1992) Retinal regeneration. Trends Neurosci 15:103–108

Hitchcock PF, Vanderyt JT (1994) Regeneration of the dopamine-cell mosaic in the retina of the goldfish. Vis Neurosci 11:209–217

Hitchcock PF, Lindsey Myhr KJ, Easter SS, Mangione-Smith R, Jones DD (1992) Local regeneration in the retina of the goldfish. J Neurobiol 23:187–203

Hitchcock PF, Macdonald RE, VanDeRyt JT, Wilson SW (1996) Antibodies against pax6 immunostain amacrine and ganglion cells and neuronal progenitors, but not rod precursors, in the normal and regenerating retina of the goldfish. J Neurobiol 29:399–413

Humphrey MF, Constable IJ, Chu Y, Wiffen S (1993) A quantitative study of the lateral spread of Müller cell responses to retinal lesions in the rabbit. J Comp Neurol 334:545–558

Humphrey MF, Chu Y, Mann K, Rakoczy P (1997) Retinal GFAP and bFGF expression after multiple argon laser photocoagulation injuries assessed by both immunoreactivity and mRNA levels. Exp Eye Res 64:361–369

Julian D, Ennis K, Korenbrot JI (1998) Birth and fate of proliferative cells in the inner nuclear layer of the mature fish retina. J Comp Neurol 394:271–282

Kajiwara K, Okano H, Sagara H, Arizumi T, Asashima M, Shooter E (1999) Transitory mixed phenotype of retinal pigmented epithelium and neural retina precursors in transdifferentiation process during retinal regeneration. Invest Ophthal Vis Sci 40:960

Kaneko Y, Saito T (1992) Appearance and maturation of voltage-dependent conductances in solitary spiking cells during retinal regeneration in the adult newt. J Comp Physiol Ser A 170:411–425

Kaplan MS, Hinds JW (1981) Neurogenesis in the 3-month-old rat visual cortex. J Comp Neurol 195:323–338

Kaplan MS, McNelly NA, Hinds JW (1985) Population dynamics of adult-formed granule neurons of the rat olfactory bulb. J Comp Neurol 239:117–125

Kästner R, Wolburg H (1982) Functional regeneration of the visual system in teleosts. Comparative investigations after optic nerve crush and damage of the retina. Z Naturforsch 37:1274–1280

Keefe J (1973a) An analysis of urodelian retinal regeneration. I. Studies of the cellular source of retinal regeneration in *Notophthalamus viridescens* utilizing 3H-thymidine and colchicine. J Exp Zool 184:185, 206

Keefe J (1973b) An analysis of urodelian retinal regeneration. IV. Studies of the cellular source of retinal regeneration in *Triturus cristatus carnifex* using 3H-thymidine. J Exp Zool 184:239, 258

Kirsche W (1960) Zur Frage der Regeneration des Mittelhirnes der Teleostei [On the question of regeneration of the midbrain in teleost fish] [in German]. Verh Anat Gesell 56:259–270

Kirsche W (1967) Über postembryonale Matrixzonen im Gehirn verschiedener Vertebraten und deren Beziehung zur Hirnbauplanlehre [Postembryonic matrix zones in the brains of various vertebrates and their relationship to the study of brain organization] [in German]. Z Mikrosk Anat Forsch 77:313–406

Kirsche W, Kirsche K (1961) Experimentelle Untersuchungen zur Frage der Regeneration und Funktion des Tectum Opticum von *Carassius auratus* [Experimental investigations on the question of regeneration and function of the optic tectum in goldfish.] [in German]. Z Mikrosk Anat Forsch 67:140–182

Klein LR, MacLeish PR, Wiesel TN (1990) Immunolabelling by a newt retinal pigment epithelium antibody during retinal development and regeneration. J Comp Neurol 293:331–339

Knight J, Raymond P (1994) Retinal pigmented epithelium does not transdifferentiate in adult goldfish. J Neurobiol 27:447–456

Kodama R, Eguchi G (1995) From lens regeneration in the newt to in vitro transdifferentiation of vertebrate pigmented epithelial cells. Semin Cell Biol 6:143–149

Kurz-Isler G, Wolburg H (1982) Morphological study on the regeneration of the retina in the rainbow trout after ouabain-induced damage: evidence for dedifferentiation of photoreceptors. Cell Tissue Res 225:165–178

Lam K (1977) Electroretinogram of the newt during retinal regeneration. Brain Res 136:148–153

Levine EM, Hitchcock PF, Glasgow E, Schechter N (1994) Restricted expression of a new paired-class homeobox gene in normal and regenerating adult goldfish retina. J Comp Neurol 348:596–606

Levine R (1975) Regeneration of the retina in the adult newt, *Triturus cristatus*, following surgical division of the eye by a limbal incision. J Exp Zool 192:363–380

Levine R (1977) Regeneration of the retina in the adult newt, *Triturus cristatus*, following surgical division of the eye by a post-limbal incision. J Exp Zool 200:41–54

Levine RL (1981) La régénérescence de la rétine chez *Xenopus laevis* [Regeneration of the retina in *Xenopus laevis*] [in French]. Rev Can Biol 40:19-27

Lewis J (1991) Rules for the production of sensory cells. In: Bock G and Whelan J (eds) Regeneration of vertebrate sensory receptor cells. Ciba Foundation Symposium, vol 160. John Wiley, Chichester, pp 25-39

Lombardo F (1968) La rigenerazione della retina negli adulti di un Teleosteo. [Regeneration of the retina in an adult teleost] [in Italian]. Accad Lincei-Rendiconti Scienze Fis Mat Nat Ser 8, 45:631-635

Lombardo F (1972) Andamento e localizzazione della mitosi durante la rigenerazione della retina di un Teleosteo adulto. [Time course and localization of mitoses during regeneration of the retina in an adult teleost.] [in Italian]. Accad Lincei-Rendiconti Scienze Fis Mat Nat Ser 8, 53:323-327

Lopashov GZ, Sologub AA (1972) Artificial metaplasia of pigmented epithelium into retina in tadpoles and adult frogs. J Embryol Exp Morphol 28:521, 547

Maier W, Wolburg H (1979) Regeneration of the goldfish retina after exposure to different doses of ouabain. Cell Tissue Res 202:99-118

McKay R (1997) Stem cells in the central nervous system. Science 276:66-71

Mensinger A, Powers M (1999) Visual function in regenerating teleost retina following cytotoxic lesioning. Vis Neurosci (in press)

Minghetti L, Levi G (1998) Microglia as effector cells in brain damage and repair: focus on prostanoids and nitric oxide. Prog Neurobiol 54:99-125

Mitashov VI (1996) Mechanisms of retina regeneration in urodeles. Int J Dev Biol 40:833-844

Mitashov VI (1997) Retinal regeneration in amphibians. Int J Dev Biol 41:893-905

Morshead CM, Reynolds BA, Craig CG, McBurney MW, Staines WA, Morassutti D, Weiss S, Van der Kooy D (1994) Neural stem cells in the adult mammalian forebrain: a relatively quiescent subpopulation of subependymal cells. Neuron 13:1071-1082

Nagy T, Reh TA (1994) Inhibition of retinal regeneration in larval *Rana* by an antibody directed against a laminin-heparan sulfate proteoglycan. Dev Brain Res 81:131-134

Negishi K, Shinagawa S (1993) Fibroblast growth factor induces proliferating cell nuclear antigen-immunoreactive cells in goldfish retina. Neurosci Res 18:143-156

Negishi K, Teranishi T, Kato S, Nakamura Y (1987) Paradoxical induction of dopaminergic cells following intravitreal injection of high doses of 6-hydroxydopamine in juvenile carp retina. Dev Brain Res 33:67-79

Negishi K, Teranishi T, Kato S, Nakamura Y (1988) Immunohistochemical and autoradiographic studies on retinal regeneration in teleost fish. Neurosci Res Suppl 8:S43-57

Negishi K, Stell WK, Teranishi T, Karkhanis A, Owusu-Yaw V, Takasaki Y (1991a) Induction of proliferating cell nuclear antigen (PCNA)-immunoreactive cells in goldfish retina following intravitreal injection with 6-hydroxydopamine. Cell Mol Neurobiol 11:639-659

Negishi K, Sugawara K, Shinagawa S, Teranishi T, Kuo CH, Takasaki Y (1991b) Induction of immunoreactive proliferating cell nuclear antigen (PCNA) in goldfish retina following intravitreal injection with tunicamycin. Dev Brain Res 63:71-83

Negishi K, Shinagawa S, Ushijima M, Kaneko Y, Saito T (1992) An immunohistochemical study of regenerating newt retinas. Dev Brain Res 68:255-264

Nottebohm F, Alvarez-Buylla A, Cynx J, Kirn J, Ling CY, Nottebohm M, Suter R, Tolles A, Williams H (1990) Song learning in birds: the relation between perception and production. Philos Trans R Soc Lond Ser B Biol Sci 329:115-124

Nye JS, Kopan R (1995) Developmental signaling: vertebrate ligands for *Notch*. Curr Biol 5:966-969

Okada TS (1980) Cellular metaplasia or transdifferentiaton as a model for retinal cell differentiation. Curr Top Dev Biol 16:349-380

Opas M (1994) Substratum mechanics and cell differentiation. Int Rev Cytol 150:119-138

Opas M, Dziak E (1994) bFGF-induced transdifferentiation of RPE to neuronal progenitors is regulated by the mechanical properties of the substratum. Dev Biol 161:440-454

Ortiz JR, Vigny M, Courtois Y, Jeanny JC (1992) Immunocytochemical study of extracellular matrix components during lens and neural retina regeneration in the adult newt. Exp Eye Res 54:861-870

Owusu-Yaw V, Kyle A, Stell W (1992) Effects of lesions of the optic nerve, optic tectum and nervus terminalis on rod precursor proliferation in the goldfish retina. Brain Res 576: 220-230

Park CM, Hollenberg MJ (1989) Basic fibroblast growth factor induces retinal regeneration. Dev Biol 134:201-205

Park CM, Hollenberg MJ (1991) Induction of retinal regeneration in vivo by growth factors. Dev Biol 148:322-333

Park CM, Hollenberg MJ (1993) Growth factor-induced retinal regeneration in vivo. Int Rev Cytol 146:49-74

Pennypacker K (1997) Transcription factors in brain injury. Histol Histopathol 12:1125-1133

Perron M, Harris W (1997) Relationships between neurogenic and proneural genes in *Xenopus* retinogenesis. Dev Biol 186:237-247

Pittack C, Jones M, Reh TA (1991) Basic fibroblast growth factor induces retinal pigment epithelium to generate neural retina in vitro. Development 113:9011-9023

Pittack C, Grunwald GB, Reh TA (1997) Fibroblast growth factors are necessary for neural retina but not pigmented epithelium differentiation in chick embryos. Development 124:805-816

Powers M, Darst J, Palmer A, Pospichal M (1998) Retinal regeneration and vision: correlation between retinal structure and visual function. Soc Neurosci Abst 24:310

Raivich G, Bluethmann H, Kreutzberg GW (1996) Signaling molecules and neuroglia activation in the injured central nervous system. Keio J Med 45:239-247

Raymond PA (1985) The unique origin of rod photoreceptors in the teleost retina. Trends Neurosci 8:12-17

Raymond PA (1991) Retinal regeneration in teleost fish. In: Bock G, Whelan J (eds) Regeneration of vertebrate sensory receptor cells. Ciba Foundation Symposium, vol 160. John Wiley, Chichester, pp 171-186

Raymond PA, Hitchcock PF (1997) Retinal regeneration: common principles but a diversity of mechanisms. Adv Neurol 72:171-184

Raymond PA, Rivlin PK (1987) Germinal cells in the goldfish retina that produce rod photoreceptors. Dev Biol 122:120-138

Raymond PA, Reifler MJ, Rivlin PK (1988a) Regeneration of goldfish retina: rod precursors are a likely source of regenerated cells. J Neurobiol 19:431-463

Raymond PA, Hitchcock PF, Palopoli MF (1988b) Neuronal cell proliferation and ocular enlargement in Black Moor goldfish. J Comp Neurol 276:231-238

Raymond Johns P (1977) Growth of the adult goldfish eye. III. Source of the new retinal cells. J Comp Neurol 176:343-358

Reh T (1991) Common mechanisms of retinal regeneration in the larval frog and embryonic chick. In: Bock G, Whelan J (eds) Regeneration of vertebrate sensory receptor cells. Ciba Foundation Symposium, vol 160. John Wiley, Chichester, pp 192-208

Reh T, Levine E (1998) Multipotent stem cells and progenitors in the vertebrate retina. J Neurobiol 36:206-220

Reh TA, Nagy T (1987) A possible role for the vascular membrane in retinal regeneration in *Rana catesbienna* tadpoles. Dev Biol 122:471-482

Reh TA, Pittack C (1995) Transdifferentiation and retinal regeneration. Semin Cell Biol 6: 137-142

Reh TA, Nagy T, Gretton H (1987) Retinal pigmented epithelial cells induced to transdifferentiate to neurons by laminin. Nature 330:68-71

Reyer RW (1977) The amphibian eye: development and regeneration. In: Crescitelli F (ed) The visual system in vertebrates. Handbook of Sensory Physiology, vol 7(5). Springer, Berlin Heidelberg New York, pp 309-390

Reynolds BA, Tetzlaff W, Weiss S (1992) A multipotent EGF-responsive striatal embryonic progenitor cell produces neurons and astrocytes. J Neurosci 12:4565-4574

Richter W, Kranz D (1981a) Autoradiographic investigation on postnatal proliferative activity of the telencephalic and diencephalic matrix-zones in the axolotl (Ambystoma mexicanum), with special reference to the olfactory organ. Z Mikrosk Anat Forsch 95:883-904

Richter W, Kranz D (1981b) Autoradiographic investigations on postnatal proliferative activity of the matrix-zones of the brain in the trout (Salmo irideus). Z Mikrosk Anat Forsch 95:491-520

Ridet JL, Malhotra SK, Privat A, Gage FH (1997) Reactive astrocytes: cellular and molecular cues to biological function. Trends Neurosci 20:570-577

Robey E (1997) Notch in vertebrates. Curr Opin Genet Dev 7:551-557

Saito T, Kaneko Y, Maruo F, Niino M, Sakaki Y (1994) Study of the regenerating newt retina by electrophysiology and immunohistochemistry (bipolar- and cone-specific antigen localization). J Exp Zool 270:491-500

Sakaguchi DS, Janick LM, Reh TA (1997) Basic fibroblast growth factor (FGF-2) induced transdifferentiation of retinal pigment epithelium: generation of retinal neurons and glia. Dev Dyn 209:387-389

Sarthy VJ, Lam DMK (1983) Retinal regeneration in the adult newt, Notophthalmus viridescens: appearance of neurotransmitter synthesis and the electroretinogram. Dev Brain Res 6:99-105

Segaar J (1965) Behavioural aspects of degeneration and regeneration in fish brain: a comparison with higher vertebrates, degeneration patterns in the nervous system. In: Singer M, Schade JP (eds) Degeneration patterns in the nervous system. Prog Brain Res, vol 14. Elsevier, Amsterdam, pp 143-231

Sologub A (1975) Differentiation of the pigmented epithelium and stimulation of its metaplasia in teleost. Ontogenez 6:39-46

Sologub AA (1977) Mechanisms of repression and derepression of artificial transformation of pigmented epithelium into retina in Xenopus laevis. Wilhelm Roux's Arch Dev Biol 182:277-291

Squier CA (1980) The stretching of mouse skin in vivo: effect on epidermal proliferation and thickness. Invest Dermatol 74:68-71

Stone LS (1950a) Neural retina degeneration followed by regeneration from surviving pigment cells in grafted adult salamander eyes. Anat Rec 106:89-110

Stone LS (1950b) The role of retinal pigment cells in regenerating neural retinae of adult salamander eyes. J Exp Zool 113:9-31

Streit WJ (1996) The role of microglia in brain injury. NeuroToxicology 17:671-678

Stroeva OG, Mitashov VI (1983) Retinal pigment epithelium: proliferation and differentiation during development and regeneration. Int Rev Cytol 83:221-293

Stuermer CAO, Niepenberg A, Wolburg H (1985) Aberrant axonal paths in regenerated goldfish retina and tectum opticum following intraocular injection of ouabain. Neurosci Lett 58:333-338

Sullivan SA, Barthel LK, Largent BL, Raymond PA (1997) A goldfish Notch-3 homologue is expressed in neurogenic regions of embryonic, adult, and regenerating brain and retina. Dev Genet 20:208-223

Taylor JSH, Jack JL, Easter SS (1989) Is the capacity for optic nerve regeneration related to continued retinal ganglion cell production in the frog? A test of the hypothesis that neurogenesis and axon regeneration are obligatorily linked. Eur J Neurosci 1:626-638

Tcheng M, Fuhrmann G, Hartmann MP, Courtois Y, Jeanny JC (1994) Spatial and temporal expression patterns of FGF receptor genes type 1 and type 2 in the developing chick retina. Exp Eye Res 58:351-358

Vaney DI (1991) Many diverse types of retinal neurons show tracer coupling when injected with biocytin or neurobiotin. Neurosci Lett 125:187–190

Wanaka A, Milbrandt J, Johnson EM (1991) Expression of FGF receptor gene in rat development. Development 111:455–468

Wässle H, Riemann H (1978) The mosaic of nerve cells in the mammalian retina. Proc R Soc Lond Ser B Biol Sci 200:441–461

Weiss S, Reynolds BA, Vescovi AL, Morshead C, Craig CG, Van der Kooy D (1996) Is there a neural stem cell in the mammalian forebrain? Trends Neurosci 19:387–393

Zhao S, Thornquist SC, Barnstable CJ (1995) In vitro transdifferentiation of embryonic rat retinal pigment epithelium to neural retina. Brain Res 677:300–310

Zhao SL, Rizzolo LJ, Barnstable CJ (1997) Differentiation and transdifferentiation of the retinal pigment epithelium. Int Rev Cytol 171:225–266

Mouse Mutants for Eye Development

Jochen Graw

1
Introduction

Much of our knowledge on the function of genes in mammalian development derived from the molecular analysis of spontaneous or induced mutations. Mutations affecting the eye can be easily identified, and, therefore, a remarkable number of such mutants have been described. A systematic evaluation of large mouse populations for mutations affecting the eye lens was initiated in 1979, when Kratochvilova and Ehling described for the first time the systematic screening for murine dominant cataract mutants in the F_1 generation after paternal radiation treatment. The systematic screening for eye mutants was extended to the use of ethylnitrosourea (ENU) as mutagenic agent (Ehling et al. 1985; West and Fisher 1986). Some of the mutants were grouped into allelic series (Kratochvilova and Favor 1992; Everett et al. 1994; Favor 1995). All the mutant phenotypes were caused by single dominant gene mutations. Together with further mutants described in the literature, a mosaic can be built up reflecting important steps during lens development and differentiation. However, to understand the mechanisms of eye development in detail, the isolation of the corresponding genes and the characterization of the mutations at the molecular level are important. A prerequisite for the molecular analysis is the chromosomal localization of the gene. Additionally, the development of the technology of homologous recombination leads to a rapidly increasing series of new null alleles of a variety of genes and to important ocular phenotypes. A general overview for the genes and mutations which are involved in eye development and mapped at a particular chromosome is given in Fig. 1. Corresponding mutants will be discussed in this chapter according to the embryological time scale, reflecting also the genetic hierarchy of the genes.

GSF National Research Center for Environment and Health, Institute of Mammalian Genetics, Laboratory of Molecular Eye Development, Ingolstädter Landstr. 1, D-85764 Neuherberg, Germany

Results and Problems in Cell Differentiation, Vol. 31
M. E. Fini (Ed.): Vertebrate Eye Development
© Springer-Verlag Berlin Heidelberg 2000

Mouse Chromosome No:

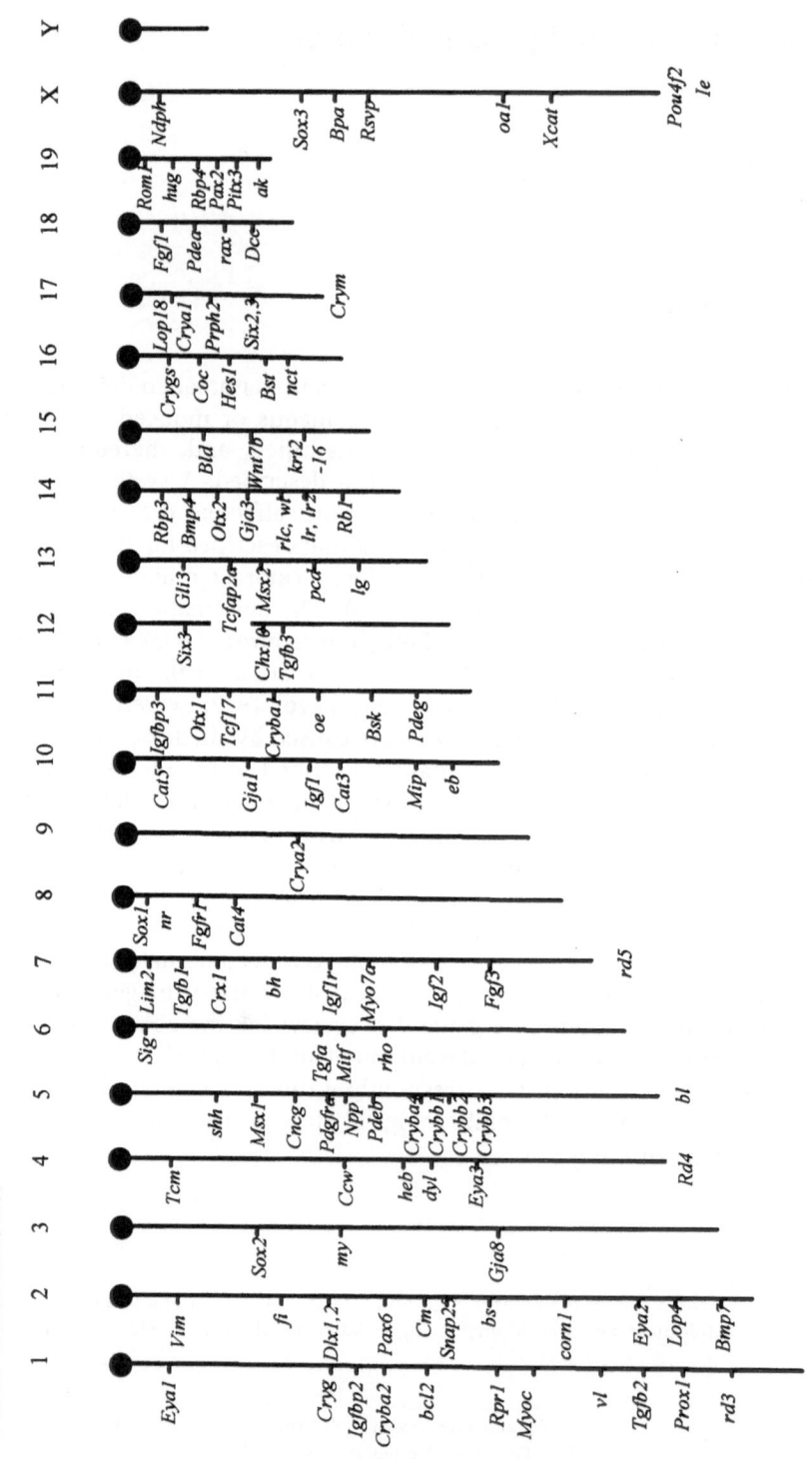

2
Mutations Affecting Early Eye Development

2.1
Mutations Affecting the Formation of the Lens Placode

One of the most important genes in eye development is the paired-box gene *Pax6*, which was recognized as being affected in the various alleles of the mouse and rat *Small eye* (*Sey*) mutants (Hill et al. 1991; Matsuo et al. 1993). In homozygous *Sey* mice, eyes and nasal cavities do not develop; they die soon after birth. The failure in lens development is attributed to a defect in the inductive interaction between the optic vesicle and the overlying ectoderm, since these tissues fail to make discrete contacts (Hogan et al. 1986). *Sey/Pax6* maps at mouse chromosome 2. The molecular analysis of *Sey* mutations revealed a point mutation leading to a TGA stop codon in the *Sey* allele and a transversion at a splice site of the *Pax6* gene in the *Sey^{1Neu}* allele (Hill et al. 1991). Homozygous *Pax6$^{-/-}$* mice (replacing *Pax6* by *lacZ*) had a phenotype similar to homozygous *Sey* mutants (St-Onge et al. 1997).

The important function of *Pax6* for eye development is indicated not only by the existence of severely affected mutants, but also by its high conservation among the animal kingdom. *Pax6* is the mammalian homologue of the *Drosophila* gene *eyeless* and the paradigm to look for *Drosophila* genes in vertebrate eye development. The murine *Pax6* gene is functional even in *Drosophila*; its ectopic expression leads to the formation of eyes at antennae, wings, or legs. Although the structure of the ommatidial insect eye is quite different from that of mammals, the *Pax6*-induced new eyes are functional from an electrophysiological point of view (Halder et al. 1995).

Pax6 transcripts are first detected in the presumptive fore- and hindbrain of 8-day old mouse embryos. Further, *Pax6* is expressed at E8.5 in the *sulcus opticus*, the lateral evagination at the basis of the forebrain. Later, Pax6 is present in the epithelial layer of optic vesicle, the optic stalk, and the surface ectoderm, which will give rise to the lens. Between E10 and E12, *Pax6* is

---

Fig. 1. Mapped genes involved in eye development. Gene symbols which are not mentioned in the text: *bh* brain hernia; *bl* blebbed; *Bld* blind; *Bsk* bareskin; *Cncg* cyclic nucleotide gated channel (cGMP- gated); *Coc* coralliform cataract; *Crym* µ-crystallin; *eb* eye blebs; *Ie* eye-ear reduction; *Igf* insulin-like growth factor; *Krt2-16* keratin complex 2 gene 16; *Lop* lens opacity; *Myo7a* myosin VIIa; *Myoc* myocilin; *Ndph* Norrie disease homologue; *nr* nervous; *oa1* ocular albinism 1; *pcd* Purkinje cell degeneration; *Pdea Pdeg* phosphodiesterase-α or -γ; *Rbp3 Rbp4* retinol binding protein 3 or 4 (interstitial); *Rpr1* photoreceptor protein 1; *Rsvp* red sensitive visual pigment; *Sig* sightless; *Tcf17* transcription factor 17; *Wnt7b* wingless-related MMTV integration site 7B. The figure is not to scale. For mapping details refer to the actual list at the Mouse Genome Informatics database: http://www.informatics.jax.org/ locus/html or the annual chromosome reports at http://www.informatics.jax.org/bin/ccr/

strongly expressed in the inner layer of the optic cup, in the lens, and in the prospective cornea. At E15.5, *Pax6* is expressed in the two layers of the neural retina, the anterior epithelium of the cornea, and in the lens. Besides in the eye, *Pax6* occurs in specific regions of the brain, the ventral neural tube, and the olfactory epithelium. These domains of expression are consistent with the phenotypes of homozygous *Sey* mutants (Walther and Gruss 1991; Hanson and van Heyningen 1995).

In addition, the function of three other genes will be discussed here briefly describing corresponding knockout mutations. The analysis of the complex and very severe phenotype of the deletion of *sonic hedgehog* (*Shh*) demonstrated that one of its functions is the definition of a midline axis during early development. This is also important for the paired formation of the eye anlagen. In $Shh^{-/-}$ mutant embryos the optic vesicles are fused at the midline and the optic stalks are deficient or absent. There is no invagination to form the characteristic double-layered optic cups, and the fused eye tissue at the midline forms a pigmented epithelium with no apparent differentiation of retinal tissue. In the cyclopic mutant eye structure, which is constituted solely of pigmented epithelium, *Pax6* is also expressed. In contrast, *Pax2*, whose expression is normally restricted to the optic stalk connecting the eye primordia to the brain (see Sect. 2.3), is not present in $Shh^{-/-}$ mutant embryos, consistent with the histologically apparent absence of the optic stalk (Chiang et al. 1996). *Shh* was mapped to mouse chromosome 5 (Marigo et al. 1995).

Homozygous mouse embryos carrying a null allele of the *Rx* gene coding for a conserved vertebrate homeobox gene have no visible eye structures, whereas mice heterozygous for the loss of *Rx* are apparently normal. The abnormal phenotype of the homozygous mutants is visible as early as E9.0 and E10.5 by a failure to form the indentation (*sulcus opticus*) that gives rise to the optic cup. It demonstrates clearly that the *Rx* gene function is required for eye formation from its initial stage. From overexpression studies it was concluded that the *Rx* gene family plays an important role in the establishment and/or proliferation of retinal progenitor cells (Mathers et al. 1997). The *Rx* gene (also referred to as *rax*: retina and anterior neural fold homeobox) is mapped to mouse chromosome 18 (Furukawa et al. 1997a).

Recently, the knockout of *Lhx2*, a LIM-homeobox gene, was reported. LIM domains do not interact directly with DNA, but they are involved in protein-protein interactions. LIM-homeobox genes are expressed widely in the developing central nervous system, and also in the region of the optic vesicle. Mice homozygous for the *Lhx2* deletion are anophthalmic. Eye development is arrested after the formation of the optic vesicle, but prior to the formation of the optic cup. Eye development did not proceed further in the mutant embryos, and by E10.5 only a small remnant of the optic vesicle was observed. At E13.5, no lens, no globe, and no retina were seen. *Lhx2* might influence the expression of *Pax6* at this region, since in the $Lhx2^{-/-}$ embryos no *Pax6* expression was observed in the lens placode (Porter et al. 1997). The chromosomal localization is as yet unknown.

The mouse mutants described above are excellent models for comparable human inherited diseases. In human, mutations in *PAX6* are responsible for human aniridia and have also been found in patients with Peter's anomaly, congenital cataracts, autosomal dominant keratitis, and isolated foveal hypoplasia (for recent review see Prosser and van Heyningen 1998).

Mutations in the *SHH* gene cause holoprosencephaly (HPE), a common developmental defect of the forebrain. HPE has a prevalence of 1:250 during embryogenesis and 1:16.000 newborn infants. Alobar HPE, the most severe form, which is usually incompatible with postnatal life, involves complete failure of division of the forebrain into the two hemispheres, and is characteristically associated with facial anomalies including cyclopia, a primitive nasal structure (proboscis) and/or midfacial clefting (Roessler et al. 1996).

2.2
Mutations Affecting the Optic Cup as the Prospective Retina

Since the first discovery of mouse microphthalmia (*Mi*) mutation (Hertwig 1942), at least 17 mutant alleles have been identified and genetically characterized at chromosome 6 (overview in Steingrímsson et al. 1994). The affected gene encodes a basic-helix-loop-helix leucine zipper family of transcription factors, referred to as *Mitf* (microphthalmia-associated transcription factor). The phenotype of these mouse mutants varies tremendously depending on the site and size of the mutation, and may be either dominant or recessive. In particular alleles (*mi/mi* and Mi^{wh}/Mi^{wh}) it has been shown that the outer layer of the optic cup (normally the future pigment layer of the retina) thickens abnormally at embryonic day 10–11, leading to a failure of the choroid fissure to close and to colobomatous microphthalmia (Müller 1950; Packer 1967; Hero et al. 1991). In addition, other cell types are also affected, like the neural-crest-derived melanocytes, which in severe cases leads to deafness owing to the lack of inner ear melanocytes. Recently, mutations in the rat mutant *mibA* (Opdecamp et al. 1998), the Syrian hamster mutant *anophthalmic white* (Hodgkinson et al. 1998), or in the *silver* homozygote (*B/B*) of the Japanese quail (Mochii et al. 1998) were characterized as mutations within the *Mitf* gene. Moreover, also mutations within the human homologue *MITF* were found in patients suffering from Waardenburg-Syndrome type 2 (Tassabehji et al. 1994) or Tietz syndrome (Amiel et al. 1998).

Ocular retardation (or) is a recessive mouse mutant with abnormal eye development. Homozygous *or* mice are blind with obvious microphthalmia, a cataractous lens, a thin retina that is morphologically poorly differentiated, and no optic nerve (Truslove 1962; Robb et al. 1978; Theiler et al. 1976; Silver and Robb 1979). The mutation is mapped to chromosome 12 (Hawes and Roderick 1990). Burmeister et al. (1996) demonstrated that the allele *or^J* has a premature stop codon in the homeobox of the *Chx10* gene. In the developing wild-type mouse, *Chx10* is expressed throughout the anterior optic vesicle and all neuroblasts of the optic cup. In the mature retina, the Chx10 protein is restricted to the inner nuclear layer. Besides the eye, *Chx10* transcripts were also detected in the developing thalamus, hindbrain, and ventral spinal cord (Liu et al. 1994). No Chx10 protein was detected in the retinal neuroepithelium of *or^J* homozygotes, leading to reduced proliferation of retinal progenitors and to a specific absence of differentiated bipolar cells (Burmeister et al. 1996).

2.3
Mutations Affecting the Optic Stalk as the Prospective Optic Nerve

Pax2 is the second *Pax* gene besides *Pax6* which is expressed in the eye. Recently, a mutation was described in the mouse, *Pax2^{1Neu}*, which exhibits defects in the optic nerve development, the retinal layer of the eye, and several defects of the kidney and brain. The mutation was mapped to mouse chromosome 19 and further characterized by a frameshift in the 5'-region of *Pax2* and a stop codon 26 amino acids downstream. The mutant gene is predicted to code for a nonfunctional protein lacking almost the entire paired domain and the complete homeodomain (Favor et al. 1996).

At the same time, when the characterization of *Pax2^{1Neu}* was published, a *Pax2* null mutant was reported (Torres et al. 1996). The phenotype of this *Pax2* knockout mouse is similar to *Pax2^{1Neu}*: extension of the pigmented retina into the optic stalk, failure of the optic fissure to close resulting in coloboma, and the ipsilateral formation of the optic tracts without formation of the *chiasma opticum*. Some malformation at the inner ear have been observed in addition to the ocular dysmorphology.

Pax2 expression during optic nerve development can be divided into two phases: the morphogenesis of the optic cup and stalk, prior to axon growth, and secondly, the period of axogenesis (Nornes et al. 1990). *Pax2* transcripts are first detected in the most distal region of the optic vesicle (opposed to the surface ectoderm), when it is making contact with the surface ectoderm. Later, *Pax2* is expressed over the ventral 2/3 of the invaginating epithelium and the ventral region of the optic stalk. The expression ends at the border of the diencephalon. During axogenesis, *Pax2* is absent from the neuroblastic layer of the retina; however, there is a strong expression in the optic disk and in the entire optic nerve (Rothenspieler and Dressler 1993).

In human, a mutational analysis of *Pax2* in a family with renal-coloboma syndrome was conducted, leading to the detection of a single nucleotide insertion. The mutation causes a frameshift in the 5'-region of *Pax2* and a stop codon 26 amino acids downstream and is identical to that observed in the *Pax2^{1Neu}* mouse mutant (Favor et al. 1996). The two affected siblings from this family exhibit optic nerve colobomas (Sanyanusin et al. 1995).

2.4
Mutations Affecting the Lens Vesicle

The formation of the lens vesicle is important for the formation of the lens and the anterior eye segment, but also for the right proportion of the retina. This key role is addressed by some mouse mutants like *aphakia* (*ak*), *eye lens aplasia* (*elap*), *eyeless*, $Gli3^{Xt}$, *head blebs* or *myelencephalic blebs*. The *eyeless* mutant *ey1* was defined as being responsible for anophthalmia (Chase 1944); its chromosomal localization is as yet unknown. In homozygous mutants, lens invagination at E10 is abnormal, the lens is smaller than normal, and often improperly centered in the optic cup. The optic vesicle of *ey1* mutants has an abnormal contact to the presumptive lens ectoderm (Webster et al. 1984).

In mouse mutants referred to as *myelencephalic blebs* (*my*), the first morphological alteration becomes visible at E9.5. A large area of extracellular matrix is formed between the optic vesicle and the overlying presumptive lens ectoderm, whereas in the wild type almost no extracellular matrix was observed. At E12, the lens capsule is ruptured, and the inner limiting membrane of the presumptive neural retina is affected. At E14, the cornea and other structures of the eye cannot be identified. In addition to the eye defects, the *my* mice suffer also from kidney anomalies (Center and Polizotto 1992). The mutation is located on mouse chromosome 3 (Davisson et al. 1976). Phenotypically similar to *my*, but mapped to mouse chromosome 4, is the *head blebs* mutation (*heb*). *Heb* mice produce abnormal eyes (sometimes also anophthalmia) due to prenatal blebs, usually on the head. Additionally, sometimes fetal death, open eyelids, and folded retinas at birth were observed (Varnum and Fox 1981).

Mice homozygous for the dominant mutation *Extra-toes* die perinatally with multiple malformations involving the eye and other organs. Concerning the ocular development, three classes of the homozygotes can be identified in early midgestation: (1) an apparently normal optic cup and an apparently normal lens vesicle is formed; (2) the optic cup is distorted and the lens vesicle is small. This group goes on to develop small eyes and coloboma because the optic fissure of the optic nerve does not close. (3) the optic cup and lens placode are not formed. A defective optic vesicle might be the primary cause for these eye malformations (Franz and Besecke 1991). The mutation has been mapped to chromosome 13 (Lyon et al. 1967) and has been shown to be a deletion within the gene *Gli3*, an oncogene encoding a zinc-finger transcription factor (Hui and Joyner 1993); the mutation is now referred to as $Gli3^{Xt}$. *Gli3* is expressed during normal mouse development in various tissues (Walterhouse et al. 1993). In human Greig syndrome families, *Gli3* is interrupted by translocation (Vortkamp et al. 1991).

The homozygous mouse mutant *aphakia* (*ak*) was characterized by bilaterally aphakic eyes without a pupil (Varnum and Stevens 1968). The abnormality in eye development of homozygous *ak* mice is first observable at the early lens vesicle stage. The irregular lens development is arrested at the lens stalk stage, which is usually present transiently during detaching of the lens vesicle from the surface epithelium (details of the ocular development have been reviewed recently by Graw 1999). All other malformations observed at later stages are most likely consequences of these initial defects during formation of the lens vesicle. After birth, the sac-like lenticular structure disappears, and most of the other ocular structures are strongly affected. In particular, the pigmented cells cover the entire anterior part of the eye, and the growing retina fills the space of the whole eye globe.

The mutation *ak* was mapped to mouse chromosome 19 (Varnum and Steven 1975). In a very detailed mapping study, Grimm et al. (1998) reported recently the position of *ak* 11 cM proximal to *Pax2*. On the basis of the location of *ak* to this defined region of chromosome 19 several genes have been considered as candidates for *ak*. However, none of them was confirmed as target of the mutation.

A phenotype similar to that in the *ak* mutants was observed in the *eye lens aplasia* mice (*elap*; formerly *lap*). Observation of fetal eye development from day 9 to 17 of gestation indicated that the eye developed normally until the start of invagination of the lens placode at day 10. However, formation of the lens vesicle progressed abnormally to form a mass of cells without a cavity at day E11. As a result of apoptotic processes, this mass was reduced in size at day E12 and had vanished by day E13/E14. Also abnormal development was observed in the cornea, vitreous body, or retina after E12. Complementary mating between mice homozygous for *elap* and *ak* produced no F_1 newborns with abnormal eyes, indicating that these mutations affect different genes (Aso et al. 1995, 1998). The chromosomal localization of *elap* is not yet known.

Congenital aphakia in man is an ocular disease based upon the absence of the eye lens. It can be subdivided into primary and secondary forms (Vermeij-Keers 1975). Primary aphakia is characterized by the total absence of the lens or the lens primordium, the iris, and the anterior chamber. In contrast, secondary congenital aphakia results from disturbances during lens development at a later stage (e.g., by rubella virus infection).

Besides this, naturally occurring mutations mice deficient in the bone morphogenetic proteins 4 and 7 (Bmp4, Bmp7) revealed abnormalities at early stages of eye development. The BMPs belong to the TGF-β superfamily of secretory signaling molecules. In *Bmp4* homozygous null mutant embryos, lens induction is absent, but the process can be rescued by exogenous BMP4 protein applied to the optic vesicle. BMP4 appears to regulate the expression of a putative downstream gene, *Msx2*, in the optic vesicle (Furuta and Hogan 1998). *Bmp4* (previously known as *Bmp-2b*) is mapped to mouse chromosome 14 (Dickinson et al. 1990).

The *Bmp7* null mutants die shortly after birth due to kidney defects. Concerning eye development, at E11, when the optic cup has developed and the invagination of the lens vesicle is visible, the lens vesicle of the mutants was smaller or even absent. In addition, the entire eye development was delayed; the lens vesicle remains connected with the surface ectoderm (Luo et al. 1995). This is consistent with the observation that BMP7 is present in the optic vesicle and the surrounding head ectoderm as early as E9.5 (Lyons et al. 1995). The *Bmp7* gene is mapped in the mouse to chromosome 2 (Marker et al. 1995).

Recently, West-Mays et al. (1999) reported the characterization of mice without the gene coding for the transcription factor AP-2α (*Tcfap2a*). These *Tcfap2a*-null embryos exhibited ocular phenotypes ranging from a complete lack of eyes (anophthalmia) to defects in the developing lens involving a persistent adhesion of the lens to the overlaying surface ectoderm (like in the *ak/ak* mutants). $Tcfap2a^{-/-}$ mutants also exhibited defects in the optic cup consisting of transdifferentiation of the dorsal retinal pigmented epithelium and the absence of a defined ganglion cell layer. *Tcfap2a* is mapped to mouse chromosome 13 (Warren et al. 1996).

2.5
Further Genes Important for Early Eye Development

From the present data it seems obvious that at least four genes, *Shh, Pax2, Pax6,* and *Rx (rax)*, are necessary to induce ocular development. However, there is growing evidence for additional control genes, including members of the *Six* and *Eya* gene families (Oliver et al. 1996; Chen et al. 1997a), *dachshund (dach*; Hammond et al. 1998; Caubit et al. 1999), *Msx1* or *Msx2* (Foerst-Potts and Sadler 1997), *Pitx3* (Semina et al. 1997), or *Prox1* (Tomarev et al. 1998). From these data, it is now becoming clear that an entire genetic cassette may be used for early eye development in vertebrates (Desplan 1997). Unfortunately, for these newly discovered genes no mutants are yet available.

3
Maturation of the Eye (1): Lens Development

3.1
Cataracts: Inherited Anomalies of the Lens

Cataracts as inherited lens opacities reflect disturbances during lenticular development and differentiation, when the lens vesicle is already formed. Because some of the mutants can be recognized by the naked eye, a variety of mouse mutants is available and will be discussed in some detail.

One of the first detected cataract mutations is the *Cataract Fraser* (Cat^{Fr}; Fraser and Schabtach 1962). Later, it was shown that Cat^{Fr} is allelic with *Lop*. The two *Cat* alleles, Cat^{Lop} and Cat^{Fr}, were mapped 20 cM distal to *Steel* (*Sl*) at chromosome 10 (Lyon et al. 1981; Muggleton-Harris et al. 1987). A candidate gene for the *Cat* locus encodes the membrane intrinsic protein (gene symbol: *Mip*; Griffin and Shiels 1992). Sequence analysis revealed that the Cat^{Fr} mutation is due to a transposon-induced splicing error leading to a truncated form of *Mip* transcripts. However, the mutation in the Cat^{LoP} leads to a single amino-acid substitution, which inhibits targeting of Mip to the cell membrane (Shiels and Griffin 1993; Shiels and Bassnett 1996). Mip forms specialized junctions between the fiber cells and can be first detected in the primary fiber cells of the early lens vesicle. In the Cat^{Fr} mutant, beginning at E14, the cell nuclei in the deep cortex become abnormally pycnotic; degeneration of cytoplasm and destruction of the lenticular nucleus follow (Zwaan and Williams 1969).

At mouse chromosome 10, another group of cataract mutations is localized 3.2 cM proximal to *Sl* (Löster et al. 1997). It consists of two alleles and is provisionally referred to as *Cat3* (Kratochvilova and Favor 1992). $Cat3^{vl}$ (vacuolated lens; Kratochvilova 1981) and $Cat3^{vao}$ (cataract with anterior opacity; Graw et al. 1986) arose independently in the F_1 generation after paternal γ-irradiation. On the basis of the location of *Cat3* to this defined region of chromosome 10, several genes have been considered as candidates for *Cat3*, but none of them was confirmed as target of the mutations (Löster et al. 1997). To the region of conserved synteny at human chromosome 12q21-24 the human eye disorder *cornea plana congenita* has been mapped. Clinical signs include opaque thickening of the cornea and adhesions between the iris and cornea (Tahvanainen et al. 1996).

Slit lamp observation revealed that the lenses of the $Cat3^{vl}$ mutants are filled up with small and large vacuoles. Pupillary dilatation is limited. The anterior chamber is flattened, and opacified corneas are frequently observed. The $Cat3^{vao}$ mutants show an opaque area immediately beneath the lenticular capsule forming multiple disk-like opacities around the optical axis. The two alleles differ mainly in the appearance of their opacities (Löster et al. 1997).

The histological analysis identified an aberrant cell layer between the anterior epithelium and the primary lens fibers at embryonic day E12.5, when the lens vesicle is filled by the primary lens fibers. It leads to a maldevelopment of the anterior lens epithelium and degeneration of the fibers. After birth, the lens capsule ruptures at the equatorial region, and synechiae with the iris occur (Graw et al. 1998). Biochemical examination of the cataractous $Cat3^{vao}$ lenses demonstrated the presence of oxidative and osmotic stress (Graw et al. 1989a, 1990b).

The dominant X-linked cataract mutation *Xcat* was recovered after parental radiation (Favor and Pretsch 1990). Histological analysis during the embryonic development revealed that in the affected embryos the primary fiber cells are irregularly arranged and show small foci of cellular disintegration. Progressive degeneration of fibers occurs. However, the lens epithelium and the newly differentiated fibers show no evident abnormality indicating that the mutation affects the differentiation of the primary lens fiber cells after their initial elongation. Analysis of crystallin and cytoskeleton proteins of postnatal cataractous lenses revealed no significant abnormalities when compared to the normal lens (Grimes et al. 1993). Detailed genetical analysis placed the *Xcat* mutation to the distal end of the mouse X chromosome. It suggests that this locus should map to a conserved block at Xp22.1-p22.3 in human. To this region, the Nance Horan Syndrome has been mapped (Stambolian et al. 1994).

Another X-linked mouse mutant with affected lenses is referred to as *bare patches* (*Bpa*). Since hemizygous males die before birth, heterozygous females have patches of bare skin. Lens cortical "frost figure" opacities are present. *Bpa* is considered as a model for human X-linked syndrome chondrodysplasia punctata (Happle et al. 1983; Angel et al. 1993).

The *total opacity To3* is placed on chromosome 7 (Kerscher et al. 1996). Mice heterozygous or homozygous for the *To3* mutation exhibit a total opacity of the lens with a dense cataract. Additionally, homozygotes exhibit microphthalmia and abnormally small eyes. Histological analysis revealed vacuolization of the lens and gross disorganization of the fibers; posterior lens rupture can be observed only in homozygotes. The *To3* mutation was characterized as a single G→T transversion within the first exon of the *Lim2* gene coding for a lens-specific integral membrane protein, MP19. It was predicted that this DNA change results in a nonconservative substitution of a valin for the normally encoded glycine at amino acid #15 of the MP19 protein (Steele et al. 1997).

The *Gja8* gene encodes the lens-specific gap junction membrane channel protein α8, which is also referred to as connexin 50 or MP70. It maps to mouse chromosome 3 (Kerscher et al. 1995) and was demonstrated recently to be affected by a single A→C transversion within codon 47 of the *No2* (nuclear opacity 2) mouse cataract. The sequence alteration is predicted to result in the nonconservative substitution of Ala for the normally encoded Asp (Steele et al. 1998). A similar phenotype (microphthalmia and nuclear cataract) was observed in *Cx50* null mice (White et al. 1998). A mutation in the corresponding human gene leads to a zonular pulverulent cataract (Shiels et al. 1998).

A knockout mutation of another type of connexin, the gene coding for connexin46 (or gap junction protein α3; gene symbol *Gja3*), exhibits nuclear cataract, which was associated with the proteolysis of crystallins. Obviously, there is no influence on the early stages of lens formation (Gong et al. 1997). *Gja3* is mapped to mouse chromosome 14 (Haeflinger et al. 1992). Similarly, disruption of the gene coding for connexin43 (gene symbol: *Gja1* for gap junction membrane channel protein α1) demonstrated normal development of the lens and differentiation of the fiber cells at the bow region. The lenses of the *Gja1*$^{-/-}$ mice exhibit grossly dilated extracellular spaces and intracellular vacuoles. These changes suggest that the osmotic balance within these cells is markedly altered (Gao and Spray 1998). *Gja1* is mapped to mouse chromosome 10 (Hsieh et al. 1991).

The *Patch* (*Ph*) mutation in mice is of spontaneous origin and was first characterized by Gruneberg and Truslove (1960). The *Ph* mutation maps to chromosome 5 and appears to be a deletion encompassing the *Pdgfra* gene coding for the platelet-derived growth factor receptor-α-subunit. However, the deletion may affect also other genes. In *Ph/Ph* embryos the number of neural-crest-derived corneal fibroblasts is reduced as well as the neural-crest-derived periocular mesenchyme. The thickness of the cornea in E16 *Ph/Ph* embryos is about half that of a normal cornea (Morrison-Graham et al. 1992). Additionally, empty spaces in the lens matrix are seen in the *Patch* mutants (Schatteman et al. 1992).

The mutant *blind-steril* (*bs*) is characterized by bilateral nuclear cataracts, microphthalmia and glossy coats. The cataracts are detectable at E16. Females are fertile, but males are sterile. The mutation was mapped to mouse chromosome 2 (Varnum 1983). The mutation *vacuolated lens* (*vl*) is mapped to mouse chromosome 1 and leads to opaque white lenses. Additionally, the mutants are characterized by a white belly spot and *spina bifida*. Small lens vacuoles are present at birth (Dickie 1967, 1969).

The so-called *rupture of lens cataract* (*rlc*) was mapped to chromosome 14 (Matsushima et al. 1996) and, recently, a similar form, *lr2* (*lens rupture 2*) was mapped to a close position (Song et al. 1997). The opacity in the *rlc/rlc* mice becomes apparent at 35–60 days of age; there are no developmental changes reported (Iida et al. 1997). Other forms of cataracts, which are formed postnatally without observed developmental alterations, are the *Nakano cataract* (*nct*; Takehana 1990, Wada et al. 1991). The mutation was mapped recently to chromosome 16 (Hiai et al. 1998). The *Tcm* mutation (total cataract with microphthalmia), a cataract with iris dysplasia and coloboma (Zhou et al. 1997) and the *Ccw* mutation, *cataract and curly whiskers* (Kerscher et al. 1996), are localized on mouse chromosome 4. The *nuclear-posterior polar opacity* (*Npp*) maps to chromosome 5, and *Cat5* (*previously To2*), a total opacity, to chromosome 10 (Everett et al. 1994).

3.2
Differentiation Processes in the Developing Lens – the Crystallin Connection

Up to 90% of the soluble protein in the postmitotic lens cells consists of proteins, which are referred to as α-, β-, and γ-crystallins (Mörner 1893). The α-crystallins form high-molecular aggregates. Recent findings on the structure and function of α-crystallins demonstrated that they have chaperone activity and belong to the family of the small heat shock proteins. In contrast to αB-crystallin (gene symbol: *Crya2*), which is ubiquitously expressed, the αA-crystallin (gene symbol: *Crya1*) occurs mainly in the lens. The β/γ-crystallin superfamily exhibits a characteristic protein motif, the so-called Greek key motif, in a quadruple organization. It is considered to be essential for the extremely high protein concentration within the lens (for a recent review on crystallins see Graw 1997).

The *Cryg* genes are organized as a cluster of six very similar genes (*Cryga>Crygf*) within approximately 50 kb on mouse chromosome 1. In mice, five independent mutations have been characterized as mutations within the *Cryg* gene cluster: first of all, the *Elo* mutant (*Eye lens obsolescence*) was characterized to carry a single nucleotide deletion in the γE-crystallin gene. The mutation destroys the reading frame of the gene, and at the protein level one of the Greek key motifs is affected (Cartier et al. 1992). Therefore, the new allele symbol is $Cryge^{elo}$. Further, three members of the allelic series of the Neuherberg *Cat2* mutants (Kratochvilova and Favor 1992) have been characterized by mutations within the *Cryg* genes (Klopp et al. 1998).

The mutation *ENU-436* was induced by ethylnitrosourea (Favor 1983; 1984), and an A→G transition was found at position #230 in exon 2 of *Cryga*. The new allele symbol is suggested as $Cryga^{1Neu}$. The deduced replacement of Asp by Gly at amino acid position 77 affects the connecting peptide between the second and third Greek key motif (Klopp et al. 1998). The lenses of the $Cryga^{1Neu}$ mutants were characterized as a small nuclear opacity (Favor 1983).

The cataract mutation $Cat2^{nop}$ arose spontaneously (Graw et al. 1984). The molecular characterization revealed that in the $Cat2^{nop}$ the 3rd exon of the *Crygb* gene is affected. The small deletion of 11 bp starting after position #416 and the insertion of 4-bp lead to a frame shift and create finally a new stop codon. The new allele symbol is suggested as $Crygb^{nop}$. The corresponding γB-crystallin protein is predicted to be truncated after 144 amino acids; the last six amino acids are different from the wild-type γB-crystallin. Western blot analysis demonstrated the stable expression of the wrong protein (Klopp et al. 1998).

Using in situ hybridization techniques with a probe detecting all *Cryg* transcripts in embryonic sections, a lower extent of *Cryg* transcripts was detected in the $Crygb^{nop}$ mutants beginning from embryonic day 13.5. The first morphological abnormality in the mutant lenses was observed as swelling of lens fibers at embryonic day 15.5 (Santhiya et al. 1995). Histological investigations at the age of 3 weeks confirmed the characterization as a nuclear opacity. The nuclei of the cortical cells could also be detected in the area of the lens nucleus of the $Crygb^{nop}$ lenses (Graw et al. 1990c). Biochemical investigations demonstrated an increase in the concentration of oxidized glutathione in the $Crygb^{nop}$ lenses over the wild type, which is not due to a corresponding decrease of an enzyme related to this metabolite (Graw et al. 1985, 1989a, 1990c).

The $Cat2^{t}$ mutant was discovered in an experiment using X-rays as mutagenic agent (Graw et al. 1986). The final result of DNA sequencing demonstrated that in this particular mutant the third exon of the γE-crystallin gene is affected. The C→G exchange at pos. #432 creates a stop codon predicting a truncated protein after amino acid 143. The new allele symbol is suggested as $Cryge^t$ (Klopp et al. 1998). In the lenses of heterozygous and homozygous $Cryge^t$ mutants, the epithelial and fiber cells were swollen, the lens capsule was ruptured, and the cellular organization of the lens was completely destroyed (Graw et al. 1990a).

The fifth mutation analyzed in this context is referred to as $Cat2^{ns}$ (previously as $Scat$; Graw et al. 1989b) and occurred spontaneously in the GSF breeding colony. Molecular analysis revealed a large deletion (>2 kB) within the 3rd exon of $Crgye$; the exact breakpoints have not yet been determined (N. Klopp and J. Graw, unpubl.). The mutation should be referred to as $Cryge^{ns}$, and causes an anterior suture opacity in heterozygotes and microphthalmia with vacuolated lenses in homozygotes. Exposing the weakly affected heterozygous mutants to UV radiation, rupture of the posterior lens capsule could be induced (Forker et al. 1997), as observed for untreated mice only in homozygous mutants (Graw et al. 1989b).

All these $Cryg$ mutations affect only the lens cells and no other part of the eye; however, the size of the entire eye is always smaller than the wild type. A common feature in all three mutants is the inhibition of a Mg^{2+}-dependent DNase in the lens. The decrease of DNase activity followed the same directionality ($Cryge^{ns} > Crygb^{nop} > Cryge^t$) as the decrease in the relative content of water-soluble lens protein, which might be used as a rough indicator for the severity of cataractogenesis (Graw and Liebstein 1993). This finding was the first experimental evidence for the stop of the lenticular differentiation process in these cataractous lenses.

From a developmental point of view, the $Cryge^{elo}$ mutants have been characterized in most details up to now. In this particular mutant line, the first change was detected at E12.5 by impaired elongation of the central fibers at the basal cytoplasm (Oda et al. 1980). Later, the lens fiber cells are morphologically abnormal, and the cavity of the lens still remain. The nuclei of the poorly elongated fiber cells are located in the posterior region of the cytoplasm, and dense bodies were noted at nuclear poles. Around birth, the lens capsule of the $Cryge^{elo}$ mice is ruptured in the posterior region. The developmental analysis of the $Cryge^{elo}$ mutants suggested that the primary effect of the deletion in the $Cryge$ gene may be specific to the fiber cell differentiation rather than to the cell proliferation and to inhibit fiber cell elongation (Yoshiki et al. 1991).

One of the genes which are important for the regulation of the lens-specific expression of the $Cryg$ genes is $Sox1$. A targeted deletion of $Sox1$ in mice caused microphthalmia and cataract. Mutant lens fiber cells fail to elongate, probably as a result of an almost complete absence of $Cryg$ transcripts (Nishiguchi et al. 1998). The phenotype of the homozygous $Sox1$ deletion mutant is very similar to the most severe $Cryg$ mutation, $Cryge^t$. $Sox1$ is mapped on mouse chromosome 8 (Malas et al. 1996).

Some of the inherited cataracts in man are also related to mutations in the crystallin-encoding genes. The locus for the hereditary human Coppock-like cataract (CCL) is closely linked to the $CRYG$ gene cluster, which is localized at

2q33-35. Detailed molecular analysis identified a missense mutation in a highly conserved segment of exon 2 of *CRYGC*. This mutation was not seen in a large control population and is the first evidence of an involvement of *CRYGC* in human cataract formation (Héon et al. 1999). Another type of human autosomal dominant cataract, the polymorphic congenital cataract (*PCC*), was mapped also to the position 2q33-35. A tri-nucleotide microsatellite marker for the *CRYGB* gene was found to cosegregate with *PCC*, suggesting strongly a mutation within this locus (Rogaev et al. 1996).

An intermediate member of the β/γ-crystallin superfamily is the γS-crystallin, previously also referred to as βS-crystallin. The corresponding gene *Crygs* was mapped recently to mouse chromosome 16 (Sinha et al. 1998). The cataract mutation *opj* (*opacity due to poor juntions*) (Everett et al. 1994; Kerscher et al. 1996) was mapped close to *Crygs*. Sequence analysis of *Crygs* from *Opj* mice revealed a mutation coding for a key residue of the core of the N-domain of the protein (Wistow et al. 1998).

A further mouse model affecting a crystallin-encoding gene is the *Philly* mouse, which was demonstrated to be an in-frame deletion of 12 bp in the βB2-crystallin encoding gene (*Crybb2*), resulting in a loss of four amino acids (Chambers and Russell 1991). The region in which the deletion occurs is close to the carboxy-terminus and essential for the formation of the tertiary structure of the βB2-crystallin. The increasing severity of the phenotype is temporally correlated to the expression of the *Crybb2* gene (Carper et al. 1982); *Crybb2* is mapped to mouse chromosome 5 (Kerscher et al. 1995). After the 1st postnatal week, the characteristic bow configuration of the nuclei in the lens cortex was replaced by a fan-shaped configuration, and swelling of the lens fibers occurred (Uga et al. 1980). Faint anterior opacities seen at postnatal day 15 are followed by sutural cataracts at day 25, nuclear cataract at 30 days, lamellar perinuclear opacities at 35 days, and total nuclear with anterior and posterior polar cataracts at 45 days. Cataractogenesis is associated with an intralenticular increase in water, sodium, and calcium, and a decrease in potassium, reduced glutathione, and ATP. An altered membrane permeability is the cause of an increased outward leak (Kador et al. 1980).

Also for this mouse mutation a human homologous disease has been described, which is referred to as cerulean cataract (*CCA2*: congenital cataract of cerulean type 2). This particular cataract is characterized by peripheral bluish and white opacifications in concentric layers with occasional central lesions arranged radially. Recently, Litt et al. (1997) mapped this particular type of cataract to a region of human chromosome 22 containing three genes coding for different β-crystallins. Sequence analysis revealed that a chain-termination mutation in *CRYBB2* is associated with this particular type of cataract in this family.

A further mouse model, a knockout of the αA-crystallin-encoding gene *Crya1*, was published recently (Brady et al. 1997). Initially, αA-crystallin-deficient lenses appear structurally normal, but they are smaller than the lenses of the wild-type littermates; $Cryal^{-/-}$ lenses develop an opacification that starts in the nucleus and progresses to a general opacification with age. Cataract formation is finally caused by insolubility of the αB-crystallin. $Cryal$ is mapped to mouse chromosome 17 close to the cataract mutation $lop18$ (Chang et al. 1996). In contrast, knockout mice of $Crya2$ encoding αB-crystallin are cataract-free, but they die prematurely because of a variety of organ defects (Wawrousek and Brady 1998). $Crya2$ is mapped on mouse chromosome 9 (Xia et al. 1996).

In human, one of the most common familial forms of congenital cataracts is referred to as the autosomal dominant congenital cataract ($ADCC$). An $ADCC$ locus was mapped recently to human chromosome 21q22.3 near the αA-crystallin encoding gene, $CRYAA$. By sequencing this candidate gene, Litt et al. (1998) found a missense mutation leading to an Arg→Cys exchange at the amino acid position 116, which is associated with $ADCC$ in one family.

3.3
Senile Cataracts at the End of Development

Cataract formation in man is not only restricted to congenital forms, but a major part of cataracts develop during aging. Even for this type of disease a variety of causes might be considered (e.g., diet, environment), but also a genetic predisposition should be kept in mind. To analyze this aspect of cataract development, two mouse mutants are well known in the literature, the $Emory$ and SAM (senescence accelerated mouse) mutants.

The $Emory$ mouse cataract is a dense, nuclear lens opacity appearing at the age of 6–8 months with a dominant mode of inheritance (Kuck et al. 1981). The cataract formation is accompanied by a variety of biochemical alterations (e.g., oxidative processes, changes in the crystallin composition, membrane proteins). These investigations have not uncovered a single metabolic lesion marked enough to be considered as an important cause of this cataract (Kuck 1990). Recent publications demonstrated the delay of cataract formation by iodide (Buchberger et al. 1991) or the acceleration by menadione (Bhuyan et al. 1997). Both results are consistent with the hypothesis that oxidative damage of lens membranes is important for the formation of this particular cataract. There is no chromosomal localization reported so far.

The SAM (senescence accelerated mouse) mutants are characterized by cataracts, periophthalmic lesions, opacity and ulcer of the cornea at the age of 4–24 months (Hosokawa et al. 1984). Interestingly, further brother-sister matings of the mutants lead to an earlier onset of cataract formation at the age of 10 weeks (Hosokawa et al. 1988). Further histological investigations demonstrated the persistence of the hyaloid vascular system in the SAM cataracts as well as an increased Ca^{2+} concentration and cross-linking transglutaminase activity (Hosokawa et al. 1993; Ashida et al. 1994). Also for this mutation, there is no chromosomal localization reported so far.

3.4
Transgenic Mice to Study Cataract Formation

To analyze the effect of particular gene products on the development of the lens a rapidly increasing number of transgenic animals were obtained by fusing the gene of interest either to the promotor of the αA- or the γF-crystallin gene. Both genes are commonly accepted to be expressed in a lens-specific manner. If the αA-crystallin promotor was fused to the highly cytotoxic *Diphteria toxin* gene, no lens is formed. This is combined with a marked reduction in eye size, structural abnormalities of the cornea, thickening of the iris, increased retinal cell density, and extensive whorling of the retinal fiber layers (Kaur 1989). Morphological alterations can be observed only from E12.5 onwards. The cells in the central posterior embryonic lens appeared to be vacuolated and undergo necrosis (Key et al. 1992). If the *Diphteria toxin* gene is fused to the γF-crystallin promotor, a considerable heterogeneity, varying from the reduced size of the eye to deficiency in nuclear fiber cells, was observed (Breitman et al. 1987, 1989).

In contrast to the total loss of the lens, the fusion of the γ-interferon gene to the αA-crystallin promoter leads to a normal development of the optic cup and lens vesicle at E12. Anomalies started at E14, when the transgene lens fails to form a well-defined lens bow because of the disorganization of lens cells at the posterior pole. At E18, lens and retinal differentiation program are completely disrupted (Egwuagu et al. 1994).

Transgenic mice expressing a retinoic acid receptor α (*rara*, fused with β-galactosidase under the control of the αA-crystallin promotor) develop cataracts and microphthalmia. Cataractous eyes are also characterized by abnormal positioning of the hyaloid artery and protrusion of lens material outside the lens capsule (Balkan et al. 1992). *Rara* is mapped to mouse chromosome 11 (Mattei et al. 1991).

Transgenic mice which are able to express and secrete acidic FGF (FGF-1; gene symbol *Fgf1*) into the extracellular space between the anterior epithelial cells revealed marked abnormalities of these cells including their elongation. Finally, the secreted FGF-1 led to microphakia and associated microphthalmia (Robinson et al. 1995b). Moreover, recent studies analyzing the function of a truncated FGF receptor (*Fgfr1*) in transgenic mice exhibited defective lens development characterized by cataracts and microphthalmia, whereas other ocular structures remained normal (Robinson et al. 1995a). An additional member of the FGF family, *Fgf3* (originally named *int-2*) was investigated under the control of the αA-crystallin promotor to target expression of *Fgf3* to the developing lens in trangenic mice. The expression of *Fgf3* in the lens rapidly induced epithelial cells throughout the lens to elongate and to express fiber cell-specific proteins including MIP and β-crystallins. This premature differentiation of the lens epithelium was followed by the degeneration of the entire lens. Ectopic expression of *Fgf3* in the lens also resulted in developmental alterations of the eyelids, cornea, and retina. It is suggested that Ffg3 activates the same receptor as Fgf1 (Robinson et al. 1998). The experiments cited above led to the conclusion that FGF molecules and their receptors are necessary for lens fiber cell differentiation.

The cytoskeleton of the lens cells exhibits a particular structure, the beaded filaments, which are believed to be specific for the lens. Vimentin is one of the main cytoskeletal proteins synthesized during lens development; the corresponding gene (*Vim*) is located on mouse chromosome 2. Overexpression of vimentin in the lenses of the transgenic mice interfered with the normal differentiation of lens fibers: cell denucleation and elongation processes were impaired and the animals developed cataracts at 6–12 weeks of age (Capetanaki et al. 1989).

When the SV40 large T antigen is targeted to the lens, the transgenic animals develop lens tumors. At E13, no elongation of primary lens fiber cells takes place but the lumen of the transgenic lens vesicle is filled by rounded cells. In juvenile transgenic mice, the lens was replaced by a disorganized mass that had ruptured the lens capsule and become infiltrated by blood vessels (Mahon et al. 1987; Bryce et al. 1993). However, if the gene coding for the large T antigen of the polyoma virus is fused to the αA-crystallin promotor, no evidence for tumor development was observed in vivo, but the expression of this T antigen resulted in microphthalmia, impairment of cell elongation, denucleation, and mitotic senescence (Griep et al. 1989). The different effects of the SV40 T-antigen and the polyoma T-antigen have been explained by their differential affinity for the *Rb1* gene product, which is responsible for the suppression of retinoblastoma. The effect of the transgene SV40 T-antigen was demonstrated to be due to the interaction with Rb1, since a modified version of the SV40 T-antigen, mutated in the binding domain for the Rb family, did not lead to any alterations in the lens (Fromm et al. 1994).

The function of the Rb gene product in lens development was analyzed in more detail using gene targeting to inactivate *Rb* in the mouse. Homozygous $Rb^{-/-}$ mice are lethal and die at E13-E15. The loss of *Rb* function is associated with unchecked proliferation of epithelial cells, impaired expression of differentiation markers, and inappropriate apoptosis in lens fiber cells. The increased apoptosis in the *Rb*-deficient lenses is dependent on the action of another tumor suppressor gene, *p53*. This effect was demonstrated in mice embryos deficient in both genes (Morgenbesser et al. 1994). Similar results were obtained by inactivation of the Rb gene product by the viral protein E7, which binds efficiently to the phosphorylated Rb protein (Pan and Griep 1994). The system is much more refined if the *bcl2* oncogene is targeted to the lens in transgenic mice by the αA-crystallin promotor. Expression of *bcl2* in the lens induces microphthalmia and cataracts. The terminal differentiation of lens fiber cells appears to be inhibited. Further, *bcl2* transgenic mice are mated to transgenic mice expressing a truncated SV40 T antigen in the lens, which inactivates Rb and activates thereby p53-dependent apoptosis. In those lenses where both transgenes (*bcl2* and truncated SV40 T antigen) are expressed, the apoptosis was found to be substantially reduced. It is concluded that *bcl2* can protect lens fiber cells from the p53-dependent apoptosis, which occurs after Rb inactivation (Fromm and Overbeek 1997).

4
Maturation of the Eye (2): Cornea, Iris, and Ciliary Body

4.1
Mouse Mutants with Lens-Corneal Adhesion

The formation of the cornea is the result of the last series of major inductive events in eye development, with the lens vesicle interacting with the overlying surface ectoderm (Hay 1979). Therefore, a variety of mutations exist resulting in lens-corneal adhesions and microphthalmia in the mouse.

The semidominant mutation *Coloboma* (*Cm*) was mapped at chromosome 2 (Theiler and Varnum 1981) and shows delayed detachment of the lens vesicle and microphthalmia; homozygotes die early in pregnancy. Until E10, heterozygous *Cm* mice develop nearly normally. At E11.5, the *Cm*/+ lenses were not detached from the corneal ectoderm, and there was epithelial continuity. At E13, the *Cm*/+ embryos exhibit a small epithelial stalk connecting the anterior lens epithelium with the epithelium of the cornea. At E14, most of the mutants had an adhesion of the anterior pole of the lens with the thinned cornea; the continuity between the anterior lens epithelium and the corneal epithelium remains present. In some cases, an anterior plug of lens fibers perforated the cornea. At E18, variation of the expressivity was noted, ranging from lens fibers penetrating through a hole in the cornea into the conjunctival sac between the cornea and the eyelid to a small depression in the cornea and a slight thickening of the anterior lens epithelium. In contrast to the *ak*/*ak* mutants, the anterior chamber and the vitreous body of the eye are present in the *Cm*/+ mutants; no folding of the retina was observed (Theiler and Varnum 1981). The radiation-induced mutation was characterized as a large deletion encompassing 1.1 to 2.2 cM and including the genes *Snap25* (synaptosomal-associated protein, 25 kDa), *Plcb1* (Phospholipase Cβ) and several simple sequence repeats (Hess et al. 1994).

A further example might be the mutant carrying the gene responsible for *dysgenetic lens* (*dyl*). This recessive mutation is mapped on mouse chromosome 4 (Sanyal et al. 1986). The defect is first recognizable at E10, when the lens vesicle of affected embryos fails to separate from the ectoderm. At E13, the lens is smaller than normal; at E14/15 vacuolar structures appear in the lens, and at E16 the lens is cataractous. Part of the lens protrudes to the exterior through the persistant ectodermal connection. Material from the lens is expelled, resulting in the greatly reduced and distorted lens seen in young animals. The pupil is markedly smaller than normal and irregular in outline; the cornea is opaque in varying degrees. In sections of the eye, the lens is much reduced in size and irregular in shape, and there is a persistent connection between the lens and the corneal epithelium (Sanyal and Hawkins 1979). Crystallins are produced in normal amounts in the early stages before degeneration becomes severe (Brahma and Sanyal 1984).

The radiation-induced mutation $Cat4^a$ was mapped at chromosome 8 and described as expressing an anterior pyramidal opacity with corneal adhesions, microphthalmia and closed eyes. The histological analysis revealed a fusion of the anterior pole of the lens to the cornea, which is associated with a defect in the corneal stroma and endothelium, which is filled with epithelial cells continuous with those covering the conically projecting anterior surface of the lens. The lens itself is also affected, but more severe in the homozygotes (small clusters of round fiber-like cells, irregular strands of lens epithelium and capsule). The anterior chamber and the vitreous body are absent; the extensively folded retina fills the small eye globe (Favor et al. 1997). Because $Cat4^a$ shows phenotypical similarities to several other independent mouse mutations including Sey, the authors speculate whether $Cat4^a$ might be one of several genes involved in a common developmental pathway and part of the $Pax6$-regulated gene cascade governing eye morphogenesis (Grimes et al. 1998).

A further mutant with persistant lens-cornea attachment is referred to as $fidget$ (fi). The mutation was mapped to mouse chromosome 2 and leads to a cataractous lens. Additionally, abnormalities in the bony labyrinth were observed, making the mice run in circles (Carter and Grüneberg 1950; Truslove 1956).

Moreover, a mouse model exists also for corneal surface disease and neovascularization. The mutation at the responsible gene, $corn1$, leads to early irregular thickening of the corneal epithelium, development of stromal neovascularization by 20 days of age, and cataract by 48 days of age. $Corn1$ is mapped at mouse chromosome 2 (Smith et al. 1996).

4.2
Mutations Affecting the Anterior Eye Development

The development of the anterior segment of the eye is affected either by the lens and/or by the retina. There are at least two mutations which affect the iris and the ciliary process in a more specific manner. One of them is the $Otx1$ knockout mutation. The $Otx1$ gene is one of the two mammalian homologues to the $Drosophila$ $orthodenticle$ homeobox-containing gene. The $Otx1$ knockout mutant exhibits severe defects in brain development, but also affects the eye. The thickness of the iris is reduced, the ciliary process and the lacrymal and Harderian glands are absent (Acampora et al. 1996). $Otx1$ is mapped to mouse chromosome 11 (DeGregorio et al. 1996).

The recessive mutation $waved-1$ $(wa1)$ was recognized because of the curly whiskers and waved coat. Many homozygotes have open eyelids at birth, leading additionally to corneal damage (Bennet and Gresham 1956). In complementation tests it was demonstrated that $wa1$ is an allele of $Tgfa$ encoding the transforming growth factor α (Luetteke et al. 1993). Targeted mutation produced eye abnormalities including anterior segment dysgenesis, lens and retinal defects, as well as corneal scarring and inflammation; some mutants are also born with open eyelids (Luetteke et al. 1993; Mann et al. 1993).

Overexpression of *Tgfa* in transgenic mice under control of the mouse *Cryaa* promotor leads to multiple eye defects including corneal opacities, cataracts, and microphthalmia. At early embryonic stages, TGFα induced the perioptic mesenchymal cells to migrate abnormally into the eye and to accumulate around the lens. The eye defects of the *Tgfa*-transgenic mice were significantly abated if they were bred to *Egfr* mutant mice *waved-2* (*wa2*) to a homozygous *wa2* background. Because the *Egfr* mutation in the *wa2* mice is located in the receptor kinase domain, this result indicates that the receptor tyrosin kinase activity is critical for signaling the migratory response. These studies by Reneker et al. (1995) demonstrated that *Tgfa* is capable of altering the migratory decisions and behavior of perioptic mesenchyme during eye development. *Tgfa* is mapped to mouse chromosome 6 (Fowler et al. 1993).

The *Tgfa* signaling events are also strengthened by the *fos-related antigen-2* (*fra2*) as is obvious from the overexpression of *fra2* in transgenic mice. These mice suffer from corneal abnormalities as early as E15.5 and a failure in eyelid fusion. Adult eyes were characterized by generalized anterior segment dysgenesis with extensive adherence of iris tissue to the posterior surface of the cornea obliterating the anterior chamber. These features are similiar to that previously reported in transgenic mice overexpressing *Tgfa*. Interaction between fra2 and Tgfa was confirmed in additional in vitro experiments (McHenry et al. 1998).

The eyelid fusion is also affected by the mutation *lid gap* (*lg*). Homozygous carriers of this mutation exhibit an extensive vacuolization of the lenses, which is first seen at E14. Later in development, defects of the cornea and retina occur, and at birth, the eyelids are open. Administration of thyroxine or cortisone at E10-11 or E14, respectively prevents the phenotype, suggesting an essential function for these two hormones at particular stages of ocular development (Stein et al. 1967; Harris et al. 1984; Juriloff 1985; Juriloff and Harris 1993). The gene *lg* is mapped to mouse chromosome 13 (Juriloff et al. 1983), but there is evidence that further loci are acting in the same pathway.

5
Maturation of the Eye (3): The Retina

5.1
Mutations Affecting the Formation of the Retina

The retina is already designed in the two layers of the optic cup: the outer layer will form the pigmented epithelium, and the inner layer, also referred to as neural layer, will be differentiated into the photoreceptor layer (consisting of rod and cone cells next to the pigmented epithelium), and into the cerebral layer (consisting of bipolar, ganglion, horizontal and amacrine cells as well as Müller cells). Therefore, all mutations which lead to phenotypes at the optic cup stage affect also the further development and differentiation of the retina (see Sect. 2.2: *Mitf, or/Chx10, Lhx2*).

Additionally, a few genes obviously affect later stages of retinal development as demonstrated by the mainly targeted disruptions of the corresponding genes. *Pou4f2* a member of the *Brn3* family of POU-domain transcription factors (previously referred to as *Brn3b*) is expressed in presumptive ganglion cell precursors as they begin to migrate from the zone of dividing neuroblasts to the future ganglion cell layer. The targeted disruption of the *Pou4f2* gene leads to the selective loss of 70% of retinal ganglion cells in homozygous deficient mice. The postmitotic ganglion cells precursors fail to properly differentiate and appear to be degenerated by apoptosis during the perinatal and early postnatal period (Xiang 1998, Xiang et al. 1996). Using FISH analysis, *Pou4f2* was assigned to the X-chromosome (Theil et al. 1994).

In contrast to the two mouse models described above, the effect on retinal development caused by the *Mash1* null mutation is not as dramatic. *Mash1* encodes a basic helix-loop-helix transcription factor, which is the mammalian homologue to the *Drosophila achaete-scute* proneural gene. A null mutation of *Mash1* delays differentiation of retinal neurons. Controversely, a null mutation of *Hes1* upregulates *Mash1* expression, accelerates retinal differentiation, and rod and horizontal cells appeared prematurely and formed abnormal rosette-like structures. In the mutant retina, bipolar cells died extensively and finally disappeared; lens and cornea development was also severely disturbed (Tomita et al. 1996; Kageyama et al. 1997). *Hes1* is the mouse homologue of *Drosophila hairy and Enhancer of split*; it is mapped to mouse chromosome 16 (Takebahashi et al. 1994). The chromosomal localization of *Mash1* is as yet unknown.

Additionally, genes are known to be important also for retina development, but no mutation has been reported up to now. The murine homologue to the *Drosophila homeobox-containing* gene *distal-less* (*Dlx1*, mapped to mouse chromosome 2) is expressed in the prospective deep retina at E11.5 to E14.5. In situ hybridization has revealed strong signals which are restricted to the neural retina layer (Dollé et al. 1992). The chicken homeobox-containing gene *GH6* has a broader spectrum of expression in the eye; besides in the neural retina it can be found in the lens epithelium and the optic nerve. *GH6* shows remarkable homology to the human homeobox gene *H6* (Stadler and Solursh 1994). *Crx1* (cone rod homeobox), a novel *Otx*-like gene, is obviously important for the photoreceptor differentiation. *Crx1* expression is restricted to developing and mature photoreceptor cells; it is obviously involved in the regulation of several photoreceptor-specific genes including *rhodopsin* (Furukawa et al. 1997b). The mouse *Crx1* gene is mapped to chromosome 7 (Chen et al. 1997b). Mutations in the human *CRX* gene lead to a variety of clinical phenotypes including Leber congenital amaurosis, cone-rod-dystrophy (CORDII), and late-onset retinitis pigmentosa (Sohocki et al. 1998).

5.2
Mutations Leading to Retinal Degeneration

Two mouse mutations have been studied very intensively which lead to retinal degeneration of distinct velocity. Therefore, they are referred to as *rd* and *rds* (retinal degeneration slow). Mice homozygous for the *rd* mutation display degeneration of the retinal rod photoreceptor cells beginning at about 1 week after birth, and by 4 weeks no photoreceptors are left. Degeneration is preceded by accumulation of cyclic GMP-phosphodiesterase. Within a 67-bp exon of the gene encoding the phosphodiesterase-β subunit (*Pdeb*), a nonsense ocher mutation was observed, which truncates the gene product and eliminates more than one half of the peptide chain including the catalytic domain (Bowes et al. 1990; Pittler and Baehr 1991). The *Pdeb* gene is located on mouse chromosome 5 (Sidman and Green 1965). Recently, the rescue of the photoreceptor function by virus-mediated gene transfer was reported for this particular mutation (Jomary et al. 1997). This might be important for the well known related human diseases (McLaughlin et al. 1993; Pittler et al. 1993).

The other mutation, *rds*, leads to abnormal development of rod and cone photoreceptors, followed by their slow degeneration. The molecular analysis demonstrated that *rds* is caused by the insertion of a 9.2 kb repetitive element into exon 2 of the *peripherin* (*Prph2*) gene. The entire insertion is included in the RNA products of the mutant locus at mouse chromosome 17 (van Nie et al. 1978). There is some evidence that *rds* mice represent a null allele (Ma et al. 1995). Even in this case, a complete rescue of the photoreceptor dysplasia and degeneration was reported (Travis et al. 1992). There are also several mutations already characterized in the human *PRPH2* gene leading to inherited retinal degeneration (Keen and Inglehearn 1996).

In addition to these well-characterized mutations, *rd* and *rds*, further mouse mutants with retinal degenerations are described. In *rd3/rd3* mice, retinal development proceeds normally through the second postnatal week. Thereafter, photoreceptor and outer nuclear layers begin to degenerate, and by 8 weeks, no photoreceptor cells remain. The mutation was mapped to mouse chromosome 1 (Chang et al. 1993). The autosomal dominant retinal degeneration *Rd4* was found in a stock carrying an inversion, which encompasses nearly all of mouse chromosome 4. In heterozygotes, the retinal outer nuclear and plexiform layers begin to reduce at 10 days of age, showing total loss at 6 weeks. The histological observations are confirmed by electroretinograms. Retinal vessel attenuation, pigment spots, and optic atrophy appeared in the fundus at 4 weeks of age. The homozygotes are lethal (Roderick et al. 1997). Additionally, the *tubby* mouse mutants (*tub*) suffer from retinal degeneration and progressive hearing loss as well as from obesity. The retinal degeneration aspects of the *tub* mutant have been presented as the effect of a separate gene mutation, *rd5*, but it appears likely that *tub* and *rd5* might be allelic (Ohlemiller et al. 1995, Noben-Trauth et al. 1996). The *tub* gene was mapped to mouse chromosome 7 (Coleman and Eicher 1990).

A further naturally occurring mouse mutation leading to degenerative retinal abnormalities was reported recently: *hugger* (*hug*) is mapped to mouse chromosome 19, and the abnormal retinal phenotype can be recognized at birth, when some retinal ganglion cells already lie in abnormal positions in the inner plexiform layer. By postnatal day 18, the number of neurons is reduced in all three cellular layers of the retina. Rod photoreceptor cells develop only rudimentary outer segments, and by 9 months of age, about 75% of the photoreceptor cells have completely disappeared. The genetic lesion is not yet characterized at the molecular level; *Rom1* (coding for the rod outer segment membrane protein 1) might be excluded as a candidate gene because of its distinct position on mouse chromosome 19 (Sidman et al. 1997).

Retinitis pigmentosa is one of the most common inherited retinal disease in man; this topic is addressed also by a variety of animal models (for a detailed overview see Kohler et al. 1997; Peterson-Jones 1998). Retinitis pigmentosa leads to degeneration of the photoreceptors and is caused by mutations affecting several genes. A major impact has the *rhodopsin* gene (*rho*) coding for the rod photopigment in the disk membranes of the photoreceptor cells; *rho* is mapped to mouse chromosome 6 (Elliott et al. 1990). The role of *rho* mutations will be clarified by several transgenic mice carrying mutations at site, which are known from human patients to be important for retinal degeneration, e.g., P23H. The expression of this particular mutation in the retina leads to disorganization of the basal disks of the outer segment, indicating defective disk membrane morphogenesis. It appears that this defective disk membrane morphogenesis results in the formation of fewer mature disks, thus accounting for observed gradual shortening of the photoreceptor outer segments with age (Liu et al. 1997). Additionally, a transgenic mouse was constructed containing three mutations near the N-terminus of opsin (P23H; V20G; P27L). These animals exhibit a slowly progressive degeneration of the rod photoreceptors (Wu et al. 1998).

Transgenic mice carrying mutated *rho* genes are also used for studies concerning treatment protocols. A successful approach uses the P347S mutation; vitamin A supplementation slowed the rate of photoreceptor degeneration, which is also caused by the P347S mutation (Li et al. 1998). The effect of vitamin A on the correct development and normal function is also demonstrated by knockout mice missing the genes coding for both the retinoic acid receptors, β2 and γ2 (Grondona et al. 1996).

5.3
Mutations Affecting the Optic Nerve

In mouse, the first axons from the retinal ganglion cells enter the optic stalk on E12.5. The axons further extend to the ventral wall of the optic stalk and reach the primary optic center of the brain. This question is addressed by knockout mice deficient in *netrin-1* (*Ntn1*) and *DCC. Ntn1* was characterized as a gene coding for an axon-guidance molecule specifically on neuroepithelial cells at the disk surrounding exiting retinal ganglion axons. These cells also express the netrin receptor, *DCC* (= deleted in colorectal cancer). In *Ntn1*- and *DCC*-deficient embryos, the axons from the retinal ganglion cells fail to exit into the optic cup, resulting in optic nerve hypoplasia (Deiner et al. 1997). Another transgenic mouse model affecting the function of the optic nerve is the *ankyrinB*-deficient mouse. Besides some effects on myelination in the brain, the optic nerve axons of *ankyrinB*-deficient mice become dilated with diameters up to eightfold greater than normal, and they degenerate by day 20 (Scotland et al. 1998). The chromosomal localization of *Ntn1* is as yet unknown, whereas *DCC* is mapped to mouse chromosome 18 (Justice et al. 1992).

In addition to these recently developed transgenic mice, several "classical" mutants have been described affecting the formation and the function of the optic nerve. In the *wabbler-lethal* (*wl*) mutants, most optic nerve axons appeared normal; however, degenerating axons surrounded by normal myelin and axons with thickened myelin sheaths were prevalent in *wl/wl* mice. Later, it leads to axonal degeneration (Carroll et al. 1992). A subpopulation of anophthalmic mice (*ZRDCT-AN*) demonstrated that their optic fissure did not involute into the optic stalk. As a consequence, the optic nerve fibers failed to exit from the eye in their appropriate position. Secondary changes in the retina were near-total loss of ganglion cells and variable attenuation of the other nuclear and plexiform layers (Silver et al. 1984). The *wl* gene is mapped to mouse chromosome 14 (Lane and Dickie 1961).

The autosomal semidominant mutation *Bst* (*belly spot and tail*) is often associated with small and atrophic optic nerves in adult mice. Histological analysis of eyes from E12 mice revealed a delayed closure of the optic fissure. At E15, disorganization of the retinal neuroepithelium can be observed, and ganglion cell axons are found between pigmented and normal retina. At birth, optic nerves of affected mice are smaller than those of wild types, and ectopic axons are found within the eyes (Rice et al. 1997). The *Bst* mutation is mapped to mouse chromosome 16 (Rice et al. 1995).

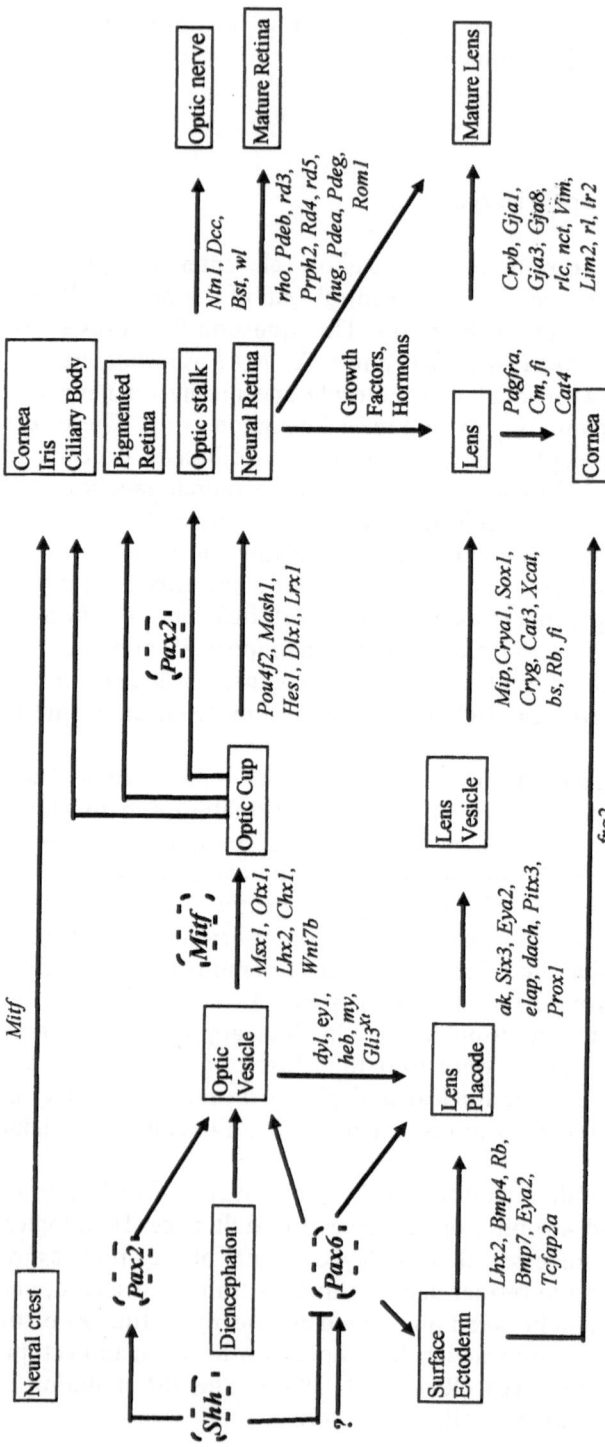

Fig. 2. Flow chart of eye development. A simplified flow chart illustrates the major events and tissue differentiation during eye development. Genes and mutations mentioned in the text are included into the scheme. Genes in open boxes are discussed as major control genes. For abbreviations refer to the main text

6
Conclusion

A great variety of mouse mutants affecting ocular development is available. They arose spontaneously, or were recovered after parental treatment by chemical mutagens or radiation. Recent development of molecular methods is leading to a rapidly increasing number of transgenic mice overexpressing particular genes in distinct tissues or representing null alleles by targeted disruption of the genes. The molecular analysis of the affected genes together with the detailed phenotypical analysis, will allow us to understand the pathways of ocular development. A simplified scheme is given in Fig. 2. Additionally, the analysis of the genes and their mutations enables us to describe precisely the mechanisms leading to the pathological situation. Together with modern techniques in biochemistry, it might become possible in the near future to treat an increasing number of inherited ocular diseases either by novel drugs or by somatic gene therapy. For each of these steps, mouse mutants are an indispensible tool.

References

Acampora D, Mazan S, Avantaggio V, Barone P, Tuorto F, Lallemand Y, Brulet P, Simeone A (1996) Epilepsy and brain abnormalities in mice lacking the *Otx1* gene. Nat Genet 14: 218–222

Amiel J, Watkin PM, Tassabehji M, Read AP, Winter RM (1998) Mutation in the *MITF* gene in albinism-deafness syndrome (Tietz syndrome). Clin Dysmorphol 7:17–20

Angel TA, Faust CJ, Gonzales JC, Kenwrick S, Lewis RA, Herman GE (1993) Genetic mapping of the X-linked dominant mutations *striated* (*Str*) and *bare patches* (*Bpa*) to a 600-kb region of the mouse X-chromosome: implications for mapping human disorders in Xq28. Mamm Genome 4:171–176

Ashida Y, Takeda T, Hosokawa M (1994) Protein alterations in age-related cataract associated with a persistent hyaloid vascular system in senescence-accelerated mouse (SAM). Exp Eye Res 59:467–473

Aso S, Horiwaki SI, Noda S (1995) Lens aplasia: a new mutation producing lens abnormality in the mouse. Lab Anim Sci 45:41–46

Aso S, Tashiro M, Baba R, Sawaki M, Noda S, Fujita M (1998) Apoptosis in the lens anlage of the heritable lens aplastic mouse (*lap* mouse). Teratology 58:44–53

Balkan W, Klintworth GK, Bock CB, Linney E (1992) Transgenic mice expressing a constitutively active retinoic acid receptor in the lens exhibit ocular defects. Dev Biol 151:622–625

Bennett JH, Gresham GA (1956) A gene for eyelids open at birth in the house mouse. Nature 178:272–273

Bhuyan DK, Huang X, Kuriakose G, Garner WH, Bhuyan KC (1997) Menadione-induced oxidative stress accelerates onset of Emory mouse cataract in vivo. Curr Eye Res 16:519–526

Bowes C, Li T, Danciger M, Baxter LC, Applebury ML, Farber DB (1990) Retinal degeneration in the *rd* mouse is caused by a defect in the β-subunit of rod cGMP-phosphodiesterase. Nature 347:677–680

Brady JP, Garland D, Duglass-Tabor Y, Robison WGJr, Groome A, Wawrousek EF (1997) Targeted disruption of the mouse αA-crystallin gene induces cataract and cytoplasmic inclusion bodies containing the small heat shock protein αB-crystallin. Proc Natl Acad Sci USA 94:884–889

Brahma SK, Sanyal S (1984) Immunohistochemical studies of lens crystallins in the dysgenetic lens (*dyl*) mutant mice. Exp Eye Res 38:305–311

Breitman ML, Clapoff S, Rossant J, Tsui LC, Glode LM, Maxwell IH, Bernstein A (1987) Genetic ablation: targeted expression of a toxin gene causes microphthalmia in transgenic mice. Science 238:1563–1565

Breitman ML, Bryce DM, Giddens E, Clapoff S, Goring D, Tsui LC, Klintworth GK, Bernstein A (1989) Analysis of lens cell fate and eye morphogenesis in transgenic mice ablated for cells of the lens lineage. Development 106:457–463

Bryce DM, Liu Q, Khoo W, Tsui LC, Breitman ML (1993) Progressive and regressive fate of lens tumors correlates with subtle differences in transgene expression in γF-crystallin-SV40 antigen transgenic mice. Oncogene 8:1611–1620

Buchberger W, Winkler R, Moser M, Rieger G (1991) Influence if iodide on cataractogenesis in Emory mice. Ophthalmic Res 23:303–308

Burmeister M, Novak J, Liang MY, Basu S, Ploder L, Hawes NL, Vidgen D, Hoover F, Goldman D, Kalnins VI, Roderick TH, Taylor BA, Hankin MH, McInnes RR (1996) Ocular retardation mouse caused by *Chx10* homeobox null allele: impaired retinal progenitor proliferation and bipolar cell differentiation. Nat Genet 12:376–384

Capetanaki Y, Smith S, Heath JP (1989) Overexpression of the vimentin gene in transgenic mice inhibits normal lens cell differentiation. J Cell Biol 109:1653–1664

Carper D, Shinohara T, Piatigorsky J, Kinoshita JH (1982) Deficiency of functional messenger RNA for a developmentally regulated β-crystallin polypeptide in a hereditary cataract. Science 217:463–464

Carrol EW, Curtis RL, Sullivan DA, Melvin JL (1992) Wallerian degeneration in the optic nerve of the *wabbler-lethal* (*wl/wl*) mouse. Brain Res Bull 29:411–418

Carter TC, Grüneberg H (1950) Linkage between *fidget* and *agouti* in the house mouse. Heredity 4:373–376

Cartier M, Breitman ML, Tsui LC (1992) A frameshift mutation in the γE-crystallin gene of the *Elo* mouse. Nat Genet 2:42–45

Caubit X, Thangarajah R, Theil T, Wirth J, Nothwang HG, Ruther U, Krauss (1999) Mouse *Dac*, a novel nuclear factor with homology to *Drosophila dachshund* shows a dynamic expression in the neural crest, the eye, the neocortex, and the limb bud. Dev Dyn 214:66–80

Center EM, Polizzotto RS (1992) Etiology of the developing eye in myelencephalic blebs (*my*) mice. Histol Histopathol 7:231–236

Chambers C, Russell P (1991) Deletion mutation in an eye lens β-crystallin. J Biol Chem 266:6742–6746

Chang B, Heckenlively JR, Hawes NL, Roderick TH (1993) New mouse primary retinal degeneration (*rd-3*). Genomics 16:45–49

Chang B, Hawes NL, Smith RS, Heckenlively JR, Davisson MT, Roderick TH (1996) Chromosomal localization of a new mouse *lens opacity* gene. Genomics 36:171–173

Chase HB (1944) Studies on an anophthalmic strain of mice. IV. A second major gene for anophthalmia. Genetics 29:264–269

Chen R, Amoui M, Zhang Z, Mardon G (1997a) Dachshund and eyes absent proteins form a complex and function synergistically to induce ectopic eye development in Drosophila. Cell 91:893–903

Chen S, Wang QL, Nie Z, Sun H, Lennon G, Copeland NG, Gilbert DJ, Jenkins NA, Zack DJ (1997b) Crx, a novel Otx-like paired-homeodomain protein, binds to and transactivates photoreceptor-cell specific genes. Neuron 19:1017–1030

Chiang C, Litingtung Y, Lee E, Young KE, Cordon JL, Westphal H, Beachy PA (1996) Cyclopia and defective axial patterning in mice lacking Sonic hedgehog gene function. Nature 383:407–413

Coleman DL, Eicher EM (1990) Fat (fat) and tubby (tub): two autosomal recessive mutatrions causing obesity syndromes in the mouse. J Hered 81:424–427

Davisson MT, Eicher EM, Green MC (1976) Genes on chromosomes 3 of the mouse. J Hered 67:155–156

DeGregorio L, Manenti G, Simeone A, Dragani TA (1996) Genetic mapping of the homeobox-containing gene Otx1 on mouse chromosome 11. Mamm Genome 7:241

Deiner MS, Kennedy TE, Fazeli A, Serafini T, Tessier-Lavigne M, Sretavan DW (1997) Netrin-1 and DCC mediate axon guidance locally at the optic disc: loss of function leads to optic nerve hypoplasia. Neuron 19:575–589

Desplan C (1997) Eye development: governed by a dictator or a junta? Cell 91:861–864

Dickie MM (1967) Vacuolated lens. Mouse News Lett 36:39–40

Dickie MM (1969): Vacuolated lens (vl). Mouse News Lett 40:29

Dickinson ME, Kobrin MS, Silan CM, et al. (1990) Chromosomal localization of seven members of the murine TGF-β superfamily suggests close linkage to several morphogenetic mutant loci. Genomics 6:505–520

Dollé P, Price M, Duboule D (1992) Expression of the murine Dlx-1 homeobox gene during facial, ocular and limb development. Differentiation 49:93–99

Egwuagu CE, Sztein J, Chan CC, Mahdi R, Nussenblatt RB, Chepelinsky AB (1994) γInterferon expression disrupts lens and retinal differentiation in transgenic mice. Dev Biol 166:557–568

Ehling UH, Charles DJ, Favor J, Graw J, Kratochvilova J, Neuhäuser-Klaus A, Pretsch W (1985) Induction of gene mutations in mice: the multiple endpoint approach. Mutat Res 150:393–401

Elliott RW, Sparkes RS, Mohandas T, Grant SG, McGinnis JF (1990) Localization of the rhodopsin gene to the fistal half of mouse chromosome 6. Genomics 6:635–644

Everett CA, Glenister PH, Taylor DM, Lyon MF, Kratochvilova-Löster J, Favor J (1994) Mapping of six dominant cataract genes in the mouse. Genomics 20:429–434

Favor J (1983) A comparison of the dominant cataract and recessive specific-locus mutation rates induced by treatment of male mice with ethylnitrosourea. Mutat Res 110:367–382

Favor J (1984) Characterization of dominant cataract mutations in mice: penetrance, fertility and homozygous viability of mutations recovered after 250 mg/kg ethylnitrosourea paternal treatment. Genet Res Camb 44:183–197

Favor J (1995) Mutagenesis and human genetic disease: dominant mutation frequencies and a characterization of mutational events in mice and humans. Environ Mol Mutagen 25 (S26):81–87

Favor J, Pretsch W (1990) Genetic localization and phenotypic expression of X-linked cataract (Xcat) in Mus musculus. Genet Res Camb 56:157–162

Favor J, Sandulache R, Neuhäuser-Klaus A, Pretsch W, Chatterjee B, Senft E, Wurst W, Blanquet V, Grimes P, Spörle R, Schughart K (1996) The mouse Pax2^{1Neu} mutation is identical to a human PAX2 mutation in a family with renal-coloboma syndrome and results in developmental defects of the brain, ear, eye, and kidney. Proc Natl Acad Sci USA 93:13870–13875

Favor J, Grimes P, Neuhäuser-Klaus A, Pretsch W, Stambolian D (1997) The mouse Cat4 locus maps to chromosome 8 and mutants express lens-corneal adhesion. Mamm Genome 8:403–406

Foerst-Potts L, Sadler TW (1997) Disruption of Msx-1 and Msx-2 reveals roles for these genes in craniofacial, eye, and axial development. Dev Dyn 209:70–84

Forker C, Wegener A, Graw J (1997) Effects of UV-B radiation on a hereditary suture cataract in mice. Exp Eye Res 64:405–411

Fowler KJ, Mann GB, Dunn AR (1993) Linkage of the murine transforming growth factor α gene with Igk, Ly-2, and Fabp1 on chromosome 6. Genomics 16:782–784

Franz T, Besecke A (1991) The development of the eye in homozygotes of the mouse mutant Extra-toes. Anat Embryol 184:355–361

Fraser FC, Schabtach G (1962) "Shrivelled": a hereditary degeneration of the lens in the house mouse. Genet Res Camb 3:383–387

Fromm L, Overbeek PA (1997) Inhibition of cell death by lens-specific overexpression of *bcl-2* in transgenic mice. Dev Genet 20:276–287

Fromm L, Shawlot W, Gunning K, Butel JS, Overbeek PA (1994) The retinoblastoma protein-binding region of simian virus 40 large T antigen alters cell cycle regulation in lenses of transgenic mice. Mol Cell Biol 14:6743–6754

Furukawa T, Kozak CA, Cepko CL (1997a) *Rax*, a novel paired-type homeobox gene, shows expression in the anterior neural fold and developing retina. Proc Natl Acad Sci USA 94:3088–3093

Furukawa T, Morrow EM, Cepko CL (1997b) *Crx*, a novel *otx*-like homeobox gene, shows photoreceptor-specific expression and regulates photoreceptor differentiation. Cell 91:531–541

Furuta Y, Hogan BLM (1998) BMP4 is essential for lens induction in the mouse embryo. Genes Dev 12:3764–3775

Gao Y, Spray DC (1998) Structural changes in lenses of mice lacking the gap junction protein connexin43. Invest Ophthalmol Visual Sci 39:1198–1209

Gong X, Li E, Klier G, Huang Q, Wu Y, Lei H, Kumar NM, Horwitz J, Gilula NB (1997) Disruption of α3 connexin gene leads to proteolysis and cataractogenesis in mice. Cell 91:833–843

Graw J (1997) The crystallins: genes, proteins and diseases. Biol Chem 378:1331–1348

Graw J (1999) Cataract mutations and lens development. Prog Retina Eye Res 18:235–267

Graw J, Liebstein A (1993) DNase activity in murine lenses: implications for cataractogenesis. Graefe's Arch Clin Exp Ophthalmol 231:354–358

Graw J, Kratochvilova J, Summer KH (1984) Genetical and biochemical studies of a dominant cataract mutant in mice. Exp Eye Res 39:37–45

Graw J, Summer KH, Michel C, Bors W (1985) Catalase and superoxide dismutase activities in lenses of cataractous *NOP*-mice. Exp Eye Res 41:577–579

Graw J, Favor J, Neuhäuser-Klaus A, Ehling UH (1986) Dominant cataract and recessive specific locus mutations in offspring of X-irradiated male mice. Mutat Res 159:47–54

Graw J, Bors W, Michel C, Reitmeir P, Summer KH, Wulff A (1989a) Oxidative stress and inherited cataracts in mice. Ophthalmic Res 21:414–419

Graw J, Kratochvilova J, Löbke A, Reitmeir P, Schäffer E, Wulff A (1989b) Characterization of *Scat* (Suture Cataract), a dominant cataract mutation in mice. Exp Eye Res 49:469–477

Graw J, Bors W, Gopinath PM, Merkle S, Michel C, Reitmeir P, Schäffer E, Summer KH, Wulff A (1990a) Characterization of *Cat-2t*, a radiation-induced dominant cataract mutation in mice. Invest Ophthalmol Visual Sci 31:1353–1361

Graw J, Reitmeir P, Wulff A (1990b) Osmotic state of lenses in three dominant murine cataract mutants. Graefe's Arch Clin Exp Ophthalmol 228:252–254

Graw J, Werner T, Merkle S, Reitmaier P, Schäffer E, Wulff A (1990c) Histological and biochemical characterization of the murine cataract mutant *Nop*. Exp Eye Res 50:449–456

Graw J, Immervoll T, Grimm C, Löster J (1998) Developmental and genetical analysis of the *Cat3* cataract mutants in mice. Invest Ophthalmol Visual Sci 39:S523

Griep AE, Kuwabara T, Lee EJ, Westphal H (1989) Perturbed development of the mouse lens by polyomavirus large T antigen does not lead to tumor formation. Genes Dev 3:1075–1085

Griffin CS, Shiels A (1992) Localisation of the gene for the major intrinsic protein of eye-lens-fibre cell membranes to mouse Chromosome 10 by in situ hybridisation. Cytogenet Cell Genet 59:300–302

Grimes PA, Favor J, Koeberlein B, Silvers WK, Fitzgerald PG, Stambolian D (1993) Lens development in a dominant X-linked congenital cataract of the mouse. Exp Eye Res 57:587–594

Grimes PA, Koeberlein B, Favor J, Neuhäuser-Klaus A, Stambolian D (1998) Abnormal eye development associated with *Cat4a*, a dominant mouse cataract mutation on chromosome 8. Invest Ophthalmol Visual Sci 39:1863–1869

Grimm C, Chatterjee B, Favor J, Immervoll T, Löster J, Klopp N, Sandulache R, Graw J (1998) *Aphakia (ak)*, a mouse mutation affecting early eye development: fine mapping, exclusion of candidate genes and altered *Pax6* and *Six3* expression. Dev Genet 23:299-316

Grondona JM, Kastner P, Gansmuller A, Decimo D, Chambon P, Mark M (1996) Retinal dysplasia and degeneration in RARβ2/RARγ2 compound mutant mice. Development 122: 2173-2188

Grüneberg H, Truslove GM (1960) Two closely linked genes in the mouse. Genet Res Camb 1:69-90

Haeflinger JA, Bruzzone R, Jenkins NA, Gilbert DJ, Copeland NG, Paul DL (1992) Four novel members of the connexin family of gap junction proteins. Molecular cloning, expression, and chromosome mapping. J Biol Chem 267:2057-2064

Halder G, Callaerts P, Gehring WJ (1995) Induction of ectopic eyes by targeted expression of the eyeless gene in *Drosophila*. Science 267:1788-1792

Hammond KL, Hanson IM, Brown AG, Lettice LA, Hill RE (1998) Mammalian and *Drosophila* dachshund genes are related to the *Ski* proto-oncogene and are expressed in eye and limb. Mech Dev 74:121-131

Hanson I, van Heyningen V (1995) Pax6: more than meets the eye. Trends Genet 11:268-272

Happle R, Phillips RJ, Roessner A, Junemann G (1983) Homologous genes for X-linked *chondrodysplasia punctata* in man and mouse. Hum Genet 63:24-27

Harris MJ, Juriloff DM, Biddle FG (1984) Cortisone cure of the lidgap defect in fetal mice: a dose-response and time-response study. Teratology 29:287-295

Hawes NL, Roderick TH (1990) Linkage of ocular retardation (*or*). Mouse Genome 87:93

Hay ED (1979) Development of the vertebrate cornea. Intern Rev Cytol 63:263-322

Héon E, Priston M, Schorderet DF, Billingsley GD, Girard PO, Lubsen N, Munier FL (1999) The γ-crystallins and human cataracts: a puzzle made clearer. Am J Hum Genet 65:1261-1267

Hero I, Farjah M, Scholtz CL (1991) The prenatal development of the optic fissure in colobomatous microphthalmia. Invest Ophthalmol Visual Sci 32:2622-2635

Hertwig P (1942) Neue Mutationen und Koppelungsgruppen bei der Hausmaus. Z Indukt Abstammungs-Vererbungsl 80:220-246

Hess EJ, Collins KA, Copeland NG, Jenkins NA, Wilson MC (1994) Deletion map of the coloboma (*Cm*) locus on mouse chromosome 2. Genomics 21:257-261

Hiai H, Kato S, Horiuchi Y, Shimada R, Tsuruyama T, Watanabe T, Matsuzawa A (1998) Mapping of the Nakano cataract gene *nct* on mouse chromosome 16. Genomics 50:119-120

Hill RE, Favor J, Hogan BLM, Ton CCT, Saunders GF, Hanson IM, Prosser J, Jordan T, Hastie ND, van Heyningen V (1991) Mouse *Small eye* results from mutations in a paired-like homeobox-containing gene. Nature 354:522-525

Hodgkinson CA, Nakayama A, Li H, Swenson LB, Opdecamp K, Asher JHJr, Arnheiter H, Glaser T (1998) Mutation at the *anophthalmic white* locus in Syrian hamsters: haploinsufficiency in the *Mitf* gene mimics human Waardenburg syndrome type 2. Hum Mol Genet 7:703-708

Hogan BLM, Horsburgh G, Cohen J, Hetherington CM, Fisher G, Lyon MF (1986) *Small eyes* (*Sey*): a homozygous lethal mutation on chromosome 2 which affects the differentiation of both lens and nasal placodes in the mouse. J Embryol Exp Morphol 97:95-110

Hosokawa M, Takeshita S, Higuchi K, Shimizu K, Irino M, Toda K, Honma A, Matsumura A, Yasuhira K, Takeda T (1984) Cataract and other ophthalmic lesions in senescence accelerated mouse (SAM). Morphology and incidence of senescence associated ophthalmic changes in mice. Exp Eye Res 38:105-114

Hosokawa M, Ashida Y, Tsuboyama T, Chen WH, Takeda T (1988) Cataract in senescence accelerated mouse (SAM). 2. Development of a new strain of mouse with late-appearing cataract. Exp Eye Res 47:629-640

Hosokawa M, Ashida Y, Matsushita T, Takahashi K, Takeda T (1993) Persistent hyaloid vascular system in age-related cataract in a SAM strain of mouse. Exp Eye Res 57:427-434

Hsieh CL, Kumar NM, Gilula NB, Francke U (1991) Distribution of genes for gap junction membrane channel proteins on human and mouse chromosomes. Somat Cell Mol Genet 17:1991-2000

Hui CC, Joyner AL (1993) A mouse model of greig cephalopolysyndactyly syndrome: the extra-toes^j mutation contains an intragenic deletion of the Gli3 gene. Nat Genet 3:241-246

Iida F, Matsushima Y, Hiai H, Uga S, Honda Y (1997) Rupture of lens cataract: a novel hereditary recessive cataract model in the mouse. Exp Eye Res 64:107-113

Jomary C, Vincent KA, Grist J, Neal MJ, Jones SE (1997) Rescue of photoreceptor function by AAV-mediated gene transfer in a mouse model of inherited retinal degeneration. Gene Ther 4:683-690

Juriloff DM (1985) Prevention of the eye closure defect in lg^{Ml}/lg^{Ml} fetal mice by thyroxine. Teratology 32:73-86

Juriloff DM, Harris MJ (1993) Retinoic acid, cortisone, or thyroxine suppresses the mutant phenotype of the eyelid development mutation, lg^{Ml}, in mice. J Exp Zool 265:144-152

Juriloff DM, Harris MJ, Miller JR (1983) The lidgap defect in mice: update and hypotheses. Can J Genet Cytol 25:246-254

Justice MJ, Gilbert DJ, Kinzler KW, et al. (1992) A molecular genetic linkage map of mouse chromosome 18 reveals extensive linkage conservation with human chromosomes 5 and 18. Genomics 13:1281-1288

Kador PF, Fukui HN, Fujushi S, Jernigan HM Jr, Kinoshita JH (1980) Philly mouse: a new model of hereditary cataract. Exp Eye Res 30:59-68

Kageyama R, Ishibashi M, Takebayashi K, Tomita K (1997) bHLH transcription factors and mammalian neuronal differentiation. Int J Biochem Cell Biol 29:1389-1399

Kaur S, Key B, Stock J, McNeish JD, Akeson R, Potter SS (1989) Targeted ablation of α-crystallin-synthesizing cells produces lens-deficient eyes in transgenic mice. Development 105:613-619

Keen TJ, Inglehearn CF (1996) Mutations and polymorphisms in the human peripherin-RDS gene and their involvement in the inherited retinal degeneration. Hum Mutat 8:297-303

Kerscher S, Church RL, Boyd Y, Lyon MF (1995) Mapping of four mouse genes encoding eye lens-specific structural, gap junction, and integral membrane proteins: Crybal (crystal-linβA3/A1), Crybb2 (crystallinβB2), Gja8 (MP70), and Lim2 (MP19). Genomics 29:445-450

Kerscher S, Glenister PH, Favor J, Lyon MF (1996) Two new cataract loci, Ccw and To3, and further mapping of the Npp and Opj cataracts in the mouse. Genomics 36:17-21

Key B, Liu L, Potter SS, Kaur S, Akeson R (1992) Lens structures exist transiently in development of transgenic mice carrying an α-crystallin-diphtheria toxin hybrid gene. Exp Eye Res 55:357-367

Klopp N, Favor J, Löster J, Lutz RB, Neuhäuser-Klaus A, Prescott A, Pretsch W, Quinlan RA, Sandilands A, Vrensen GJFM, Graw J (1998) Three murine cataract mutants (Cat2) are defective in different γ-crystallin genes. Genomics 52:152-158

Kohler K, Guenther E, Zrenner E (1997) Tiermodelle in der Retinitis-pigmentosa-Forschung. Klin Monatsbl Augenheilkd 211:84-93

Kratochvilova J (1981) Dominant cataract mutations detected in offspring of gamma-irradiated male mice. J Hered 72:302-307

Kratochvilova J, Ehling UH (1979) Dominant cataract mutations induced by γ-irradiation of male mice. Mutat Res 63:221-223

Kratochvilova J, Favor J (1992) Allelism tests of 15 dominant cataract mutations in mice. Genet Res Camb 59:199-203

Kuck JF (1990) Late onset hereditary cataract of the Emory mouse. A model for human senile cataract. Exp Eye Res 50:659-664

Kuck JF, Kuwabara T, Kuck KD (1981-82) The Emory mouse cataract: an animal model for human senile cataract. Curr Eye Res 1:643-649

Lane PW, Dickie MM (1961) Linkage of wabbler-lethal and hairless in the mouse. J Hered 52:159-160

Li T, Sandberg MA, Pawlyk BS, Rosner B, Hayes KC, Dryja TP, Berson EL (1998) Effect of vitamin A supplementation on *rhodopsin* mutants threonine-17→methionine and proline-347→serine in transgenic mice and in cell cultures. Proc Natl Acad Sci USA 95:11933–11938

Litt M, Carrero-Valenzuela R, LaMorticella D, Schultz DW, Mitchell TN, Kramer P, Maumenee IH (1997) Autosomal dominant cerulean cataract is associated with a chain termination mutation in the human β-crystallin gene CRYBB2. Hum Mol Genet 6:665–668

Litt M, Kramer P, LaMorticella DM, Murphey W, Lovrien EW, Weleber RG (1998) Autosomal-dominant congenital cataract associated with a missense mutation in the human alpha-crystallin gene *CRYAA*. Hum Mol Genet 7:471–474

Liu ISC, Chen JD, Ploder L, Vidgen D, van der Kooy D, Kalnins VI, McInnes RR (1994) Developmental expression of a novel murine homeobox gene (*Chx10*): evidence for roles in determination of the neuroretina and inner nuclear layer. Neuron 13:377–393

Liu X, Wu TH, Stowe S, Matsushita A, Arikawa K, Naash MI, Williams DS (1997) Defective phototransductive disk membrane morphogenesis in transgenic mice expressing opsin with a mutated N-terminal domain. J Cell Sci 110:2589–2597

Löster J, Immervoll T, Schmitt-John T, Graw J (1997) *Cat3^{vl}* and *Cat3^{vao}*, cataract mutations on mouse chromosome 10: phenotypic characterization, linkage studies and analysis of candidate genes. Mol Gen Genet 257:97–102

Luetteke NC, Qui TH, Pfeiffer RL, Oliver P, Smithies O, Lee DC (1993) TGFα deficiency results in hair follicle and eye abnormalities in targeted and waved-1 mice. Cell 73:263–278

Luo G, Hofmann C, Bronckers AL, Sohocki M, Bradley A, Karsenty G (1995) BMP-7 is an inducer of nephrogenesis, and is also required for eye development and skeletal patterning. Genes Dev. 9:2808–2820

Lyon MF, Morris T, Searle AG, Butler J (1967) Occurences and linkage relations of the mutant "*extra-toes*" in the mouse. Genet Res 9:383–385

Lyon MF, Jarvis SE, Sayers I, Holmes RS (1981) Lens opacity: a new gene for congenital cataract on chromosome 10 of the mouse. Genet Res Camb 38:337–341

Lyons KM, Hogan BLM, Robertson EJ (1995) Colocalization of BMP7 and BMP2 RNAs suggests that these factors cooperatively mediate tissue interactions during murine development. Mech. Dev. 50:71–83

Ma J, Norton JC, Allen AC, Burns JB, Hasel KW, Burns JL, Sutcliffe JG, Travis GH (1995) Retinal degeneration slow (*rds*) in mouse results from simple insertion of a haplotype-specific element into protein-coding exon II. Genomics 28:212–219

Mahon KA, Chepelinsky AB, Khillan JS, Overbeek PA, Piatigorsky J, Westphal H (1987) Oncogenesis of the lens in transgenic mice. Science 235:1622–1628

Malas S, Sartor M, Duthie S, Hadjantonakis K, Lovell-Badge R, Episkopou V (1996) Genetic and physical mapping of the murine *Sox1* gene. Mamm Genome 7:620–621

Mann GB, Fowler KJ, Gabriel A, Nice EC, Williams RL, Dunn AR (1993) Mice with a null mutation of the *TGFα* gene have abnormal skin architecture, wavy hair, and curly whiskes and often develop corneal inflammation. Cell 73:249–261

Marigo V, Roberts DJ, Lee SM, et al. (1995) Cloning, expression, and chromosomal location of SHH and IHH: two human homologues of the *Drosophila* segment polarity gene *hedgehog*. Genomics 28:44–51

Marker PC, King JA, Copeland NG, Jenkins NA, Kingsley DM (1995) Chromosomal localization, embryonic expression, and imprinting tests for *Bmp7* on distal mouse chromosome 2. Genomics 28:576–580

Mathers PH, Grinberg A, Mahon KA, Jamrich M (1997) The *Rx* homeobox gene is essential for vertebrate eye development. Nature 387:603–607

Matsuo T, Osumi-Yamashita N, Noji S, Ohuchi H, Koyama E, Myokai F, Matsuo N, Taniguchi S, Doi H, Iseki S, Ninomiya Y, Fujiwara M, Watanabe T, Eto K (1993) A mutation in the Pax-6 gene in rat *small eye* is associated with impaired migration of midbrain crest cells. Nat Genet 3:229–304

Matsushima Y, Kamoto T, Iida F, Abujang P, Honda Y, Hiai H (1996) Mapping of *rupture of lens cataract (rlc)* on mouse chromosome 14. Genomics 36:553-554

Mattei MG, Riviere M, Krust A, Ingvarsson S, Vennstrom B, Islam MQ, Levan G, Kautner P, Zelent A, Chambon P, Szpirer J, Szpirer C (1991) Chromosomal assignment of retinoic acid receptor *(RAR)* genes in the human, mouse, and rat genomes. Genomics 10:1061-1069

McHenry JZ, Leon A, Matthaei KI, Cohen DR (1998) Overexpession of *fra-2* in transgenic mice perturbs normal eye development. Oncogene 17:1131-1140

McLaughlin ME, Sandberg MA, Berson EL, Dryja TP (1993) Recessive mutations in the gene encoding the β-subunit of rod phosphodiesterase in patients with retinitis pigmentosa. Nat Genet 4:130-134

Mochii M, Ono T, Matsubara Y, Eguchi G (1998) Spontaneous transdifferentiation of quail pigmented epithelial cell is accompanied by a mutation in the *Mitf* gene. Dev Biol 196:145-159

Morgenbesser SD, Williams BO, Jacks T, dePinho RA (1994) p53-dependent apoptosis produced by *Rb*-deficiency in the developing mouse lens. Nature 317:72-74

Mörner CT (1893) Untersuchungen der Proteinsubstanzen in den lichtbrechenden Medien des Auges. Z Physiol Chem 18:61-106

Morrison-Graham K, Schatteman GC, GC, Bork T, Bowen-Pope DF, Weston JA (1992) A PDGF-receptor mutation in the mouse *(Patch)* perturbs the development of a non-neuronal subset of neural crest-derived cells. Development 115:133-142

Muggleton-Harris AL, Festing MFW, Hall M (1987) A gene location for the inheritance of the Cataract Fraser *(Cat^{Fr})* mouse congenital cataract. Genet Res Camb 49:235-238

Müller G (1950) Eine entwicklungsgeschichtliche Untersuchung über das erbliche Kolobom mit Mikrophthalmus bei der Hausmaus. Z Mikrosk Anat Forsch 56:520-558

Nishiguchi S, Wood H, Kondoh H, Lovell-Badge R, Episkopou V (1998) *Sox1* directly regulates the γ-crystallin gene and is essential for lens development in mice. Genes Dev 12:776-781

Noben-Trauth K, Naggert JK, North MA, Nishina PM (1996) A candidate gene for the mouse mutation *tubby*. Nature 380:534-538

Nornes HO, Dressler GR, Knapik EW, Deutsch U, Gruss P (1990) Spatially and temporally restricted expression of Pax2 during murine neurogenesis. Development 109:797-809

Oda SI, Watanabe K, Fujisawa H, Kameyama Y (1980) Impaired development of lens fibers in genetic microphthalmia, eye lens obsolescence, *Elo*, of the mouse. Exp Eye Res 31:673-681

Ohlemiller KK, Hughes RM, Mosinger-Ogilvie J, Speck JD, Grosof DH, Silverman MS (1995) Cochlear and retinal degeneration in the *tubby* mouse. Neuroreport 6:845-849

Oliver G, Loosli F, Köster R, Wittbrodt J, Gruss P (1996) Ectopic lens induction in fish in response to the murine homeobox gene *Six3*. Mech Dev 60:233-239

Opdecamp K, Vanvooren P, Riviere M, Arnheiter H, Motta R, Szpirer J, Spirer C (1998) The rat microphthalmia-associated transcription factor gene *(Mitf)* maps at 4q34-q41 and is mutated in the *mib* rats. Mamm Genome 9:617-621

Packer SO (1967) The eye and skeletal effects of two mutant alleles at the *microphthalmia* locus of *Mus musculus*. J Exp Zool 165:21-45

Pan H, Griep AE (1994): Altered cell cycle regulation in the lens of HPV-16 E6 or E7 transgenic mice: implications for tumor suppressor gene function in development. *Genes Dev.* 8:1285-1299

Petersen-Jones SM (1998) Animal models of human retinal dystrophies. Eye 12:566-570

Pittler SJ, Baehr W (1991) Identification of a nonsense mutation in the rod photoreceptor cGMP phosphodiesterase β-subunit gene of the *rd* mouse. Proc Natl Acad Sci USA 88:8322-8326

Pittler SJ, Keeler CE, Sidma RL, Baehr W (1993) PCR analysis of DNA from 70-year-old sections of rodless retina demonstrates identity with the mouse *rd* defect. Proc Natl Acad Sci USA 90:9616-9619

Porter FD, Drago J, Xu Y, Cheema SS, Wassif C, Huang SP, Lee E, Grinberg A, Massalas JS, Bodine D, Alt F, Westphal H (1997) *Lhx2*, a LIM homeobox gene, is required for eye, forebrain, and definitive erythrocyte development. Development 124:2935-2944

Prosser J, van Heyningen V (1998) *PAX6* mutations reviewed. Hum Mutat 11:93–108

Reneker LW, Silversides DW, Patel K, Overbeek PA (1995) TGFa can act as a chemoattractant to perioptic mesenchymal cells in developing mouse eyes. Development 121:1669–1680

Rice DS, Williams RW, Ward-Bailey P, Johnson KR, Harris BS, Davisson MT, Goldowitz D (1995) Mapping the *Bst* mutation on mouse chromosome 16: a model for human optic atrophy. Mamm Genome 6:546–548

Rice DS, Tang Q, Williams RW, Harris BS, Davisson MT, Goldowitz D (1997) Decreased retinal ganglion cell number and misdirected axon growth associated with fissure defects in *Bst/+* mutant mice. Invest Ophthalmol Visual Sci 38:2112–2124

Robb RM, Silver J, Sullivan RT (1978) Ocular retardation (*or*) in the mouse. Invest Ophthalmol Visual Sci 17:468–473

Robinson ML, MacMillan-Crow LA, Thompson JA, Overbeek PA (1995a) Expression of a truncated FGF receptor results in defective lens development in transgenic mice. Development 121:3959–3967

Robinson ML, Overbeek PA, Verran DJ, Grizzle WE, Stockards CR, Friesel R, Maciag T, Thompson JA (1995b) Extracellular FGF-1 acts as a lens differentiation factor in transgenic mice. Development 121:505–514

Robinson ML, Ohtaka-Maruyama C, Chan CC, Jamieson S, Dickson C, Overbeek PA, Chepelinsky AB (1998) Disregulation of ocular morphogenesis by lens-specific expression of *FGF-3/int-2* in transgenic mice. Dev Biol 198:13–31

Roderick TH, Chang B, Hawes NL, Heckenlively JR (1997) A new dominant *retinal degeneration* (*Rd4*) associated with a chromosomal inversion in the mouse. Genomics 42:393–396

Roessler E, Belloni E, Gaudenz K, Jay P, Berta P, Scherer SW, Tsui LC, Muenke M (1996) Mutations in the human *Sonic Hedgehog* gene cause holoprosencephaly. Nat Genet 14:357–360

Rogaev EI, Rogaeva EA, Korovaitseva GI, Farrer LA, Petrin AN, Keryanov SA, Turaeva S, Chumakov I, St. George-Hyslop P, Ginter EK (1996) Linkage of polymorphic congenital cataract to the γ-crystallin gene locus on human chromosome 2q33-35. Hum Mol Genet 5:699–703

Rothenspieler UW, Dressler GR (1993) *Pax-2* is required for mesenchyme-to-epithelium conversion during kidney development. Development 119:711–720

Santhiya ST, Abd-alla SM, Löster J, Graw J (1995) Reduced levels of γ-crystallin transcripts during embryonic development of murine *Cat2^{nop}* mutant lenses. Graefe's Arch Clin Exp Ophthalmol 233:795–800

Sanyal S, Hawkins RK (1979) *Dysgenetic lens* (*dyl*) – a new gene in the mouse. Invest Opthalmol Visual Sci 18:642–645

Sanyal S, van Nie R, de Moes J, Hawkins RK (1986) Map position of dysgenetic lens (*dyl*) locus on chromosome 4 in the mouse. Genet Res Camb 48:199–200

Sanyanusin P, Schimmenti LA, McNoe LA, Ward TA, Pierpont MEM, Sullivan MJ, Dobyns WB, Eccles MR (1995) Mutation of the *PAX2* gene in a family with optic nerve colobomas, renal anomalies and vesicoureteral reflux. Nat Genet 9:358–363

Schatteman GC, Morrison-Graham K, van Koppen A, Weston JA, Bowen-Pope DF (1992) Regulation and role of PDGF receptor α-subunit expression during embryogenesis. Development 115:123–131

Scotland P, Zhou D, Benveniste H, Bennett V (1998) Nervous system defects of *AnkyrinB* (-/-) mice suggest functional overlap between the cell adhesion molecule L1 and 440-kd AnkyrinB in premyelinated axons. J Cell Biol 143:1305–1315

Semina EV, Reiter RS, Murray JC (1997) Isolation of a new homeobox gene belonging to the *Pitx/Rieg* family: expression during lens development and mapping to the *aphakia* region on mouse chromosome 19. Hum Mol Genet 6:2109–2116

Shiels A, Bassnett S (1996) Mutations in the founder of the MIP gene family underlie cataract development in the mouse. Nat Genet 12:212–215

Shiels A, Griffin CS (1993) Aberrant expression of the gene for lens major intrinsic protein in the CAT mouse. Curr Eye Res 12:913–921

Shiels A, Mackay D, Ionides A, Berry V, Moore A, Bhattacharya S (1998) A missense mutation in the human *connexin50* gene (*GJA8*) underlies autosomal dominant "*zonular pulverulent*" cataract, on chromosome 1q. Am J Hum Genet 62:526–532

Sidman RL, Green MC (1965) Retinal degeneration in the mouse: location of the *rd* locus in linkage group XVII. J Hered 56:23–29

Sidman RL, Tang M, Kosoras B, Phillips SJ, Taylor BA (1997) Mapping and retinal phenotype of the *hugger* mutation in the mouse. Mamm Genome 8:399–402

Silver J, Robb RM (1979) Studies on the development of the eye cup and optic nerve in normal mice and in mutants with congenital optic nerve aplasia. Dev Biol 68:175–190

Silver J, Puck SM, Albert DM (1984) Development and aging of the eye in mice with inherited optic nerve aplasia: histopathological studies. Exp Eye Res 38:257–266

Sinha D, Esumi N, Jaworski C, Kozak CA, Pierce E, Wistow G (1998) Cloning and mapping of the mouse *Crygs* gene and non-lens expression of γS-crystallin. Mol Vis 4:8

Smith RS, Hawes NL, Kuhlmann SD, Heckenlively JR, Chang B, Roderick TH, Sundberg JP (1996) *Corn1*: a mouse model for corneal surface disease and neovascularization. Invest Ophthalmol Visual Sci 37:397–404

Sohocki MM, Sullivan LS, Mintz-Hittner HA, Birch D, Heckenlively JR, Freund CL, McInnes RR, Daiger SP (1998) A range of clinical phenotypes associated with mutations in *CRX*, a photoreceptor transcription-factor gene. Am J Hum Genet 63:1307–1315

Song CW, Okumoto M, Mori N, Kim JS, Han SS, Esaki K (1997) Mapping of new recessive cataract gene (*lr2*) in the mouse. Mamm Genome 8:927–931

Stadler HS, Solursh M (1994) Characterization of the homeobox-containing gene *GH6* identifies novel regions of homeobox gene expression in the developing chick embryo. Dev Biol 161:251–262

Stambolian D, Favor J, Silvers W, Avner P, Chapman V, Zhou E (1994) Mapping of the X-linked cataract (*Xcat*) mutation, the gene implicated in the Nance Horan Syndrome, on the mouse X chromosome. Genomics 22:377–380

Steele ECJr, Kerscher S, Lyon MF, Glenister PH, Favor J, Wang J, Church RL (1997) Identification of a mutation in the MP19 gene, *Lim2*, in the cataractous mouse mutant *To3*. Mol Vis 3:5

Steele ECJr, Lyon MF, Favor J, Guillot PV, Boyd Y, Church RL (1998) A mutation in the *connexin 50* (*Cx50*) gene is a candidate for the *No2* mouse cataract. Curr Eye Res 17:883–889

Stein KF, Norris BE, Mason J (1967) Development of an *open eyelid* mutant in *Mus musculus*. Dev Biol 16:315–330

Steingrímsson E, Moore KJ, Lamoreux ML, Ferré-D'Amaré A, Burley SK, Zimring DCS, Skow LC, Hodgkinson CA, Arnheiter H, Copeland NG, Jenkins NA (1994) Molecular basis of mouse *microphthalmia* (*mi*) mutations helps explain their developmental and phenotypic consequences. Nat Genet 8:256–263

St-Onge L, Sosa-Pineda B, Chowdhury K, Mansouri A, Gruss P (1997) *Pax6* is required for differentiation of glucagon-producing α-cells in mouse pancreas. Nature 387:406–409

Tahvanainen E, Villanueva AS, Forsus H, Salo P, de la Chapelle A (1996) Dominantly and recessively inherited cornea plana congenita map to the same small region of chromosome 12. Genome Res 6:249–254

Takebayashi K, Sasai Y, Sakai Y, Watanabe T, Nakanishi S, Kageyama R (1994) Structure, chromosomal locus, and promoter analysis of the gene encoding the mouse helix-loop-helix factor HES-1. Negative autoregulation through the multiple N box elements. J Biol Chem 269:5150–5156

Takehana M (1990) Hereditary cataract of the Nakano mouse. Exp Eye Res 50:671–676

Tassabehji M, Newton VE, Read AP (1994) Waardenburg syndrome type 2 caused by mutations in the human microphthalmia (*MITF*) gene. Nat Genet 8:251–255

Theil T, Zechner U, Klett C, Adolph S, Moroy T (1994) Chromosomal localization and sequences of the murine *Brn-3* family of developmental control genes. Cytogenet Cell Genet 66:267–271

Theiler K, Varnum DS (1981) Development of *coloboma* (*Cm*/+), a mutation with anterior lens adhesion. Anat Embryol 162:121–126

Theiler K, Varnum DS, Nadeau JH, Stevens LC, Cagianut B (1976) A new allele of *ocular retardation*: early development and morphogenetic cell death. Anat Embryol 150:85–97

Tomarev SI, Zinovieva RD, Chang B, Hawes NL (1998) Characterization of the mouse *Prox1* gene. Biochem Biophys Res Commun 248:683–689

Tomita K, Nakanishi S, Guillemot F, Kageyama R (1996) *Mash1* promotes neuronal differentiation in the retina. Genes Cells 8:765–774

Torres M, Gomez-Pardo E, Gruss P (1996) *Pax2* contributes to inner ear patterning and optic nerve trajectory. Development 122:3381–3391

Travis GH, Groshan KR, Lloyd M, Bok D (1992) Complete rescue of photoreceptor dysplasia and degeneration in transgenic *retinal degeneration slow* (*rds*) mice. Neuron 9:113–119

Truslove GM (1956) The anatomy and development of the *fidget* mouse. J Genet 54:64–86

Truslove GM (1962) A gene causing ocular retardation in the mouse. J Embryol Exp Morphol 10:652–660

Uga S, Kador PF, Kuwabara T (1980) Cytological study of Philly mouse cataract. Exp Eye Res 30:79–92

van Nie R, Ivanyi D, Demant P (1978) A new H-2-linked mutation, *rds*, causing retinal degeneration in the mouse. Tissue Antigens 12:106–108

Varnum DS (1983) *Blind-sterile*: a new mutation on chromosome 2 of the house mouse. J Hered 74:206–207

Varnum DS, Fox SC (1981) Head blebs: a new mutation on chromosome 4 of the mouse. J Hered 72:293

Varnum DS, Stevens LC (1968) Aphakia, a new mutation in the mouse. J Hered 59:147–150

Varnum DS, Stevens LC (1975) Report from the Jackson Lab. Mouse News Lett 53:35

Vermeji-Keers C (1975) Primary congenital aphakia and the rubella syndrome. Teratology 11:257–266

Vortkamp A, Gessler M, Grzeschik KH (1991) *Gli3* zinc-finger gene interrupted by translocations in Greig syndrome families. Nature 352:539–540

Wada E, Koyama-Ito H, Matsuzawa A (1991) Biochemical evidence for conversion to milder form of hereditary mouse cataract by different genetic background. Exp Eye Res 52:501–506

Walterhouse D, Ahmed M, Slusarski D, Kalamaras J, Boucher D, Holmgren R, Iannaccone P (1993) *Gli*, a zinc finger transcription factor and oncogene, is expressed during normal mouse development. Dev Dyn 196:91–102

Walther C, Gruss P (1991) *Pax-6*, a murine paired box gene, is expressed in the developing CNS. Development 113:1435–1449

Warren G, Gordon M, Siracusa LD, Buchberg AM, Williams T (1996) Physical and genetic localization of the gene encoding the AP-2 transcription factor to mouse chromosome 13. Genomics 31:234–237

Wawrousek EF, Brady JP (1998) αB-Crystallin gene knockout mice develop a severe, fatal phenotype late in life. Invest Ophthalmol Visual Sci 39:S523

Webster EHJr, Silver AF, Consalves NI (1984) The extracellular matrix between the optic vesicle and presumptive lens during lens morphogenesis in an anophthalmic strain of mice. Dev Biol 103:142–150

West JD, Fisher G (1986) Further experience of the mouse dominant cataract mutation test from an experiment with ethylnitrosourea. Mutat Res 164:127–136

West-Mays JA, Zhang J, Nottoli T, Hagopian-Donaldson S, Libby D, Strissel KJ, Williams T (1999) AP-2a transcription factor is required for early morphogenesis of the lens vesicle. Dev Biol 206:46–62

White TW, Goodenough DA, Paul DL (1998) Targeted ablation of connexin50 in mice results in microphthalmia and zonular pulverulent cataracts. J Cell Biol 143:815–825

Wistow G, Sinha D, Lyon M, Kozak C, Pierce E, Esumi N, Jaworski C (1998) γs-Crystallin in lens, retina and *Opj* cataract. Invest Ophthalmol Visual Sci 39:S523

Wu TH, Ting TD, Okajima TI, Pepperberg DR, Ho YK, Ripps H, Naash MI (1998) Opsin localization and rhodopsin photochemistry in a transgenic mouse model of retinitis pigmentosa. Neuroscience 87:709–717

Xia Y, Welch CL, Warden CH, Lange E, Fukao T, Lusis AJ, Gatti RA (1996) Assignment of the mouse *ataxia-telangiectasia* gene (*Atm*) to mouse chromosome 9. Mamm Genome 7:554–555

Xiang M (1998) Requirement for *Brn-3b* in early differentiation of postmitotic retinal ganglion cell precursors. Dev Biol 197:155–169

Xiang M, Gan L, Zhou L, Klein WH, Nathans J (1996) Targeted deletion of the mouse POU domain gene *Brn-3a* causes selective loss of neurons in the brainstem and trigeminal ganglion, uncoordinated limb development, and impaired suckling. Proc Natl Acad Sci USA 93:11950–11955

Yoshiki A, Hanazono M, Oda SI, Wakasugi N, Sakakura T, Kusakabe M (1991) Developmental analysis of the lens obsolescence (*Elo*) gene in the mouse: cell proliferation and *Elo* gene expression in the aggregation chimera. Development 113:1293–1304

Zhou E, Grimes P, Favor J, Koeberlein B, Pretsch W, Neuhäuser-Klaus A, Sidjanin D, Stambolian D (1997) Genetic mapping of a mouse ocular malformation locus, *Tcm*, to chromosome 4. Mamm Genome 8:178–181

Zwaan J, Williams RM (1969) Cataracts and abnormal proliferation of the lens epithelium in mice carrying the *CatFr* gene. Exp Eye Res 8:161–167

Genetic Analysis of Eye Development in Zebrafish

Jarema Malicki

1
Introduction

The vertebrate eye is a remarkable biological sensor which provides an extraordinary assortment of detailed information about the surrounding world. The accurate performance of the eye requires the cooperation of many elements with fundamentally different functional and structural properties such as the lens, ocular muscles, and the retina. The embryological origins of the eye are as diverse as its structure. The neural retina and the pigmented epithelium originate as an outpocketing of the neural tube, the lens as an ectodermal placode, while the ocular muscles arise from the mesenchyme. In the past decade or so, a small aquatic vertebrate, the zebrafish, has emerged as an excellent model system to study the genetic basis of vertebrate eye development.

Several advantages speak in favor of using zebrafish as a model system for developmental phenomena in general and the retina in particular. Zebrafish adults are easy to maintain in large numbers. Females are capable of producing hundreds of offspring in a single mating. Embryos develop outside the mother and are initially unpigmented, allowing one to monitor development through the body wall. Finally, embryogenesis is exceptionally rapid, producing well-developed organ systems by 48 h postfertilization (hpf). Owing to these attractive features, the zebrafish embryo has proven useful for studies of many organs including the brain, spinal cord, notochord, somites, heart, kidney, ear, vasculature, and skeleton (Weinstein et al. 1995; Brand et al. 1996a; Brand et al. 1996b; Chen et al. 1996; Kelsh et al. 1996; Malicki et al. 1996b; Odenthal et al. 1996; Piotrowski et al. 1996; Schier et al. 1996; Schilling et al. 1996; Stemple et al. 1996; van Eeden et al. 1996; Whitfield et al. 1996; Drummond et al. 1998). The beneficial characteristics of zebrafish embryogenesis are also apparent in eye development. The optic lobe becomes clearly delineated around 12 hpf, the lens rudiment around 20 hpf, and the pigmented epithelium starts accumulating pigment granules around 24 hpf. By 28 hpf,

Dept. of Ophthalmology, Harvard Medical School/MEEI, 243 Charles St., Boston, Massachusetts 02114, USA

Results and Problems in Cell Differentiation, Vol. 31
M. E. Fini (Ed.): Vertebrate Eye Development
© Springer-Verlag Berlin Heidelberg 2000

the first ganglion cells become postmitotic, and by 60 hpf neurons of the retina form laminar pattern (Nawrocki 1985; Hu and Easter 1999). Throughout this period, growth and morphological transformations of the optic lobe can be easily monitored under a dissecting microscope. Optokinetic and startle responses appear between 60 and 80 hpf, indicating that the zebrafish visual system becomes functional (Easter and Nicola 1996). The exquisitely precise cellular patterning of the retina at 72 hpf is shown in Fig. 1A.

The early morphogenetic movements leading to the formation of the neural retina, the pigmented epithelium, and the lens have been characterized through microscopic observations and histological studies (Schmitt and Dowling 1994). By midday 2 of embryonic development, eye morphogenesis is already complete and in the following 24 h, the majority of neuroepithelial

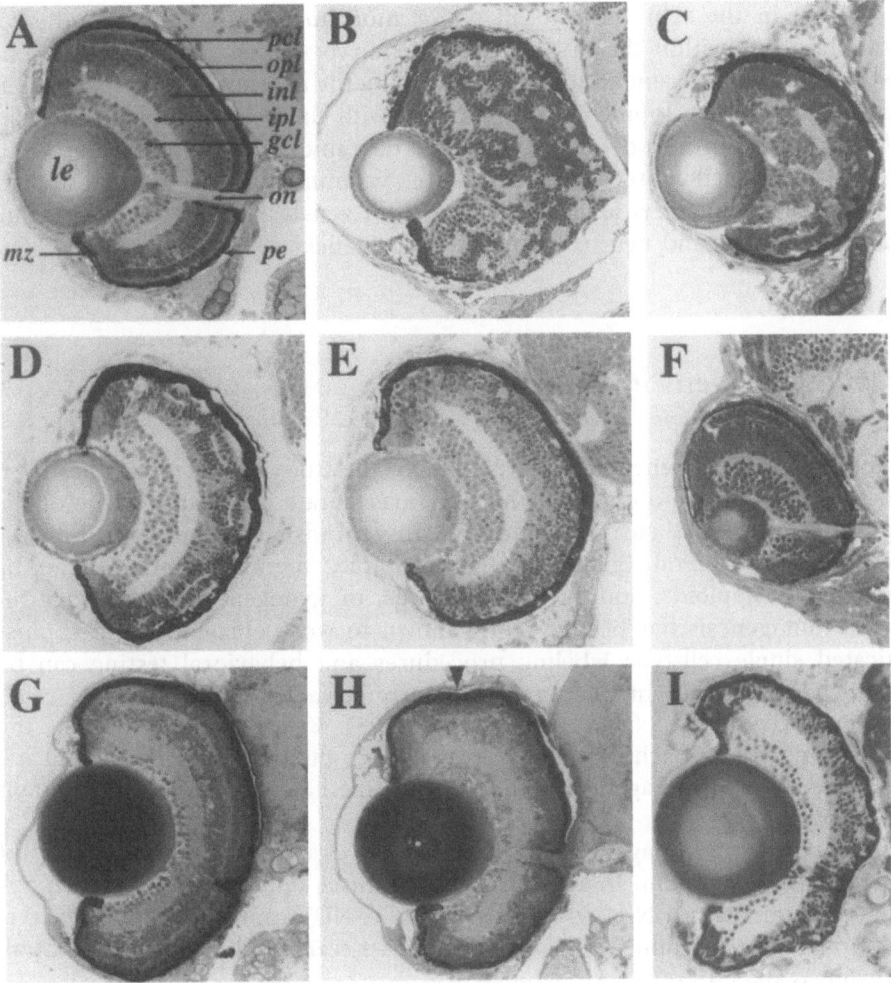

cells of the central retina differentiate into retinal neurons and glia. In all layers of the retina, neuronal differentiation begins in a small group of ventral cells (Schmitt and Dowling 1996; Passini et al. 1997). From there, it spreads into the other quadrants of the retina. As indicated by birthdating studies, the bulk of retinal neurogenesis occupies a narrow window of time between 27 and 55 hpf (Nawrocki 1985; Hu and Easter 1999). As in other vertebrates, ganglion cells are the first to be born (Altshuler et al. 1991) and the other cell types follow shortly thereafter. At the retinal margin, cells remain undifferentiated and neurogenesis continues throughout the lifetime of the organism (Marcus et al. 1999).

As in other vertebrates, the retina in zebrafish contains six major types of neurons and one type of glia (Cajal 1893; Rodieck 1973; Dowling 1987). The ganglion and the photoreceptor cells are by far the best studied cells of the zebrafish retina. The sequence of events which lead to ganglion cell differentiation has been described in considerable detail. The axons of ganglion cells become distinguishable as these cells migrate towards the vitreal surface shortly after the last mitotic division (Bodick and Levinthal 1980). The first axonal projections exit the retina at about 32–34 hpf, cross the body midline an hour later, and reach the rostral contralateral tectum at 44–46 hpf (Stuermer 1988; Burrill and Easter 1995). By 72 hpf, all tectal quadrants are innervated. In addition to the optic tectum, the ganglion cells innervate nine other brain targets. Except for one, that is innervated from both retinae, they are all targeted by contralateral projections (Burrill and Easter 1994).

Five types of photoreceptor cells are present in the zebrafish retina: rods, and four types of cones – blue, red, green, and UV. Individual photoreceptor

---

Fig. 1A–I. Histological analysis of mutations affecting the zebrafish retina. **A.** Cells of wild–type retina display precise, laminar architecture at 3 dpf. Photoreceptor cells localize to the photoreceptor cell layer (*pcl*) adjacent to the pigmented epithelium (*pe*). The inner nuclear layer (*inl*) contains horizontal, bipolar and amacrine interneurons. Ganglion cells occupy the ganglion cell layer (*gcl*) adjacent to the lens (*le*). The axonal projections of ganglion cells exit the retina through the optic nerve (*on*). Laminae composed mainly of pericaria are separated by layers of neuronal projections: the inner plexiform layer (*ipl*) and the outer plexiform layer (*opl*). By 3 dpf, the majority of neurons in the central retina are postmitotic. In contrast, cells of the marginal zone (*mz*) continue to proliferate. **B.** In the retina of *oko meduzy*, lamination is almost entirely absent. Instead of forming layers, neuronal projections aggregate in patches. **C.** The retina of *zimny* is also disorganized. Other sections of the *zimny* mutant retinae reveal that this phenotype is associated with massive cell death (not shown). **D.** Photoreceptor cells of *krenty* do not form a uniform layer and instead differentiate in patches. **E.** In *mikre oko*, few or sometimes no photoreceptors are distinguishable by morphological criteria. **F.** The retina of *out of sight* is greatly reduced compared to the wild type. **G.** In wildtype retina at 5 dpf, the photoreceptor cell layer is clearly visible. **H.** In the *photoreceptors absent* mutant retina, photoreceptor cells are missing with exception of cells located near the marginal zone (*arrowhead*). **I.** Except for sporadic cells in the central retina, most of the photoreceptor cell layer is absent in *not really finished*, a mutant induced by a retroviral insertion. All panels show transverse sections at 3 dpf (A to F) or 5 dpf (G to I). Midline is to the right

cell types are characterized by the expression of specific *opsin* genes which define their spectral sensitivities (Raymond et al. 1993; Vihtelic et al. 1999). Rod opsin is first detected at 50 hpf, whereas blue and red cone opsins are expressed by 52 hpf (Raymond et al. 1995). The elongated shapes of photoreceptor cells can be distinguished in the ventral patch by 48 hpf (Robinson et al. 1995). Within the following 6 h, the first rudiments of outer segments appear on electron micrographs (Branchek and Bremiller 1984; Schmitt and Dowling 1999). By 4 days postfertilization (dpf), UV cones can be morphologically distinguished from other cell types, and by day 12 all types of photoreceptor cells have developed distinct morphologies (Branchek and Bremiller 1984). In tangential sections through the photoreceptor cell layer in the adult retina, photoreceptors of different types are distributed in a precise spatial pattern. This spatial arrangement can be already distinguished by 72 hpf (Larison and Bremiller 1990).

Thus, in the zebrafish eye all major morphogenetic and differentiative events occur within the first 3 days of embryogenesis. This carries a substantial bonus for genetic studies by reducing the amount of time involved in screening for developmental defects even with the help of sophisticated labeling procedures and behavioral tests. Its rapid embryonic development, simple and well-characterized structure, prominent size and accessibility, make the zebrafish eye an excellent model for genetic dissection of vertebrate morphogenesis and neurogenesis (see also Malicki 1999).

2
Genetic Screens in Zebrafish

The repertoire of genetic tools available for the study of vertebrates has been historically much poorer than that for invertebrate organisms. Some of the genetic techniques currently available in zebrafish can, however, impress even an invertebrate geneticist. For example, zebrafish embryos can be studied as haploids, androgenetic diploids, or gynogenetic diploids; retroviral mutagenesis has been recently shown to work efficiently; and sophisticated single cell-type labeling procedures and behavioral testing can be used to detect mutant phenotypes in early larvae (Clark 1981; Baier et al. 1996; Corley-Smith et al. 1996; Walker 1999). Below, I briefly discuss the mutagenesis approaches available in zebrafish, the types of schemes utilized for breeding of mutagenized animals, and methods used to detect mutant phenotypes.

A diversity of mutagenesis approaches is an essential aspect of a successful genetic model system. In zebrafish, mainly three approaches have been applied: gamma rays, N-ethyl-N-nitrosourea (ENU), and, more recently, retroviral vectors (Mullins et al. 1994; Solnica-Krezel et al. 1994; Gaiano et al. 1996b; Walker 1999).

Gamma rays were the first mutagenic agent used in zebrafish (Grunwald et al. 1988; Streisinger et al. 1989). Gamma ray alleles of a desired locus can be generated with a higher frequency than ENU alleles (Walker 1999). Many represent deletions invaluable in testing whether chemically induced alleles are hypomorphic or amorphic. The most significant disadvantage of gamma-ray mutagenesis is that it often produces large chromosomal rearrangements. In the absence of other alleles, these are frequently of limited use because they simultaneously affect many genes and produce compound defects rather than single gene phenotypes (Walker 1999). ENU mutagenesis is another efficient way of inducing mutations (Mullins et al. 1994; Solnica-Krezel et al. 1994). In contrast to gamma rays, ENU mainly induces point mutations. Nucleotide sequencing of more than ten ENU-induced mutant alleles revealed that they mostly involve single base pair substitutions, leading to the appearance of premature stop codons, defective splice sites, or substitutions of conserved amino acids (Talbot et al. 1995; Brownlie et al. 1998; Lun and Brand 1998; Moens et al. 1998; Reifers et al. 1998). Since ENU was chosen for the largest mutagenesis screens performed in zebrafish so far, the majority of the currently available mutant alleles are ENU-induced.

An important recent addition to mutagenesis approaches in zebrafish is the use of retroviral vectors. Insertions of such vectors into a gene or its immediate vicinity are capable of producing mutations (Gaiano et al. 1996b). The major driving force behind retroviral mutagenesis screens is the ease of mutant gene cloning (Gaiano et al. 1996b; Becker et al. 1998). The mutated gene can be quickly identified using the retroviral sequence as a genomic tag. Retroviral mutagenesis is burdened with at least two major disadvantages. First, the mutagenesis process itself is very laborious, involving the injection of a retroviral vector into embryos one by one, followed by the selection of F1 progeny which have inherited the highest number of retroviral insertions using Southern blot analysis or an equivalent approach (Amsterdam and Hopkins 1999). Second, the efficiency of retroviral mutagenesis is 1 to 2 orders of magnitude lower than that of ENU. Nonetheless, the efficiency of retroviral mutagenesis has been improving steadily in recent years, making this approach increasingly more attractive (Lin et al. 1994; Gaiano et al. 1996a; Amsterdam and Hopkins 1999).

The choice of a breeding scheme is an important consideration in a genetic screen. When screening for recessive embryonic or larval defects in a diploid organism, one has to devote F2 generation to generating multiple carriers of a mutant allele, which then can be used to produce F3 homozygous mutant animals. If an adult phenotype is the subject of a screen, two full generations are necessary to produce adult homozygotes. This is time-consuming and, even more importantly, requires a prohibitive amount of space for raising F3 families. The zebrafish is unique in that at least two approaches can be used to generate mutant hemi- or homozygotes in the F2 generation.

The first approach is to generate haploid F2 embryos; the second, to produce parthenogenetic F2 diploids. Both methods involve some short-comings. The development of haploid embryos deviates from the wild type (Walker 1999). The body axis of haploids is shortened and their brains, eyes, and otic vesicles are morphologically abnormal. The number of defects accumulating in haploid development greatly reduces the usefulness of haploid embryos for screening after 3 dpf. In contrast to haploids, parthenogenetic diploids develop normally to adulthood (Pelegri and Schulte-Merker 1999). Even in this approach, however, a large proportion of embryos display abnormalities during the first 24 h of development and the majority die by 5 dpf. An additional complication is that the fraction of mutant homozygotes in a clutch of parthenogenetic embryos generated through the early pressure treatment varies with the distance of the mutant locus from the centromere (Beattie et al. 1999). Thus, in some cases, the few embryos which display a mutant phenotype may be difficult to identify against the background of nongenetic developmental abnormalities. Despite their shortcomings, both haploid and parthenogenetic screens have been used successfully to identify genetic defects in zebrafish (Henion et al. 1996; Guo et al. 1999).

Screens for dominant mutations offer yet another way to reduce the amount of time and space necessary to identify mutant phenotypes. Dominant phenotypes can be already detected in F1 animals. In practice, screening F1 embryos is not efficient, however, because more than half of them do not reach sexual maturity and therefore the mutations which they carry cannot be recovered (van Eeden et al. 1999). Screens for dominant defects performed on F2 larvae, on the other hand, are characterized by recovery rates approaching 100 % (van Eeden et al. 1999). Dominant screens eliminate the problem of genetic redundancy and are invaluable for isolating late–onset phenotypes. The biggest disadvantage of dominant screens is that viable dominant defects are much less frequent than recessive mutations (van Eeden et al. 1996). This drawback may be more pronounced in the case of gross morphological abnormalities. Late phenotypes confined to small cell populations may be better represented by dominant mutations.

Another major advantage of using zebrafish for genetics research is the great diversity of methods which can be used to detect mutant phenotypes. Morphological, histochemical, and behavioral criteria have been recently used to search for defects in the zebrafish visual system. Importantly, nearly all of them can be applied during the early larval period. The simplest and least time-consuming approach is the visual inspection of embryonic and larval phenotypes under a dissecting microscope. Even this simple method is not possible in other vertebrate model systems, due to intrauterine embryogenesis and low fecundity. In one of the largest recent experiments, F3 progeny of mutagenized animals were screened for eye defects using the criteria of eye size, shape, and pigmentation at four time points between 1 and 5 dpf (Malicki et al. 1996a). This simple screening strategy led to the identification of 49 mutations affecting the development of the retina.

Although this visual inspection screen produced very encouraging results, it only detected mutations resulting in gross morphological defects. How could one screen for more subtle defects such as the absence or overproduction of a rare cell type in the retina? Two recent screens have employed histochemical methods to visualize specific cell populations in the retina.

To isolate defects in retinotectal projections, Baier and colleagues used lipophilic tracers to label retinal ganglion cells (Baier et al. 1996). DiI and DiO were injected in parallel into two separate quadrants of the retina in F3 progeny of mutagenized animals. Since over 100,000 larvae were screened in this experiment, it was of the essence to minimize the time spent on dye application. This was accomplished through constructing partially automated mounting and injection devices. Through these technical improvements, the time spent on the analysis of a single larva was reduced to 1 min. This screen led to the isolation of 114 mutations affecting the formation of retinotectal projections. Although many of them result in gross morphological defects in the brain, others appear to be very specific to the retinotectal pathway, and thus are likely to involve primary defects in retinal ganglion cells (Baier et al. 1996; Karlstrom et al. 1996; Trowe et al. 1996).

Another small cell population recently targeted in a mutagenesis screen are catecholaminergic neurons, which reside in the forebrain, hindbrain, and retina. In the retina, only the interplexiform cells are catecholaminergic. Since their number is very small, an antibody against tyrosine hydroxylase was used to visualize defects in their development. Immunohistochemical screening of 700 early pressure-treated F2 egg clutches led to the identification of two mutations affecting interplexiform neurons (Guo et al. 1999). This and the retinotectal projection screen provide a good example of how zebrafish embryos can be used to analyze subtle defects in small cell populations. The great advantage of marker-guided screens is that they allow one to focus on very specific developmental processes. Taken to the extreme, marker-guided screens could be used to monitor developmental phenomena on a subcellular level, by analyzing such events as the formation of adherens junctions, the orientation of mitotic spindles, or the position of centrosomes. The position of the centrosomes, for example, could be used to analyze the formation of retinal neuroepithelium (Malicki and Driever 1999).

Behavioral screens are another attractive dimension of zebrafish genetics. Vision–dependent behavioral responses have been already described by some of the early proponents of the zebrafish model. Clark and colleagues employed both the optomotor and the optokinetic responses to analyze the visual acuity of normal and mutant zebrafish larvae (Clark 1981). They suggested that the optokinetic response is particularly valuable in mutagenesis screens because it appears early in development, does not involve the optic tectum, and is present even in developmentally retarded individuals. Interest in optokinetic screens has been recently revived. In one recent experiment, screening of 266 mutagenized genomes led to the identification of two mutations affecting retinal function (Brockerhoff et al. 1995). In addition to the optomotor and the optokinetic responses, at least three other vision-dependent behaviors have been described in zebrafish: phototaxis – the tendency of zebrafish larvae to gather in well-lit areas (Brockerhoff et al. 1995); the visual startle response – the avoidance behavior induced by a sudden decrease in light intensity (Easter and Nicola 1996); and the escape response – the escape behavior induced by a moving dark shape (Li and Dowling 1997). The escape response has been used in a dominant screen of 345 F1 individuals and has led to the identification of two nonresponsive mutant lines (Li and Dowling 1997).

Both marker-guided and behavioral screens tend to be labor-intensive. In contrast to marker-guided experiments, however, behavioral screens are characterized by low specificity because they seldom allow one to focus on a single cell population. The most significant advantage of the behavioral screening approach is that it allows one to detect very subtle developmental defects which frequently cannot be visualized using any of the available molecular markers.

3
Currently Available Eye Mutants

Screens performed in zebrafish in recent years have led to the identification of over 70 loci affecting many aspects of eye development (Malicki 1999). The classification of mutant genes based on phenotypes alone is difficult for several reasons. First, it is not clear whether the currently available mutations represent amorphic alleles. It is likely that a number of them are hypomorphic and do not reveal all aspects of gene function. Second, because individual genes frequently function in more than one tissue and at more than one developmental stage, late functions may be obscured by early ones. Finally, it is frequently difficult to distinguish primary from secondary defects. Nonetheless, attempts to classify mutants based on phenotypes are very useful. Phenotype-based classifications are, after all, the easiest way to relate genetic loci to biological processes. It has to be kept in mind, however, that current phenotype-based classifications are most likely incomplete. A more accurate classification of mutant loci will emerge as underlying genes are characterized through genetic and molecular approaches.

3.1
Specification of the Eye Field and Optic Cup Morphogenesis

During zebrafish gastrulation, the cells of the two prospective retinae form a single field. Cell fate mapping has shown that subsequent morphogenetic rearrangements separate this field into two domains, each corresponding to a separate optic lobe (Woo and Fraser 1995). Several loci identified in zebrafish mutagenesis screens clearly play a role in this process. Mutations in *cyclops* (*cyc*), *bozozok* (*boz*), *squint* (*sqt*), *one eyed pinhead* (*oep*), *uncle freddy* (*unf*), and *schmalspur* (*sur*) result in a ventral displacement and fusion of the optic lobes. In all mutants in which this has been investigated, this eye phenotype is accompanied by a reduction of the ventral brain and a reduction or absence of prechordal plate (Hatta et al. 1994; Brand et al. 1996b; Solnica-Krezel et al. 1996; Schier et al. 1997; Feldman et al. 1998).

The defect of prechordal plate in cyclopia-inducing mutants suggests that eye fusion is due to a deficiency in this tissue. This hypothesis is supported by classic embryological experiments as well as more recent studies indicating that the prechordal plate is involved in the specification of ventral forebrain and the eye field. Surgical removal of the prechordal plate in chicks and amphibians leads to the fusion of eye field, and the appearance of cyclopia (Adelmann 1930; Li et al. 1997). Moreover, displacement of the prechordal plate to the vicinity of the optic lobe results in the suppression of eye development (Li et al. 1997). These experiments suggest that the prechordal plate plays an essential role in the positioning of the eye primordia, possibly by subdividing the eye field. Four of the cyclopia-inducing loci have recently been cloned: *cyclops* (*cyc*) and *squint* (*sqt*) are nodal-related TGFβ family factors; *one eyed pinhead* (*oep*) is an EGF–related factor; and *bozozok* (*boz*) encodes a homeobox transcription factor (Feldman et al. 1998; Sampath et al. 1998; Zhang et al. 1998; Fekany et al. 1999). In blastula and early gastrula, *bozozok* and *squint* are mainly expressed at the dorsal margin of the blastoderm. Their expression ceases by the end of gastrulation (Feldman et al. 1998; Yamanaka et al. 1998). Thus, the function of these genes in the positioning of eye primordia appears to be indirect and related to their role in mesoderm formation. On the other hand, *one eyed pinhead* and *cyclops* are both expressed in the prechordal plate and in the brain (Sampath et al. 1998; Zhang et al. 1998); therefore their functions may be more directly involved in eye field specification. Mosaic analysis has revealed that *cyclops* acts cell-nonautonomously in the ventral forebrain and possibly in the prechordal plate (Hatta et al. 1994). These studies, combined with recent expression data, suggest that *cyclops* functions in both the prechordal plate and the ventral brain. This may also be true for *one eyed pinhead*. Apart from expression data, however, there is no evidence that *oep* functions in brain tissues.

Sonic hedgehog (*shh*) is another gene associated with cyclopia. The gene knockout of *shh* in the mouse results in the deletion of ventral forebrain fates (Chiang et al. 1996). Likewise, spontaneous mutations in the human *shh* gene produce a similar defect known as holoprosencephaly (Roessler et al. 1996). In contrast, deletion of the zebrafish *shh* homologue, originally identified as the *sonic-you* (*syu*) locus, does not lead to cyclopia or even a narrowing of the distance between the eyes (Schauerte et al. 1998). This has been explained by a presumptive redundancy of gene function with hedgehog paralogues *tiggy-*

winkle hedgehog (*twhh*) and *echidna hedgehog* (*ehh*). Despite the lack of a loss-of-function eye phenotype in zebrafish, evidence for the role of *sonic hedgehog* in eye morphogenesis has been provided by overexpression experiments. Injection of *sonic hedgehog* mRNA into embryonic zebrafish at the early blastula stage results in a reduction of eye size – a phenotype very similar to the one produced by prechordal plate transplants into the vicinity of the optic lobe (Macdonald et al. 1995).

Cyclopia in zebrafish is also caused by mutations in three other loci: *knypek* (*kny*), *trylobite* (*tri*), and *silberblick* (*slb*). These mutations affect convergent extension movements during gastrulation, causing a shortening and broadening of the body axis (Solnica-Krezel et al. 1996; Heisenberg and Nusslein-Volhard 1997). No gross deletions of anterior mesodermal structures, such as the prechordal plate, have been reported in these mutants. Their eye phenotypes are not fully penetrant, and range from a slight narrowing of the distance between the eyes to the appearance of a single eye. At the beginning of somitogenesis, these mutants are characterized by an abnormally large gap between the *sonic hedgehog* expression domain and the anterior edge of the neural plate as marked by *dlx-3* expression (Heisenberg and Nusslein-Volhard 1997; Marlow et al. 1998). The magnitude of this gap in *kny*, *tri*, and the *kny;tri* double mutants correlates with the degree of cyclopia. This, and the observation that *shh* affects both *Pax-6* and *Pax-2* expression (see below), has led to the hypothesis that cyclopia in this group of mutants is caused by the lack of eye field subdivision by hedgehog signaling (Heisenberg and Nusslein-Volhard 1997; Marlow et al. 1998).

Genetic analysis of eye development in the mouse has identified several loci involved in the early development of the optic vesicles. At least three of them have been characterized at the molecular level: *Rx*, *lhx2*, and *Pax-6*. The loss-of-function phenotypes of these genes have been revealed through knockout analysis and spontaneous mutations (Hill 1991; Mathers et al. 1997; Porter et al. 1997). The most striking defect in these mutants is the failure of eye development beyond the early optic vesicle stage. Eye rudiments of these mutants, when present, form in the proper positions, and the brain abnormalities associated with these mutant phenotypes are distinct from cyclopias. The mutant phenotype of *Pax-6*, *Rx*, and *lhx2* has not been observed in zebrafish so far. None of the less severe phenotypes isolated in mutagenesis screens has been shown to involve this group of genes either. The absence of the *Pax-6* phenotype among the zebrafish mutants may be due to chance or genetic redundancy. The latter possibility is supported by the discovery of two zebrafish *Pax-6* paralogues: *Pax-6.1* and *Pax-6.2* (Nornes et al. 1998). What is the relationship between the cyclopia-inducing loci and the *Pax-6* pathway? In the absence of the *cyclops* function, the *Pax-6* expression domain is expanded into the ventral forebrain. On the other hand, overexpression of *sonic hedgehog* leads to the reduction of the *Pax-6* expression and a severe reduction in eye size (Macdonald et al. 1995). Thus, the *Pax-6* gene appears to act downstream of *cyclops, sonic hedgehog,* and presumably other loci in this category.

no isthmus (noi), the zebrafish gene which encodes the *Pax-2.1* transcription factor, plays a role in yet another aspect of early eye development. During embryogenesis, the *noi* protein is present in the optic stalk and in cells surrounding the choroid fissure. In the adult fish, it is expressed in the optic nerve astrocytes (Macdonald et al. 1997). Several mutant alleles of *noi* are available (Lun and Brand 1998). The strongest of them, noi^{tu29a}, is presumably amorphic and produces a defect of choroid fissure closure, failure of glial differentiation in the optic nerve, and abnormal axonal pathfinding of optic nerve axons (Macdonald et al. 1997; Lun and Brand 1998). In the developing eye, *Pax-2* and *Pax-6* expression domains closely abut each other and are regulated in a reciprocal way. In *cyclops* mutant animals, the expansion of *Pax-6* expression parallels a reduction of *Pax-2*. The opposite effect, the expansion of *Pax-2* and the reduction of *Pax-6* expression domains, is seen in response to *shh* overexpression (Macdonald et al. 1995). The reciprocal regulation of *Pax-2* and *Pax-6* expression patterns may reflect the fact that these two genes participate in the specification of two mutually exclusive domains of the optic vesicle: the optic stalk and the retina. Like the *Pax-6* gene, *Pax-2* is also represented by two paralogues in the zebrafish genome: *Pax-2.1* and *Pax-2.2* (Pfeffer et al. 1998). No mutations are currently known to affect *Pax-2.2.*

Compared to other mutant categories affecting eye development, mutants of early eye morphogenesis are by far the best characterized at the molecular level. The molecular structure of five mutant genes has been recently determined. In the majority of cases, this was accomplished through the candidate approach (Brand et al. 1996a; Feldman et al. 1998; Sampath et al. 1998; Fekany et al. 1999). The abundance of candidate genes for this particular category of mutants partially stems from the fact that the molecular basis of early embryonic patterning is well conserved throughout the animal kingdom and has long been the focus of intensive research in both vertebrate and invertebrate model systems.

3.2
Growth of the Eye

The optic lobes of the zebrafish embryo are clearly developed by 12 hpf. From this time on, all cells of the neural retina continue to be mitotically active, until the first postmitotic neurons appear between 27 and 28 hpf, marking the beginning of neuronal differentiation (Hu and Easter 1999). By 3 dpf, nearly all cells of the central retina become postmitotic, but proliferation continues in the marginal zone. The cells of retinal margin remain pluripotent and generate all retinal cell types throughout the larval period and adulthood (Marcus et al. 1999). In addition to the retinal margin, the entire retina contains a small number of mitotically active cells which continue to generate rod photoreceptors (Johns and Fernald 1981; Marcus et al. 1999).

Mutations of at least five loci affect the size of the eye but do not result in excessive cell death or any other obvious abnormalities in the retina. These are *out of sight* (*out*), *spy eye* (*spy*), *cleopatra* (*cle*), *visionary* (*vis*), and *podgladacz* (*pod*) (Malicki et al. 1996a). With the exception of *out*, the phenotypes of these mutants are first recognizable by visual inspection between 4 and 5 dpf. The relatively late appearance of their phenotype suggests that they affect eye growth at the retinal margin. *out of sight* (Fig. 1F) stands out in this mutant category because of the early appearance of its phenotype between 30 and 36 hpf. Although retinal neuroepithelial cells continue to proliferate from the time of optic lobe formation at 12 hpf through the beginning of neuronal differentiation at 27 hpf, no difference of eye size is observed between the wild type and the *out* mutant fish until at least 30 hpf. The decrease of eye size in *out* roughly correlates with the appearance of the first postmitotic neurons and the shortening of the cell cycle. At 24 hpf, the length of the cell cycle in the fish retina is estimated at 10 h (Nawrocki 1985). Around 28 hpf, it shortens to 5 h (Hu and Easter 1999). Thus the decrease of eye size in *out* may be caused by a premature exit from the cell cycle or by an inability of the mutant cells to decrease the cell cycle length. As in other mutants in this category, no patterning defects or excessive cell death are observed in the retina. The *out* phenotype resembles *cyclin D1* knockout mice which also display a dramatic decrease of cell numbers in the retina (Sicinski et al. 1995). Unlike the *cyclin D1* knockout, however, *out* animals display normal body size.

3.3
Specification and Differentiation of Distinct Cell Populations

The vertebrate neural retina initially consists of a morphologically uniform population of neuroepithelial cells. In a narrow window of time, from 30 to 60 hpf, the vast majority of these cells exit the cell cycle and assume one of seven major fates characterized by distinct morphologies, positions, and functions. What genetic mechanisms participate in the specification and differentiation of distinct populations of retinal cells? One way to approach this problem is to identify mutations affecting the development of individual cell populations in the retina. This approach has been used successfully in the mouse. Both spontaneous mutations and targeted gene knockouts have revealed loci involved in the differentiation of retinal neurons (Burmeister et al. 1996; Gan et al. 1996).

Mutagenesis screens in zebrafish provide an effective, alternative way to isolate loci involved in the differentiation of specific retinal cell types. A large collection of mutations in this category have been isolated in the screens performed to date. Among them, photoreceptor defects are by far the most numerous due to at least two reasons. First, photoreceptor cells are very abundant in the retina, and due to their outer segments, bulkier than other cell types. Thus loss of photoreceptors results in a dramatic and readily detectable decrease in eye size. Second, compared to the majority of other neurons, photoreceptor cells are unusual in their structure and function. While the majority of neurons are likely to share developmental pathways, much of the genetic machinery responsible for photoreceptor differentiation may be unique. Thus, photoreceptor defects may tend to be specific to this cell type. In contrast, mutations of other neuronal populations of the retina may tend to be accompanied by defects in many other parts of the CNS. Such general developmental defects are frequently discarded as nonspecific.

Photoreceptor mutants display several distinct patterns of cell loss. Three types of photoreceptor cell loss were identified in a morphological screen for defects in eye development (Malicki et al. 1996a). Some of these were also represented by mutations isolated in other screens (Fadool et al. 1997). Here I will use a broadened version of this categorization to assign all currently described zebrafish photoreceptor defects to four subcategories. The first includes mutations characterized by a loss of cells throughout the photoreceptor cell layer, with the marginal cells being slightly more affected than the cells in the central retina. This category includes: *mikre oko* (*mok*), *niezerka* (*nie*), and *not really finished* (*nrf*) (Malicki et al. 1996a; Becker et al. 1998). *mok*^{m632} appears to be the earliest currently known defect of zebrafish photoreceptor cells (Fig. 1E). By late day 3, no photoreceptors are distinguishable by morphological criteria in *mok*^{m632}. Abnormally shaped cells are detectable, however, using antibodies specific to photoreceptor cells. By 5 dpf, less than 5 % of cells are distinguishable by immunohistochemical criteria. Electron microscopic analysis reveals that nearly all *mok* photoreceptor cells do not develop inner and outer segments (G. Doerre and J. Malicki, unpublished). The *niezerka* phenotype is less severe than *mikre oko*. As revealed by immunohistochemistry, some photoreceptor cells are present in this mutant until 7 dpf (G. Doerre and J. Malicki, unpublished). So far, *not really finished* (*nrf*) is the only eye mutant induced via retroviral mutagenesis which does not display a widespread degeneration of the central nervous system (Fig. 1I). Based on transcript expression, this mutation results in a complete absence of gene product. Cloning of the *nrf* gene revealed that it encodes Nuclear Respiratory Factor (NRF), a transcription factor first discovered as an activator of the cytochrome c (Virbasius et al. 1993). The majority of photoreceptor cells are absent in the *nrf* retinae by 5 dpf. Small numbers of surviving cells develop outer segments and are capable of mediating the optokinetic response (Becker et al. 1998).

The second subgroup of photoreceptor mutants produce a patchy loss of photoreceptor cells by 3 dpf. (Fig. 1D). Not all photoreceptors are affected, and foci of normal cells survive at least until 5 dpf. *krenty* (*krt*), *discontinous* (*dis*), and *sinusoida* (*sid*) belong to this category. All have slightly abnormal brains as well (Malicki et al. 1996a).

In the third subcategory, the pattern of photoreceptor loss roughly follows their order of differentiation. Thus, frequently, the photoreceptor cells first become apoptotic in the ventral portion of the retina and only later does the degeneration spread into the center. The characteristic feature of this class of photoreceptor loss is that the peripheral cells are the least affected (Fig. 1H, compare to wild type in G). Presumably, this is because in the teleost fish the retina continues to expand at the margin so that at any given time the marginal cells are the youngest. *brudas* (*bru*), *elipsa* (*eli*), *fleer* (*flr*), *photoreceptors absent* (*pca*), and *night blindness a* (*nba*) display this pattern of cell loss (Malicki et al. 1996a; Fadool et al. 1997; Li and Dowling 1997; Drummond et al. 1998). Many of these mutations also produce obvious phenotypes in other organs. For example, *elipsa* and *fleer* involve kidney defects while *brudas* affects touch response. *Night blindness a* was originally isolated as a late dominant photoreceptor loss. In the homozygous state, this mutation causes a severe embryonic phenotype involving rapid degeneration of the retina and brain. *nba* mutant homozygotes display a loss of virtually all retinal neurons by 3.5 dpf (Li and Dowling 1997).

Finally, the mutant *partial optokinetic response b* (*pob*) forms a category of its own. Only one of the five photoreceptor subtypes is affected in *pob*. The red opsin-expressing photoreceptor cells, normal at 3 dpf, are absent by 5 dpf. Other cone types appear to survive at least 3 days longer. *pob* has been shown by linkage analysis not to affect the red opsin gene – the only gene currently known to be specific to red cones (Brockerhoff et al. 1997).

Apart from photoreceptor defects, few of the mutations isolated in zebrafish so far affect specific subpopulations of retinal cells. Most likely, this is because the majority of the retina–oriented screens have been based on morphological criteria. More sophisticated screening methodologies may be required to isolate defects in relatively small populations of cells. As mentioned above, a recent histochemical screen has revealed two loci affecting retinal interplexiform cells: *motionless* (*mot*) and *foggy* (*fog*) (Guo et al. 1999). Interplexiform cells are missing in *foggy*, while their number is reduced in *motionless*. Both mutants also affect brain catecholaminergic neurons and display defects in the cardiovascular system. Since no other retinal cell types were analyzed in these mutants, it is not clear whether they affect only interplexiform cells. As cell-type-specific defects provide an important entry into the molecular analysis of retinal cell-type identity, further marker-guided screens will most likely be necessary to isolate mutations affecting individual cell populations of the retina.

3.4
Patterning of Cell Populations

The precise patterning of neurons is a striking feature of the vertebrate retina. More so than in other parts of the central nervous system, individual cell types of the retina are organized in distinct laminae (Muller 1857; Cajal 1893; Rodieck 1973; Dowling 1987). The layers occupied primarily by pericaria are separated by plexiform laminae mainly containing neuronal projections. In the day-3 zebrafish retina, four cell types can be distinguished in histological sections based on their positions relative to the plexiform layers: ganglion cells, amacrine cells, horizontal cells, and photoreceptors (Fig. 1A). Additional laminae can be distinguished through a more detailed analysis.

What is the nature of genetic mechanisms that position individual cell types in their proper cellular laminae? Since only a handful of genes are known to be involved in retinal lamination (Tomasiewicz 1993; Stumpo et al. 1995; Tomita et al. 1996; Georges-Labouesse et al. 1998), this issue remains largely unexplored. Three zebrafish loci affecting neuronal organization in the retina provide an attractive opportunity to gain insight into the genetic basis of neuronal lamination: *oko meduzy* (*ome*), *nagie oko* (*nok*) and *glass onion* (*glo*). All three produce almost complete absence of retinal lamination (Malicki et al. 1996a; Malicki and Driever 1999). Instead of forming layers, the neuronal projections congregate in patches (Fig. 1B). Among the neuronal patterning mutants, *oko meduzy* has been described in most detail. Based on staining experiments with cell–type–specific markers, at least seven cell types differentiate in *oko meduzy* by 3 dpf but are localized to abnormal positions (Malicki and Driever 1999). The *oko meduzy* defect is already detectable in the retinal neuroepithelium at 30 hpf. Mutant neuroepithelial cells do not align their termini at the apical surface of the retina and divide in ectopic locations. As evidenced by the studies of genetically mosaic animals, this phenotype is cell-nonautonomous, indicating that *oko meduzy* affects cell-cell interactions of retinal neuroepithelium (Malicki and Driever 1999). The association of a neuronal patterning phenotype with an early defect in the neuroepithelial sheet is intriguing. Since undifferentiated retinal neurons migrate to their destination in the environment of a pseudostratified neuroepithelial sheet, it is tempting to speculate that the neuroepithelial defect in *oko meduzy* leads to a disruption of the positional cues which guide postmitotic cells to their proper positions in the neuronal laminae.

The patterning phenotypes of *ome*, *nok*, and *glo* are not identical. While ectopic photoreceptor cells in *oko meduzy* and *nagie oko* appear to be scattered randomly, *glass onion* photoreceptors form clusters (J. Malicki, unpublished). The orientation of photoreceptor cells in these clusters resembles aggregates of cells, so-called rosettes, observed in tissue culture experiments (Sheffield and Moscona 1970; Fujisawa 1971). Outside of the eye, *oko meduzy* affects brain morphology, the shape of the body axis, and blood circulation. Similar but more pronounced defects are present in *nagie oko*. Despite shape abnormalities, histological analysis did not reveal any gross patterning defects in the brain of either mutant. The *glass onion* mutation is the most pleiotropic of the three. Mutant embryos display severe abnormalities in brain shape associated with a wave of degeneration between 12 and 24 hpf. Histological sections through the *glass onion* brain reveal a disorganized pattern of neurons, similar to the defect seen in the retina (J. Malicki, unpublished). In addition, the tail of *glass onion* is shortened and blood circulation impeded or missing.

In addition to neuronal patterning abnormalities, *oko meduzy*, *nagie oko*, and *glass onion* share a defect in eye pigmentation. Although body coloration remains unaltered, the normally uniform dark pigmentation of the eye appears patchy in these mutants (Malicki et al. 1996a). This coincidental appearance of defects in the retinal neuroepithelium and the pigmented epithelium may reflect the origin of these tissues – both are derived from a common evagination of the neural tube. Although weaker, a retina-specific pigmentation defect is also present in *heart and soul* (*has*) (Malicki et al. 1996a). Neuronal patterning in *has* is difficult to evaluate, however, because extensive cell death affects the eye, starting between 48 to 60 hpf.

Assigning a mutant phenotype to the cellular patterning category has to be done cautiously since a disruption of cellular pattern frequently occurs in parallel to extensive cell death in the retina (Malicki et al. 1996a). Patterning abnormalities associated with massive cell death have been observed, for example, in the mutants *turbulent* and *zimny* (Malicki et al. 1996a). Although in *zimny* mutant animals cell death is not obvious on some histological sections (Fig. 1C), thorough analysis reveals that this phenotype is associated with abundance of cell corpses in the retina. The significance of lamination defect in *zimny* is not clear. It may be due to a disruption of a patterning mechanism or to the breakdown of retinal architecture resulting from massive cell loss. Thus, *zimny* and *turbulent* eye phenotypes are likely to involve a severe cell survival deficit rather than a true patterning defect. A similar situation is present in the cyclopia-inducing loci *cyclops* and *bozozok*, which also involve some degree of neuronal disorganization associated with a relatively mild increase in cell death (Malicki et al. 1996a; Fulwiler et al. 1997). The expression patterns of these genes unambiguously show, however, that their role in retinal lamination is secondary to much earlier developmental events.

3.5
Cellular Survival

A very frequent phenotype easily detectable by morphological criteria is a reduction of eye size combined with cell death in multiple layers of the retina. The onset of this phenotype varies from 2 dpf to several months. The magnitude of cell death also varies considerably. In the least severe mutations, for example *round eye* (*rde*), cell death is transient and affects relatively few cells (Fadool et al. 1997). In other mutants, cell corpses fill the entire retina by 3 dpf. As mentioned above, this has been observed in the most extreme cases of the *zimny* phenotype (J. Malicki, unpublished) as well as in *nba* homozygotes (Li and Dowling 1997). At least two cell death patterns can be distinguished in this class of mutants. In the first case, cell death occurs throughout the retina, whereas in the second, a considerably higher density of cell corpses is observed at the retinal margin. The first category includes mutants: *pyry* (*pyr*), *turbulent* (*tub*), *zimny* (*zny*), *lichee* (*che*), *ziemniok* (*zem*), *mizerny* (*miz*), *punktata* (*pkt*), *wide eye* (*wde*), and *archie* (*arch*); the second: *piegus* (*pgu*), *marginal eye* (*mre*), and *round eye* (*rde*). Mutations which selectively affect the retinal margin suggest that some aspects of the genetic control of neurogenesis differ in this region as compared to the central retina.

Degeneration of the central nervous system including the retina belongs to the most frequent phenotypes identified in zebrafish mutagenesis screens. Approximately 15 % of the 2383 mutants isolated in a recent ENU mutagenesis screen involved nonspecific CNS degeneration (Driever et al. 1996). Similar observations were made in a retroviral mutagenesis screen (Amsterdam and Hopkins 1999). These mutant phenotypes affect many CNS regions, and therefore are likely to be caused by genes involved in housekeeping functions, such as cell metabolism, rather than in developmental events. Because of that, they are frequently not considered worthy of a thorough characterization. Indeed, cloning of *dead eye* (*dye*), a gene which causes extensive cell death in the retina and brain at 2 dpf, followed by a severe reduction of the brain, eyes, and the craniofacial skeleton by 5 dpf, demonstrated that it encodes a protein with homology to a nuclear pore component (Allende et al. 1996; Amsterdam and Hopkins 1999). The nonspecific characteristics of these mutations usually preclude their classification together with eye defects. In screens focusing on eye development, only mutants with no or relatively mild brain and craniofacial skeleton defects are characterized further (Malicki et al. 1996a).

3.6
Targeting of the Retinotectal Projections

One of the more sophisticated screening methodologies was applied to search for defects in the retinotectal projections. Injections of lipophilic tracers DiI and DiO were used to visualize retinotectal projections in the F3 generation of mutagenized animals; 114 mutations in 35 genes were discovered in the course of this screen (Baier et al. 1996; Karlstrom et al. 1996; Trowe et al. 1996). Many aspects of retinotectal navigation, possibly reflecting important pathfinding choice points along the navigation route, are affected in this group of mutants: (1) pathfinding within the eye; (2) selection of ipsilateral versus contralateral pathways; (3) midline crossing; (4) pathfinding along the anteroposterior body axis; (5) sorting of axons in the optic tract; and (6) formation of the tectal map. Some of these categories can be subdivided further based on a more thorough characterization of the mutant phenotypes, while others overlap because mutations in a single gene affect axonal pathfinding at multiple points. *bashful* (*bal*) axons, for example, make pathfinding errors within the eye, occasionally project toward the ipsilateral tectum, fail to turn dorsally after crossing the midline, and in most cases project towards the telencephalon (Karlstrom et al. 1996).

Almost certainly, many of the retinotectal phenotypes originate in the brain. For example, the majority of mutants which form abnormal ipsilateral projections display midline abnormalities as well, such as a defective floor plate (Brand et al. 1996b). These mutations are likely to affect the patterning of the ventral CNS. This prediction has been recently confirmed by the cloning of *you-too* (*yot*), which encodes the *gli2* zinc-finger transcription factor involved in *shh* signaling (Karlstrom et al. 1999). Similarly, the majority of mutations affecting antero-posterior axon guidance display abnormal brain morphology.

The site of the defect in retinotectal mutants can be evaluated by whole eye transplantations. Transplantation of *bashful* eyes into a wild-type host results in the rescue of the retinotectal pathfinding in the brain (Chien and Bonhoeffer 1997), indicating that the *bashful* retinal axon guidance phenotype in the optic nerve and tract is due to the function of this gene in the brain and not the retina.

bashful (*bal*), *chameleon* (*con*), and *blowout* (*blw*) are so far the only retinotectal mutants that display obvious defects in the retina itself (Karlstrom et al. 1996): *bashful* (*bal*) ganglion cells do not differentiate uniformly and their axons are frequently misdirected within the eye; *chameleon* (*con*) axons frequently fail to exit the eye and instead grow in anterior and posterior directions along the eye equatorial region; and *blowout* (*blw*) retinae appear to rupture the pigmented epithelium and extend into the brain. Neuronal patterning of the ectopic retinal tissue is normal in *blw*. All three mutants share the tendency to send their projections to the ipsilateral rather than contralateral optic tectum (Karlstrom et al. 1997).

3.7
Other Mutants

Several mutants do not fit into any of the above developmental categories. For example, *pandora* (*pan*) appears to selectively affect the ventral eye (Malicki et al. 1996a). In this mutant, the portion of the retina ventral to the optic nerve is missing at 3 dpf (J. Malicki, unpublished). The eye phenotype of *pan* resembles defects of retinoid signaling, in particular the ventral eye deficiency induced by citral, an inhibitor of aldehyde dehydrogenase (Marsh-Armstrong et al. 1994). Deficiencies of the ventral retina have also been reported in compound retinoic acid receptor knockouts (Mascrez et al. 1998). In addition, body size, brain shape, heart, ear, tail, and pigmentation are also affected in *pandora*.

Behavioral screening for the loss of optokinetic response has led to the isolation of five mutants: *no optokinetic response a* (*noa*); *partial optokinetic response a (poa)*; *partial optokinetic response b (pob)*; *no optokinetic response b (nrb)*; and *no optokinetic response c (nrc)* (Brockerhoff et al. 1998). Apart from *pob*, which was described above, only *nrc* displays a histological abnormality – thinning of the outer plexiform layer (Allwardt 1999). A more in-depth understanding of the *nrc* phenotype emerged from electron microscopic analysis. The postsynaptic processes of bipolar and horizontal cells in the *nrc* mutant do not invaginate properly into the photoreceptor synaptic termini. Furthermore, the photoreceptor synaptic ribbons remain unattached in the cytoplasm instead of connecting through the arciform density to the presynaptic membrane (Allwardt 1999). So far, histological analysis of the remaining three mutants has not revealed any defects. All three display abnormal electroretinograms characterized by a reduction of the b-wave (Brockerhoff et al. 1998).

Finally, two additional groups of mutants, one with reduced eye size and one with lens defects, have been briefly described (Heisenberg et al. 1996). As their eye phenotypes were not characterized by any criteria other than external appearance, it is difficult to assign them to specific phenotypic categories

4
Future Prospects of Zebrafish Genetics

Although over 100 mutations have been identified, it is safe to say that screening for mutants of the zebrafish visual system is still in its adolescence. The saturation estimates in all of the recent experiments are low, indicating that continued screening, even via the same methods, will lead to the identification of many new loci (Baier et al. 1996; Driever et al. 1996; Haffter et al. 1996; Malicki et al. 1996a). Screening criteria, however, are likely to change. Future screens will most likely focus on progressively more subtle phenotypes discernible only with the help of markers. For example, currently available procedures do not allow one to screen for defects, especially subtle ones, affecting bipolar or horizontal cells. This may change when appropriate markers become available. The appearance of zebrafish transgenics is an important step in this direction. Using cell type-specific promoters, the expression of transgenes such as GFP can be directed to specific cell populations (Scheer and Campos-Ortega 1999). Thus, the tedious labeling of target cell populations with antibodies or in situ probes may be replaced by transgene expression. Other types of screens will also become a necessity. The identification of additional alleles will greatly aid in the functional analysis of many loci. New alleles can be obtained through noncomplementation screens such as the ones already performed for *sonic you* and several other mutants (van Eeden et al. 1999). Some eye structures, for example the cornea and the iris, have not been approached in any of the mutagenesis screens performed to date. This gap can be filled relatively easily even using simple visual inspection screening protocols. Finally, as aging in humans is frequently associated with the degeneration of eye tissues, late defects of survival rather than differentiation are also likely to be the focus of future screens.

Retroviral mutagenesis represents a major advance in zebrafish genetics. To date, at least seven retrovirus-induced mutations have been characterized (Amsterdam and Hopkins 1999). One of them affects photoreceptor development. Improved methods for the delivery of retroviral vectors into zebrafish embryos are likely to increase the efficiency of retroviral mutagenesis screens in the future. New generations of retroviral vectors will be able to function not only as mutagenic agents but also as enhancer traps or gene traps, allowing for easier identification of mutagenic insertion events. Another option for future mutagenic agents in zebrafish are transposable elements. Both *C. elegans* and *Drosophila* transposons from the Tc1/*mariner* superfamily have been shown to insert into the zebrafish genome from exogenous DNA constructs injected into the embryos (Fadool et al. 1998; Raz et al. 1998). These elements behave in a Mendelian fashion, and at least the *C. elegans* Tc1 can be subsequently mobilized by transposase injections (Raz et al. 1998). Thus, at some point in the future, transposons too may be used as enhancer traps, gene traps, or mutagenic agents.

Given the multitude of genetic defects accumulated through mutagenesis screens, the obvious next step is the molecular characterization of the mutant genes. This can be accomplished either by the candidate or positional cloning approach. As of today, positional cloning in zebrafish is still laborious and technically demanding. As the field of zebrafish genomics is developing at a breathtaking pace, this may soon change. The past several years have brought many significant developments in zebrafish genome research including: simple sequence length polymorphism (SSLP)-based genetic maps, several types of large insert genomic libraries, radiation hybrid panels, and expressed sequence tag (EST) sequencing and mapping projects (Knapik et al. 1998; Kwok et al. 1998; Zhong et al. 1998). The positional cloning of the first mutant genes has already been accomplished (Brownlie et al. 1998; Zhang et al. 1998). These developments, in particular EST identification and mapping, will also benefit the candidate cloning approach. Rapid progress in zebrafish genomics encourages speculations that within a period of several years, positional cloning will become a straightforward approach to molecular characterization of mutant loci. As molecular genetic analysis is catching up with mutagenesis and screening approaches, the zebrafish is rapidly becoming an elegant and efficient genetic system to study vertebrate development.

Acknowledgments. The author is grateful to Drs. Pamela Yelick, Francesca Pignoni, Geoffrey Doerre, Tiansen Li, Brian Link, and Arindam Majumdar for commenting on this manuscript. Histological sections in Fig. 1G and H were provided by Dr. Jim Fadool while Fig. 1I was contributed by Dr. Tom Becker. This work was supported in part by the Research to Prevent Blindness Career Development Award and the March of Dimes Birth Defects Foundation.

References

Adelmann HB (1930) Experimental studies of the development of the eye. III. J Exp Zool 57:223-281

Allende ML, Amsterdam A, Becker T, Kawakami K, Gaiano N, Hopkins N (1996) Insertional mutagenesis in zebrafish identifies two novel genes, pescadillo and dead eye, essential for embryonic development. Genes Dev 10:3141-3155

Allwardt B (1999) Ultrastructural analysis of photoreceptors in wild-type and mutant zebrafish larvae. Ph.D. thesis, Harvard University, Cambridge, Massachusetts

Altshuler D, Turner D, Cepko C (1991) Specification of cell type in the vertebrate retina. In: Man-Kit Lam D and Shatz C (eds) Development of the Visual System. The MIT Press, Cambridge, Massachusetts, pp 37-58

Amsterdam A, Hopkins N (1999) Retrovirus-mediated insertional mutagenesis in zebrafish. Methods Cell Biol 60:87-98

Baier H, Klostermann S, Trowe T, Karlstrom RO, Nusslein-Volhard C, Bonhoeffer F (1996) Genetic dissection of the retinotectal projection. Development 123:415-425

Beattie CE, Raible DW, Henion PD, Eisen JS (1999) Early pressure screens. Methods Cell Biol 60:71-86

Becker TS, Burgess SM, Amsterdam AH, Allende ML, Hopkins N (1998) not really finished is crucial for development of the zebrafish outer retina and encodes a transcription factor highly homologous to human Nuclear Respiratory Factor-1 and avian Initiation Binding Repressor. Development 125:4369-4378

Bodick N, Levinthal C (1980) Growing optic nerve fibers follow neighbors during embryogenesis. Proc Natl Acad Sci USA 77:4374-4378

Branchek T, Bremiller R (1984) The Development of Photoreceptors in the Zebrafish, Brachydanio rerio. I. Structure. J Comp Neurol 224:107-115

Brand M, Heisenberg CP, Jiang YJ, Beuchle D, Lun K, Furutani-Seiki M, Granato M, Haffter P, Hammerschmidt M, Kane DA, Kelsh RN, Mullins MC, Odenthal J, van Eeden FJ, Nusslein-Volhard C (1996a) Mutations in zebrafish genes affecting the formation of the boundary between midbrain and hindbrain. Development 123:179-190

Brand M, Heisenberg CP, Warga RM, Pelegri F, Karlstrom RO, Beuchle D, Picker A, Jiang YJ, Furutani-Seiki M, van Eeden FJ, Granato M, Haffter P, Hammerschmidt M, Kane DA, Kelsh RN, Mullins MC, Odenthal J, Nusslein-Volhard C (1996b) Mutations affecting development of the midline and general body shape during zebrafish embryogenesis. Development 123:129-142

Brockerhoff SE, Dowling JE, Hurley JB (1998) Zebrafish retinal mutants. Vision Res 38:1335-1339

Brockerhoff SE, Hurley JB, Janssen-Bienhold U, Neuhauss SC, Driever W, Dowling JE (1995) A behavioral screen for isolating zebrafish mutants with visual system defects. Proc Natl Acad Sci USA 92:10545-10549

Brockerhoff SE, Hurley JB, Niemi GA, Dowling JE (1997) A new form of inherited red-blindness identified in zebrafish. J Neurosci 17:4236-4242

Brownlie A, Donovan A, Pratt SJ, Paw BH, Oates AC, Brugnara C, Witkowska HE, Sassa S, Zon LI (1998) Positional cloning of the zebrafish sauternes gene: a model for congenital sideroblastic anaemia. Nat Genet 20:244-250

Burmeister M, Novak J, Liang MY, Basu S, Ploder L, Hawes NL, Vidgen D, Hoover F, Goldman D, Kalnins VI, Roderick TH, Taylor BA, Hankin MH, McInnes RR (1996) Ocular retardation mouse caused by Chx10 homeobox null allele: impaired retinal progenitor proliferation and bipolar cell differentiation. Nat Genet 12:376-384

Burrill J, Easter S (1995) The first retinal axons and their microenvironment in zebrafish cryptic pioneers and the pretract. J Neurosci 15:2935-2947

Burrill JD, Easter SS, Jr. (1994) Development of the retinofugal projections in the embryonic and larval zebrafish (*Brachydanio rerio*). J Comp Neurol 346 : 583–600

Cajal SR (1893) La retine des vertebres. La Cellule 9 : 17–257

Chen JN, Haffter P, Odenthal J, Vogelsang E, Brand M, van Eeden FJ, Furutani-Seiki M, Granato M, Hammerschmidt M, Heisenberg CP, Jiang YJ, Kane DA, Kelsh RN, Mullins MC, Nusslein-Volhard C (1996) Mutations affecting the cardiovascular system and other internal organs in zebrafish. Development 123 : 293–302

Chiang C, Litingtung Y, Lee E, Young KE, Corden JL, Westphal H, Beachy PA (1996) Cyclopia and defective axial patterning in mice lacking Sonic hedgehog gene function. Nature 383 : 407–413

Chien C, Bonhoeffer F (1997) Zebrafish retinotectal pathfinding mutants projecting to extra-tectal targets. Soc Neurosci Abstr 23 : 604

Clark T (1981) Visual responses in developing zebrafish (*Brachydanio rerio*). PhD Thesis, University of Oregon, Eugene, Oregon

Corley-Smith GE, Lim CJ, Brandhorst BP (1996) Production of androgenetic zebrafish (*Danio rerio*). Genetics 142 : 1265–1276

Dowling J (1987) The Retina. Harvard University Press, Cambridge, Massachusetts

Driever W, Solnica-Krezel L, Schier AF, Neuhauss SC, Malicki J, Stemple DL, Stainier DY, Zwartkruis F, Abdelilah S, Rangini Z, Belak J, Boggs C (1996) A genetic screen for mutations affecting embryogenesis in zebrafish. Development 123 : 37–46

Drummond IA, Majumdar A, Hentschel H, Elger M, Solnica-Krezel L, Schier AF, Neuhauss SC, Stemple DL, Zwartkruis F, Rangini Z, Driever W, Fishman MC (1998) Early development of the zebrafish pronephros and analysis of mutations affecting pronephric function. Development 125 : 4655–4667

Easter S, Nicola G (1996) The development of vision in the zebrafish (*Danio rerio*). Dev Biol 180 : 646–663

Fadool JM, Brockerhoff SE, Hyatt GA, Dowling JE (1997) Mutations affecting eye morphology in the developing zebrafish (*Danio rerio*). Dev Genet 20 : 288–295

Fadool JM, Hartl DL, Dowling JE (1998) Transposition of the mariner element from *Drosophila mauritiana* in zebrafish. Proc Natl Acad Sci USA 95 : 5182–5186

Fekany K, Yamanaka Y, Leung T, Sirotkin HI, Topczewski J, Gates MA, Hibi M, Renucci A, Stemple D, Radbill A, Schier AF, Driever W, Hirano T, Talbot WS, Solnica-Krezel L (1999) The zebrafish bozozok locus encodes dharma, a homeodomain protein essential for in-duction of gastrula organizer and dorsoanterior embryonic structures. Development 126 : 1427–1438

Feldman B, Gates MA, Egan ES, Dougan ST, Rennebeck G, Sirotkin HI, Schier AF, Talbot WS (1998) Zebrafish organizer development and germ-layer formation require nodal-related signals. Nature 395 : 181–185

Fujisawa H (1971) A complete reconstruction of the neural retina of chick embryo grafted onto the chorio-allantoic membrane. Dev Growth Differ (Nagoya) 13 : 25–36

Fulwiler C, Schmitt EA, Kim JM, Dowling JE (1997) Retinal patterning in the zebrafish mutant cyclops. J Comp Neurol 381 : 449–460

Gaiano N, Allende M, Amsterdam A, Kawakami K, Hopkins N (1996a) Highly efficient germ-line transmission of proviral insertions in zebrafish. Proc Natl Acad Sci USA 93 : 7777–7782

Gaiano N, Amsterdam A, Kawakami K, Allende M, Becker T, Hopkins N (1996b) Insertional mutagenesis and rapid cloning of essential genes in zebrafish. Nature 383 : 829–832

Gan L, Xiang M, Zhou L, Wagner DS, Klein WH, Nathans J (1996) POU domain factor Brn-3b is required for the development of a large set of retinal ganglion cells. Proc Natl Acad Sci USA 93 : 3920–3925

Georges-Labouesse E, Mark M, Messaddeq N, Gansmuller A (1998) Essential role of alpha 6 integrins in cortical and retinal lamination. Curr Biol 8 : 983–986

Grunwald DJ, Kimmel CB, Westerfield M, Walker C, Streisinger G (1988) A neural degeneration mutation that spares primary neurons in the zebrafish. Dev Biol 126 : 115–128

Guo S, Wilson SW, Cooke S, Chitnis AB, Driever W, Rosenthal A (1999) Mutations in the zebrafish unmask shared regulatory pathways controlling the development of catecholaminergic neurons. Dev Biol 208:473–487

Haffter P, Granato M, Brand M, Mullins MC, Hammerschmidt M, Kane DA, Odenthal J, van Eeden FJ, Jiang YJ, Heisenberg CP, Kelsh RN, Furutani-Seiki M, Vogelsang E, Beuchle D, Schach U, Fabian C, Nusslein-Volhard C (1996) The identification of genes with unique and essential functions in the development of the zebrafish, Danio rerio. Development 123:1–36

Hatta K, Puschel AW, Kimmel CB (1994) Midline signaling in the primordium of the zebrafish anterior central nervous system. Proc Natl Acad Sci USA 91:2061–2065

Heisenberg CP, Brand M, Jiang YJ, Warga R, Beuchle D, Eeden F, Furutani-Seiki M, Granato M, Haffter P, Hammerschmidt M, Kane D, Kelsh R, Mullins M, Odenthal J, Nusslein-Volhard C (1996) Genes involved in forebrain development in the zebrafish, Danio rerio. Development 123:191–203

Heisenberg CP, Nusslein-Volhard C (1997) The function of silberblick in the positioning of the eye anlage in the zebrafish embryo. Dev Biol 184:85–94

Henion PD, Raible DW, Beattie CE, Stoesser KL, Weston JA, Eisen JS (1996) Screen for mutations affecting development of Zebrafish neural crest. Dev Genet 18:11–17

Hill Rea (1991) Mouse Small eye results from mutations in a paired-like homeobox-containing gene. Nature 354:522–525

Hu M, Easter SS (1999) Retinal neurogenesis: the formation of the initial central patch of postmitotic cells. Dev Biol 207:309–321

Johns PR, Fernald RD (1981) Genesis of rods in teleost fish retina. Nature 293:141–142

Karlstrom RO, Talbot WS, Schier AF (1999) Comparative synteny cloning of zebrafish you-too: mutations in the Hedgehog target gli2 affect ventral forebrain patterning. Genes Dev 13:388–393

Karlstrom RO, Trowe T, Bonhoeffer F (1997) Genetic analysis of axon guidance and mapping in the zebrafish. Trends Neurosci 20:3–8

Karlstrom RO, Trowe T, Klostermann S, Baier H, Brand M, Crawford AD, Grunewald B, Haffter P, Hoffmann H, Meyer SU, Muller BK, Richter S, van Eeden FJ, Nusslein-Volhard C, Bonhoeffer F (1996) Zebrafish mutations affecting retinotectal axon pathfinding. Development 123:427–438

Kelsh RN, Brand M, Jiang YJ, Heisenberg CP, Lin S, Haffter P, Odenthal J, Mullins MC, van Eeden FJ, Furutani-Seiki M, Granato M, Hammerschmidt M, Kane DA, Warga RM, Beuchle D, Vogelsang L, Nusslein-Volhard C (1996) Zebrafish pigmentation mutations and the processes of neural crest development. Development 123:369–389

Knapik EW, Goodman A, Ekker M, Chevrette M, Delgado J, Neuhauss S, Shimoda N, Driever W, Fishman MC, Jacob HJ (1998) A microsatellite genetic linkage map for zebrafish (Danio rerio). Nat Genet 18:338–343

Kwok C, Korn RM, Davis ME, Burt DW, Critcher R, McCarthy L, Paw BH, Zon LI, Goodfellow PN, Schmitt K (1998) Characterization of whole genome radiation hybrid mapping resources for non-mammalian vertebrates. Nucleic Acids Res 26:3562–3566

Larison K, Bremiller R (1990) Early onset of phenotype and cell patterning in the embryonic zebrafish retina. Development 109:567–576

Li H, Tierney C, Wen L, Wu JY, Rao Y (1997) A single morphogenetic field gives rise to two retina primordia under the influence of the prechordal plate. Development 124:603–615

Li L, Dowling JE (1997) A dominant form of inherited retinal degeneration caused by a non-photoreceptor cell-specific mutation. Proc Natl Acad Sci USA 94:11645–11650

Lin S, Gaiano N, Culp P, Burns JC, Friedmann T, Yee JK, Hopkins N (1994) Integration and germ-line transmission of a pseudotyped retroviral vector in zebrafish. Science 265:666–669

Lun K, Brand M (1998) A series of no isthmus (noi) alleles of the zebrafish pax2.1 gene reveals multiple signaling events in development of the midbrain-hindbrain boundary. Development 125:3049–3062

Macdonald R, Barth KA, Xu Q, Holder N, Mikkola I, Wilson SW (1995) Midline signalling is required for Pax gene regulation and patterning of the eyes. Development 121:3267-3278

Macdonald R, Scholes J, Strahle U, Brennan C, Holder N, Brand M, Wilson SW (1997) The Pax protein Noi is required for commissural axon pathway formation in the rostral forebrain. Development 124:2397-2408

Malicki J (1999) Development of the retina. Methods Cell Biol 59:273-299

Malicki J, Driever W (1999) oko meduzy mutations affect neuronal patterning in the zebrafish retina and reveal cell-cell interactions of the retinal neuroepithelial sheet. Development 126:1235-1246

Malicki J, Neuhauss SC, Schier AF, Solnica-Krezel L, Stemple DL, Stainier DY, Abdelilah S, Zwartkruis F, Rangini Z, Driever W (1996a) Mutations affecting development of the zebrafish retina. Development 123:263-273

Malicki J, Schier AF, Solnica-Krezel L, Stemple DL, Neuhauss SC, Stainier DY, Abdelilah S, Rangini Z, Zwartkruis F, Driever W (1996b) Mutations affecting development of the zebrafish ear. Development 123:275-283

Marcus RC, Delaney CL, Easter SS, Jr. (1999) Neurogenesis in the visual system of embryonic and adult zebrafish (Danio rerio). Vis Neurosci 16:417-424

Marlow F, Zwartkruis F, Malicki J, Neuhauss SC, Abbas L, Weaver M, Driever W, Solnica-Krezel L (1998) Functional interactions of genes mediating convergent extension, knypek and trilobite, during the partitioning of the eye primordium in zebrafish. Dev Biol 203:382-399

Marsh-Armstrong N, McCaffery P, Gilbert W, Dowling JE, Drager UC (1994) Retinoic acid is necessary for development of the ventral retina in zebrafish. Proc Natl Acad Sci USA 91:7286-7290

Mascrez B, Mark M, Dierich A, Ghyselinck NB, Kastner P, Chambon P (1998) The RXRalpha ligand-dependent activation function 2 (AF-2) is important for mouse development. Development 125:4691-4707

Mathers PH, Grinberg A, Mahon KA, Jamrich M (1997) The Rx homeobox gene is essential for vertebrate eye development. Nature 387:603-607

Moens CB, Cordes SP, Giorgianni MW, Barsh GS, Kimmel CB (1998) Equivalence in the genetic control of hindbrain segmentation in fish and mouse. Development 125:381-391

Muller H (1857) Anatomisch-physiologische untersuchungen uber die Retina bei Menschen und Wirbelthieren. Z Wiss Zool 8:1-122

Mullins MC, Hammerschmidt M, Haffter P, Nusslein-Volhard C (1994) Large-scale mutagenesis in the zebrafish: in search of genes controlling development in a vertebrate. Curr Biol 4: 189-202

Nawrocki W (1985) Development of the neural retina in the zebrafish, Brachydanio rerio. PhD Thesis, University of Oregon, Eugine, Oregon

Nornes S, Clarkson M, Mikkola I, Pedersen M, Bardsley A, Martinez JP, Krauss S, Johansen T (1998) Zebrafish contains two pax6 genes involved in eye development. Mech Dev 77:185-196

Odenthal J, Haffter P, Vogelsang E, Brand M, van Eeden FJ, Furutani-Seiki M, Granato M, Hammerschmidt M, Heisenberg CP, Jiang YJ, Kane DA, Kelsh RN, Mullins MC, Warga RM, Allende ML, Weinberg ES, Nusslein-Volhard C (1996) Mutations affecting the formation of the notochord in the zebrafish, Danio rerio. Development 123:103-115

Passini MA, Levine EM, Canger AK, Raymond PA, Schechter N (1997) Vsx-1 and Vsx-2: differential expression of two paired-like homeobox genes during zebrafish and goldfish retinogenesis. J Comp Neurol 388:495-505

Pelegri F, Schulte-Merker S (1999) A gynogenesis-based screen for maternal-effect genes in the zebrafish, Danio rerio. Methods Cell Biol 60:1-20

Pfeffer PL, Gerster T, Lun K, Brand M, Busslinger M (1998) Characterization of three novel members of the zebrafish Pax2/5/8 family: dependency of pax5 and pax8 expression on the pax2.1 (noi) function. Development 125:3063-3074

Piotrowski T, Schilling TF, Brand M, Jiang YJ, Heisenberg CP, Beuchle D, Grandel H, van Eeden FJ, Furutani-Seiki M, Granato M, Haffter P, Hammerschmidt M, Kane DA, Kelsh RN, Mullins

MC, Odenthal J, Warga RM, Nusslein-Volhard C (1996) Jaw and branchial arch mutants in zebrafish II: anterior arches and cartilage differentiation. Development 123:345-356

Porter FD, Drago J, Xu Y, Cheema SS, Wassif C, Huang SP, Lee E, Grinberg A, Massalas JS, Bodine D, Alt F, Westphal H (1997) Lhx2, a LIM homeobox gene, is required for eye, forebrain, and definitive erythrocyte development. Development 124:2935-2944

Raymond P, Barthel L, Curran G (1995) Developmental patterning of rod and cone photoreceptors in embryonic zebrafish. J Comp Neurol 359:537-550

Raymond P, Barthel L, Rounsifer M, Sullivan S, Knight J (1993) Expression of rod and cone visual pigments in godfish and zebrafish: A rhodopsin-like gene is expressed in cones. Neuron 10:1161-1174

Raz E, van Luenen HG, Schaerringer B, Plasterk RHA, Driever W (1998) Transposition of the nematode *Caenorhabditis elegans* Tc3 element in the zebrafish *Danio rerio*. Curr Biol 8:82-88

Reifers F, Bohli H, Walsh EC, Crossley PH, Stainier DY, Brand M (1998) Fgf8 is mutated in zebrafish acerebellar (ace) mutants and is required for maintenance of midbrain-hindbrain boundary development and somitogenesis. Development 125:2381-2395

Robinson J, Schmitt E, Dowling J (1995) Temporal and spatial patterns of opsin gene expression in zebrafish (*Danio rerio*). Visual Neuroscience 12:895-906

Rodieck RW (1973) The vertebrate retina. Principles of structure and function. W. H. Freeman & Co, San Francisco, California

Roessler E, Belloni E, Gaudenz K, Jay P, Berta P, Scherer SW, Tsui LC, Muenke M (1996) Mutations in the human Sonic Hedgehog gene cause holoprosencephaly. Nat Genet 14:357-360

Sampath K, Rubinstein AL, Cheng AM, Liang JO, Fekany K, Solnica-Krezel L, Korzh V, Halpern ME, Wright CV (1998) Induction of the zebrafish ventral brain and floorplate requires cyclops/nodal signalling. Nature 395:185-189

Schauerte HE, van Eeden FJ, Fricke C, Odenthal J, Strahle U, Haffter P (1998) Sonic hedgehog is not required for the induction of medial floor plate cells in the zebrafish. Development 125:2983-2993

Scheer N, Campos-Ortega JA (1999) Use of the Gal4-UAS technique for targeted gene expression in the zebrafish. Mech Dev 80:153-158

Schier AF, Neuhauss SC, Harvey M, Malicki J, Solnica-Krezel L, Stainier DY, Zwartkruis F, Abdelilah S, Stemple DL, Rangini Z, Yang H, Driever W (1996) Mutations affecting the development of the embryonic zebrafish brain. Development 123:165-178

Schier AF, Neuhauss SC, Helde KA, Talbot WS, Driever W (1997) The one-eyed pinhead gene functions in mesoderm and endoderm formation in zebrafish and interacts with no tail. Development 124:327-342

Schilling TF, Piotrowski T, Grandel H, Brand M, Heisenberg CP, Jiang YJ, Beuchle D, Hammerschmidt M, Kane DA, Mullins MC, van Eeden FJ, Kelsh RN, Furutani-Seiki M, Granato M, Haffter P, Odenthal J, Warga RM, Trowe T, Nusslein-Volhard C (1996) Jaw and branchial arch mutants in zebrafish I: branchial arches. Development 123:329-344

Schmitt E, Dowling J (1994) Early eye morphogenesis in the Zebrafish, *Brachydanio rerio*. J of Comp Neurol 344:532-542

Schmitt EA, Dowling JE (1996) Comparison of topographical patterns of ganglion and photoreceptor cell differentiation in the retina of the zebrafish, *Danio rerio*. J Comp Neurol 371:222-234

Sheffield J, Moscona A (1970) Electron microscopic analysis of aggregation of embryonic cells: the structure and differentiation of aggregates of neural retina cells. Dev Biol 23:36-61

Sicinski P, Donaher JL, Parker SB, Li T, Fazeli A, Gardner H, Haslam SZ, Bronson RT, Elledge SJ, Weinberg RA (1995) Cyclin D1 provides a link between development and oncogenesis in the retina and breast. Cell 82:621-630

Solnica-Krezel L, Schier A, Driever W (1994) Efficient recovery of ENU-induced mutations from the zebrafish germline. Genetics 136:1-20

Solnica-Krezel L, Stemple DL, Mountcastle-Shah E, Rangini Z, Neuhauss SC, Malicki J, Schier AF, Stainier DY, Zwartkruis F, Abdelilah S, Driever W (1996) Mutations affecting cell fates and cellular rearrangements during gastrulation in zebrafish. Development 123:67–80

Stemple DL, Solnica-Krezel L, Zwartkruis F, Neuhauss SC, Schier AF, Malicki J, Stainier DY, Abdelilah S, Rangini Z, Mountcastle-Shah E, Driever W (1996) Mutations affecting development of the notochord in zebrafish. Development 123:117–128

Streisinger G, Coale F, Taggart C, Walker C, Grunwald DJ (1989) Clonal origins of cells in the pigmented retina of the zebrafish eye. Dev Biol 131:60–69

Stuermer CA (1988) Retinotopic organization of the developing retinotectal projection in the zebrafish embryo. J Neurosci 8:4513–4530

Stumpo D, Bock C, Tuttle J, Blackshear P (1995) MARCKS deficiency in mice leads to abnormal brain development and perinatal death. Proc Natl Acad Sci USA 92:944–948

Talbot WS, Trevarrow B, Halpern ME, Melby AE, Farr G, Postlethwait JH, Jowett T, Kimmel CB, Kimelman D (1995) A homeobox gene essential for zebrafish notochord development. Nature 378:150–157

Tomasiewicz H (1993) Genetic deletion of a neural cell adhesion molecule variant (N-CAM-180) produces distinct defects in the central neuvous system. Neuron 11:1163–1174

Tomita K, Ishibashi M, Nakahara K, Ang SL, Nakanishi S, Guillemot F, Kageyama R (1996) Mammalian hairy and Enhancer of split homolog 1 regulates differentiation of retinal neurons and is essential for eye morphogenesis. Neuron 16:723–734

Trowe T, Klostermann S, Baier H, Granato M, Crawford AD, Grunewald B, Hoffmann H, Karlstrom RO, Meyer SU, Muller B, Richter S, Nusslein-Volhard C, Bonhoeffer F (1996) Mutations disrupting the ordering and topographic mapping of axons in the retinotectal projection of the zebrafish, Danio rerio. Development 123:439–450

van Eeden F, Granato M, Schach U, Brand M, Furutani-Seiki M, Haffter P, Hammerschmidt M, Heisenberg CP, Jiang YJ, Kane D, Kelsh R, Mullins M, Odenthal J, Warga R, Allende M, Weinberg E, Nusslein-Volhard C (1996) Mutations affecting somite formation and patterning in the zebrafish, Danio rerio. Dev Suppl 123:153–164

van Eeden FJ, Granato M, Odenthal J, Haffter P (1999) Developmental mutant screens in the zebrafish. Methods Cell Biol 60:21–41

Vihtelic TS, Doro CJ, Hyde DR (1999) Cloning and characterization of six zebrafish photoreceptor opsin cDNAs and immunolocalization of their corresponding proteins. Vis Neurosci 16:571–585

Virbasius CA, Virbasius JV, Scarpulla RC (1993) NRF-1, an activator involved in nuclear-mitochondrial interactions, utilizes a new DNA-binding domain conserved in a family of developmental regulators. Genes Dev 7:2431–2445

Walker C (1999) Haploid screens and gamma-ray mutagenesis. Methods Cell Biol 60:43–70

Weinstein BM, Stemple DL, Driever W, Fishman MC (1995) Gridlock, a localized heritable vascular patterning defect in the zebrafish. Nat Med 1:1143–1147

Whitfield TT, Granato M, van Eeden FJ, Schach U, Brand M, Furutani-Seiki M, Haffter P, Hammerschmidt M, Heisenberg CP, Jiang YJ, Kane DA, Kelsh RN, Mullins MC, Odenthal J, Nusslein-Volhard C (1996) Mutations affecting development of the zebrafish inner ear and lateral line. Development 123:241–254

Woo K, Fraser SE (1995) Order and coherence in the fate map of the zebrafish nervous system. Development 121:2595–2609

Yamanaka Y, Mizuno T, Sasai Y, Kishi M, Takeda H, Kim CH, Hibi M, Hirano T (1998) A novel homeobox gene, dharma, can induce the organizer in a non-cell-autonomous manner. Genes Dev 12:2345–2353

Zhang J, Talbot WS, Schier AF (1998) Positional cloning identifies zebrafish one-eyed pinhead as a permissive EGF-related ligand required during gastrulation. Cell 92:241–251

Zhong TP, Kaphingst K, Akella U, Haldi M, Lander ES, Fishman MC (1998) Zebrafish genomic library in yeast artificial chromosomes. Genomics 48:136–138

Subject Index